Lecture Notes in Computer Science   9448

Commenced Publication in 1973
Founding and Former Series Editors:
Gerhard Goos, Juris Hartmanis, and Jan van Leeuwen

More information about this series at http://www.springer.com/series/7412

Reneta P. Barneva · Bhargab B. Bhattacharya
Valentin E. Brimkov (Eds.)

Combinatorial
Image Analysis

17th International Workshop, IWCIA 2015
Kolkata, India, November 24–27, 2015
Proceedings

 Springer

Editors
Reneta P. Barneva
SUNY Fredonia
Fredonia, NY
USA

Valentin E. Brimkov
SUNY Buffalo State
Buffalo, NY
USA

Bhargab B. Bhattacharya
Indian Statistical Institute
Advanced Computing and Microelectronics
    Unit (ACMU)
Kolkata, West Bengal
India

ISSN 0302-9743   ISSN 1611-3349   (electronic)
Lecture Notes in Computer Science
ISBN 978-3-319-26144-7   ISBN 978-3-319-26145-4   (eBook)
DOI 10.1007/978-3-319-26145-4

Library of Congress Control Number: 2015953004

LNCS Sublibrary: SL6 – Image Processing, Computer Vision, Pattern Recognition, and Graphics

Springer Cham Heidelberg New York Dordrecht London

Printed on acid-free paper

Springer International Publishing AG Switzerland is part of Springer Science+Business Media
(www.springer.com)

Preface

This volume contains the articles presented at the 17th International Workshop on Combinatorial Image Analysis, IWCIA 2015, which was held at the Indian Statistical Institute in Kolkata, during November 24–27, 2015. The 16 previous meetings were held in Paris (France) 1991, Ube (Japan) 1992, Washington DC (USA) 1994, Lyon (France) 1995, Hiroshima (Japan) 1997, Madras (India) 1999, Caen (France) 2000, Philadelphia, PA (USA) 2001, Palermo (Italy) 2003, Auckland (New Zealand) 2004, Berlin (Germany) 2006, Buffalo, NY (USA) 2008, Playa del Carmen (Mexico) 2009, Madrid (Spain) 2011, Austin, TX (USA) 2012, and Brno (Czech Republic) 2014.

Combinatorial image analysis provides theoretical foundations and methods for solving problems from various areas of human practice. In contrast to traditional approaches to image analysis that implement continuous models, float arithmetic, and rounding, combinatorial image analysis features discrete models using integer arithmetic. The developed algorithms are based on studying combinatorial properties of classes of digital images, and often appear to be more efficient and accurate than those based on continuous models.

IWCIA is an exciting opportunity for scholars, graduate students, and educators across the world to meet and share information about their latest findings in the field of combinatorial image analysis, get enriched with new ideas, reflect on some open problems, learn about new applications, and reconnect with colleagues. All papers submitted to the conference were carefully reviewed as each manuscript was sent for a double-blind review to at least three highly qualified members of the international Program Committee. The submission and review process of the workshop was carried out through the professional EasyChair conference management system. After a rigorous review process, 25 papers authored by 71 researchers from 12 countries were accepted for presentation at the workshop and for inclusion in this volume.

IWCIA 2015 featured keynote talks delivered by two outstanding scholars, whose excellent presentations inspired the audience with new ideas.

Eric Andres (Head of Signal-Image-Communications Department at the Université de Poitiers, France) presented a survey of analytically defined digital geometric objects. He started from some elements of digitization and their relation to continuous geometry. Then, he explained how mathematically sound digital objects could be built from simple assumptions about the properties a digital object should have. He concluded with some open problems and challenges.

Tetsuo Asano (President of Japan Advanced Institute of Science and Technology) surveyed algorithms for image processing that use small amounts of memory. He started with in-place algorithms, where the input image matrix can be modified, and introduced small-work-space algorithms for several important basic problems on image processing, including connected components labeling. This is a problem that has been extensively studied and a number of approaches have been proposed so far under several different computational models. For example, a linear-time algorithm is known for the problem using linear work space. Asano showed that it is possible to reduce the amount of work space into the square root of n while increasing the running time from $O(n)$ to $O(n\log n)$. He also discussed other space-efficient algorithms and their effectiveness for applications to scanners.

The full paper of the keynote talk of Prof. Andres is included in the beginning of this volume.

The contributed papers are grouped into two parts. The first one includes 16 papers devoted to theoretical foundations of combinatorial image analysis, in particular studies on geometry and topology of digital curves and surfaces, the design of space-efficient algorithms, array grammars and languages for image analysis, research on picture transformations, and others. The second part includes nine papers presenting applications of combinatorial methods in image analysis.

We believe that all presented works were of high quality and the attendees benefited from the scientific program.

We would like to express our gratitude to everyone who contributed to the success of IWCIA 2015 – from the Steering to the Program and Organizing Committees. We are indebted to the Indian Statistical Institute, Kolkata, for providing space and equipment for holding the conference.

We wish to express our special thanks to the invited speakers Eric Andres and Tetsuo Asano for their remarkable talks and overall contribution to the workshop program. We thank all authors for their valuable works and hope that the reader will find them interesting and useful. We wish to thank the participants and everyone who made this workshop an enjoyable and fruitful scientific event. We appreciate the support of SUNY Buffalo State, and especially of President Katherine S. Conway-Turner, Provost Melanie L. Perreault, and Dean Mark W. Severson, for sponsoring the best student paper awards. Finally, we express our gratitude to Springer Computer Science Editorial, and especially to Alfred Hofmann and Anna Kramer, for their efficient and kind cooperation in the timely production of this book.

November 2015   Reneta P. Barneva
                                                    Bhargab B. Bhattacharya
                                                       Valentin E. Brimkov

Organization

IWCIA 2015 was held at the Indian Statistical Institute, Kolkata, India, November 24–27, 2015

General Chair

Bhargab B. Bhattacharya   Indian Statistical Institute, Kolkata, India

Program Co-chairs

Arindam Biswas   IIEST Shibpur, India
Partha Bhowmick   IIT Kharagpur, India

Publication Chair

Reneta P. Barneva   SUNY Fredonia, USA

Steering Committee

Valentin E. Brimkov   SUNY Buffalo State, USA
Gabor T. Herman   CUNY Graduate Center, USA
Kostadin Koroutchev   Universidad Autonoma de Madrid, Spain
Josef Slapal   Technical University of Brno, Czech Republic
Petra Wiederhold   CINVESTAV-IPN, Mexico

Invited Speakers

Eric Andres   Université de Poitiers, France
Tetsuo Asano   JAIST, Japan

Program Committee

Lyuba Alboul   Sheffield Hallam University, UK
Akira Asano   Kansai University, Japan
Soumen Bag   ISM Dhanbad, India
Péter Balázs   University of Szeged, Hungary
George Bebis   University of Nevada at Reno, USA
Sara Brunetti   Università degli Studi di Siena, Italy
Guillaume Damiand   LIRIS-CNRS, Université de Lyon, France
Partha Pratim Das   IIT Kharagpur, India
Mousumi Dutt   IIIT Kalyani, India
Isabelle Debled-Rennesson   Nancy University, LORIA, France
Chiou-Shann Fuh   National Taiwan University, Taiwan

Organizing Committee

Arijit Bishnu - Chair	ISI Kolkata, India
Ansuman Banerjee	ISI Kolkata, India
Ranita Biswas	IIT Kharagpur, India
Suprativ Biswas	ISI Kolkata, India
Mousumi Dutt	IIEST Shibpur, India
Sasthi Charan Ghosh	ISI Kolkata, India
Nilanjana Karmakar	IIEST Shibpur, India
Papia Mahato	IIT Kharagpur, India
Sachchidanand Mahato	ISI Kolkata, India
Debabrata Mitra	ISI Kolkata, India
Apurba Sarkar	IIEST Shibpur, India

Sponsoring Institutions

Indian Statistical Institute, Kolkata, India
SUNY Buffalo State, Buffalo, NY, USA

Contents

Invited Talk

Digital Analytical Geometry: How Do I Define a Digital Analytical Object?

Eric Andres[✉]

Laboratoire XLIM, SIC, UMR CNRS 7252, Université de Poitiers,
BP 30179, 86962 Futuroscope Chasseneuil, France
eric.andres@univ-poitiers.fr

Abstract. This paper is meant as a short survey on analytically defined digital geometric objects. We will start by giving some elements on digitizations and their relations to continuous geometry. We will then explain how, from simple assumptions about properties a digital object should have, one can build mathematically sound digital objects. We will end with open problems and challenges for the future.

Keywords: Digital analytical geometry · Digital objects

1 Introduction

Geometry is historically the field of mathematics dealing with objects and their properties: length, angle, volume, shape, position and transform. The word Geometry stems from the ancient greek words for Earth and Measure. Geometry was the science of shapes and numbers as practical tool for measuring fields, distances between far away places, volumes for commerce, etc. For centuries, properties were proven and geometric objects were constructed based on construction rules. Euclid with his manuscripts *Elements*, revolutionized geometry with his formalization of abstract reasoning in mathematics and more significantly in geometry. The second revolution was brought upon by René Descartes with the introduction of coordinates. This marked a profound change in the way geometry was considered. It established a link between Euclidean geometry and algebra: Analytical Geometry was born. Many advances were now possible in astronomy, physics, engineering, etc. Many different forms of geometries have since been proposed such as Differential geometry, Algebraic geometry, etc.

Digital Geometry is one of the most recent forms of geometry. It can be broadly defined as the geometry of digital objects and transforms in a digital space. In this paper we are mainly considering digital points with integer coordinates (points in $\mathbb{Z}^n$). Digital Geometry has the particularity of, usually, not being an independent geometry but a digital counterpart of Euclidean geometry. Digital objects are supposed to behave and look as much as possible as their continuous counterpart. This question of representing/coding the continuous world in a finite computer is, of course, not limited to digital geometry. From the beginning, when sensors went from analog to digital and when the display mode went

© Springer International Publishing Switzerland 2015
R.P. Barneva et al. (Eds.): IWCIA 2015, LNCS 9448, pp. 3–17, 2015.
DOI: 10.1007/978-3-319-26145-4_1

from continuous (vector monitor) to digital (raster graphics), the fundamental question of object and space definition has been raised. It proved more elusive than initially thought [49]. Elementary rules of topology or geometry, that seem so obvious that they have been raised to the *axiomatic status* by Euclid, have proven to be false in Digital Geometry [20]: two, non identical, parallel 2D digital straight lines can have an infinite number of intersection points while two orthogonal 2D digital straight lines may have no intersection point. Particular versions of the Jordan theorem had to be divised that are in some sense specific to digital geometry [55].

This confrontation between the digital and the continuous worlds has given birth to various theories. One way of solving this hiatus is to consider the digital information as a sampled version of continuous information. The digital world is an approximation where information has been lost. Signal Theory provides the theoretical toolkit. Although one of the most efficient approaches when it comes to handling digital information (image processing, image analysis), it does little in helping defining actual geometry. It does not really provide any tool if one wants to draw, for instance, a line on a screen. We are considering another approach that finds its origins in the question of drawing digital equivalents of continuous objects on a raster screen (or earlier on, on a plotter). Digital Geometry is, in this sense, more closely linked to computer graphics or arithmetics. As for the continuous geometry, digital geometry started out focusing on very concrete and basic questions: how can one generate a digital analog of a continuous object for visualization purposes? This algorithmic approach has prevailed for many decades, with algorithms such as the Bresenham Digital Straight line drawing algorithm or Arie Kaufman et al. that proposed many digital primitive generation algorithms [40–42,47,48,61]. The main drawback of such an algorithmic approach is that it is difficult to ensure global properties from the local construction scheme. The other problem with a definition by construction is that you can only generate finite digital objects. As an alternative, researchers tried to describe and categorize digital objects not as a result of an algorithm but as digital classes with properties, be it geometrical or, more generally, topological [34,38,44,45,51,56]. This allows to define (classes of) digital objects that are infinite and without boundaries such as planes or surfaces in general. This approach proved useful to construct object classes with desired properties but it proved difficult to ensure tightness for the classes. And, as for the continuous geometry, analytical characterization of digital objects has proven to be effective in describing objects and the related transforms. It is a bit early to claim that it will revolutionize Digital Geometry but it allowed new insight and brought new tools for the definition of digital objects, in pattern recognition and design of digital transforms. Consider this paper as a short introduction paper into digitization transforms in general and Digital Analytical Geometry in particular.

In Sect. 2, we are going to discuss different types of digitizations. In Sect. 3 we are going to focus on digital analytical objects. We will then conclude and propose some perspectives.

2 Digitization

2.1 Notations

Let us denote n the dimension of space (digital or Euclidean) in this paper. Let $\{e_1,\ldots,e_n\}$ denote the canonical basis of the n-dimensional Euclidean vector space and O the center of the associated geometric coordinate system. Let $\mathbb{Z}^n$ be the subset of $\mathbb{R}^n$ that consists of all the integer coordinate points. A *digital (resp. Euclidean) point* is an element of $\mathbb{Z}^n$ (resp. $\mathbb{R}^n$). We denote by x_i the i-th coordinate, associated to e_i, of a point or a vector x. A *digital (resp. Euclidean) geometric object* is a set of digital (resp. Euclidean) points. A *digital inequality* is an inequality with coefficients in $\mathbb{R}$ from which we retain only the integer coordinate solutions. A *digital analytical object* is a digital object defined as union and intersection of a finite set of digital inequalities. The family of sets over $\mathbb{Z}^n$ (resp. $\mathbb{R}^n$) is denoted $\mathfrak{P}\,(\mathbb{Z}^n)$ (resp. $\mathfrak{P}\,(\mathbb{R}^n)$). A digitization is a transform from sets in the Euclidean to sets in the digital world: $\Delta : \mathfrak{P}\,(\mathbb{R}^n) \to \mathfrak{P}\,(\mathbb{Z}^n)$.

For all $k \in \{0,\ldots,n-1\}$, two integer points v and w are said to be *k-adjacent* or *k-neighbors*, if for all $i \in \{1,\ldots,n\}$, $|v_i - w_i| \le 1$ and $\sum_{j=1}^{n} |v_j - w_j| \le n-k$. In the 2-dimensional plane, the 0- and 1-neighborhood notations correspond respectively to the classical 8- and 4-neighborhood notations. In the 3-dimensional space, the 0-, 1- and 2-neighborhood notations correspond respectively to the classical 26- ,18- and 6-neighborhood notations [5,6,55].

A *k-path* is a sequence of integer points such that every two consecutive points in the sequence are k-adjacent. A digital object E is *k-connected* if there exists a k-path in E between any two points of E. A maximum k-connected subset of E is called a *k-connected component*. Let us suppose that the complement of a digital object E, $\mathbb{Z}^n \setminus E$ admits exactly two k-connected components F_1 and F_2, or in other words that there exists no k-path joining integer points of F_1 and F_2, then E is said to be *k-separating* in $\mathbb{Z}^n$. If there is no path from F_1 to F_2 then E is said to be 0-separating or simply *separating*. A point v of a k-separating object E is said to be a *k-simple point* if $E \setminus \{v\}$ is still k-separating. A k-separating object that has no k-simple points is said to be *strictly k-separating*. The notion of k-separation is defined for digital surfaces without boundaries. See [24] for more general notions.

For A and B two subsets of $\mathbb{R}^n$, $A \oplus B = \{a + b : a \in A, b \in B\}$ is the Minkowski sum of A and B. Let us denote $\breve{A} = \{-a : a \in A\}$ the reflection set of A. Let us denote $\overline{A}$ the flat of smallest dimension containing A. For a distance d, then the let us denote $\mathcal{B}_d(r) = \{x \in \mathbb{R}^n : d(x,O) \le r\}$, the ball of radius r for the distance d. Let us denote d_1, d_2, d_∞ respectively the Manhattan, Euclidean and Chebychev distance. Let us denote $\|x\|_k$ the corresponding norm (with $k = 1,2,\infty$).

2.2 General Remarks on Digitizations

Let us first start with some general remarks about digitization methods. The digitization of objects is fundamentally an ill-defined problem [49]: any digital objects can be considered as the digitization of any continuous object. Usually the goal is to have digital objects that *ressemble* the continuous object. The resulting digital objects may keep some, but not all, properties of the continuous object [21,24,52,53]. See [21,53] for a more formal presentation of a link between the continuous and the digital worlds based on non-standard analysis.

A digitization is defined broadly as a transform from the family of Euclidean sets to the family of the digital sets. However, most of the literature deals with digital objects defined as digitization of specific classes of geometric objects [1,2, 9,12,13,15,26,27,30–32,40–42,47,48,61–63]: for instance, the Bresenham digital straight line segment generation algorithm [12] works only for continuous straight line segments between two digital points. In this case, the digitization transform is usually implicit. The fact that the digitization scheme is not explicitely defined is also an important problem for pattern recognition: comparing two digital circle recognition algorithm supposes that the underlying digital circles are defined in the same way or otherwise it is like comparing apples to oranges. Other digitization transforms are defined only for linear objects [5,6] and others still for all objects [7].

Let us mention some classes of digitization transforms that are important: A *general* digitization is a digitization that is defined for all continuous objects. A *coherent* digitization transform Δ verifies the following property $E \subset F \Rightarrow \Delta(E) \subset \Delta(F)$.

2.3 Morphological Digitizations

Let us build a narrative for the construction of a general, coherent digitization transform Δ. For a geometric object E, how can we build its digital counterpart $\Delta(E)$ that *ressembles* E? Simply considering that $\Delta(E) = E \cap \mathbb{Z}^n$ is not a good idea. There are no particular reasons for E to pass through digital points and we may end up with $\Delta(E) = \varnothing$. So let us consider points that are *close* to E:

$$\Delta(E) = \{p \in \mathbb{Z}^n : d(p, E) \leq r\}, \text{ where } d \text{ is a distance and } r \in \mathbb{R} \quad (1)$$

There are some important immediate properties that go with such a definition: $\Delta(E \cup F) = \Delta(E) \cup \Delta(F)$ and $E \subset F \Rightarrow \Delta(E) \subset \Delta(F)$, which is a stronger version of the coherence property. These are fundamental properties when it comes to digital modeling of complex objects. It defines a general, coherent digitization transform. There are two parameters to work with: the distance d and a *thickness* parameter r. Let us note that the parameter r can also be defined as a function. See [9,32,63] for examples of digital objects defined with a non-constant thickness. Considering the points that are close to the original continuous object seems reasonable if we want the digital object to look like the original. There are also theoretical reasons for such a choice [21,53].

If a point p verifies $d(p, E) \leq r$ then a ball $\mathcal{B}_d(r)$ of radius r, for the distance d, centered on p intersects E which leads to the following formulation:

$$\Delta(E) = \{p \in \mathbb{Z}^n : (\mathcal{B}_d(r) \oplus p) \cap E \neq \varnothing\} \quad (2)$$

This type of digitization method is part of digitization methods called *morphological digitization* [37,46,54,59,60] with $\mathcal{B}_d(r)$ as structuring element.

Classically, the distances that have been considered are the Manhattan, the Euclidean and the Chebychev distances. An interesting set of distances well adapted for digitization transforms is the set based on *adjacency norms* [63]. Every digital adjacency relationship can be associated to a norm.

Definition 1. *For an integer k, $0 \le k < n$, the k-adjacency norm $[\cdot]_k$ is defined as follows:* $\forall x \in \mathbb{R}^n$, $[x]_k = \max\left\{ \|x\|_\infty, \frac{\|x\|_1}{n-k} \right\}$.

These distances are interesting because they verify the following property [63]: Let $p, q \in \mathbb{Z}^n$, then, p and q are k-adjacent *iff* $[p - q]_k \le 1$. See Fig. 1 for adjacency distance balls.

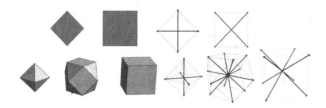

Fig. 1. 2D and 3D balls for the adjacency distances and the corresponding Flakes [63].

For morphological digitizations [37,43,46], the structuring element is not necessarily a distance ball as in formula (2). One can consider any continuous object F as structuring element and define a digitization transform of a continuous object E by [46]:

$$\Delta(E) = \left\{ p \in \mathbb{Z}^n : \left(\check{F} \oplus p \right) \cap E \neq \varnothing \right\} \qquad (3)$$

The region $\left\{ x \in \mathbb{R}^n : \left(\check{F} \oplus x \right) \cap E \neq \varnothing \right\}$ is called the *offset region*. Formulation (3) has implicitly already been used in digitizations such as the grid intersection digitization [43] with half-open structuring elements. This is also the starting point for the analytical characterization of digital objects with the analytical description of the offset region. Note that, for an arbitrary structuring element F, it is the reflection $\check{F}$ that appears in formula (3) (Fig. 2).

3   Analytical Characterization of Digital Objects

Let us first define what we understand by analytical characterization of a digital object: a digital object is defined by a set of equations (inequalities typically). A point belongs to the digital object iff it verifies the set of equations. The cardinality of the set of equations should be independent of the number of digital points of the object. The analytical characterization of digital objects has a great

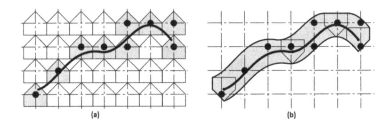

Fig. 2. This figure has been proposed in [46]. (a) $\{p \in \mathbb{Z}^2 : F \oplus p \neq \varnothing\}$ (b) $(\check{F} \oplus E) \cap \mathbb{Z}^2$. The region in gray in (b) is called the offset zone.

interest in digital geometry. A digital object is defined in *comprehension* and not as a voxel enumeration. Infinite digital objects can be represented. This was also one of the reasons for trying to define digital objects based on topology [34,38,44,45,51,56]. The key to the analytical characterization is that it allows a characterization of digital objects with *interesting* topological properties.

Since Reveilles proposed the analytical characterization of digital straight lines [52], many papers have been proposed that describe or discuss properties of analytical digital objects. Those papers can be roughly classified into two groups:

- Direct defined Analytical Digital Object: Papers that introduce an analytical definition of digital objects or classes of objects, or that analytically characterize previously known digital objects. Those objects are defined *directly* in the digital space without being explicitly associated to a digitization transform.
- Digitized Analytical Objects: papers that introduce a digitization transform that allows an analytical characterization of digital objects.

3.1 Direct Defined Analytical Digital Objects

Let us first list some of the digital objects that have been *directly* analytically defined in the digital space without an explicit reference to a digitization transform. The list is of course not exhaustive.

Digital Analytical Hyperplane: The first class of digital object that has been analytically characterized has been the digital straight 2D line [19,25]. It was J-P. Reveilles that proposed an analytical description of a Digital Straight Line (DSL) $0 \leq ax - by + c < \omega$ [52] with a thickness parameter ω that allows a parametrization of its topology. He also made an explicit link between digital straight lines, topology, quasi-affine transforms and arithmetics [10,23,39,52]. Many papers have been devoted to its study. Indeed, the structure of digital straight lines is rich, with immediate links to word theory, the Stern-Brocot tree, the Farey sequence, etc. It allows a natural extension to higher dimensions [1,27,52] with the analytical characterization of digital hyperplanes:

$$H : 0 \leq a_0 + \sum_{i=1}^{n} a_i x_i < \omega. \tag{4}$$

See [18,43] for a survey of digital linearity and planarity with interesting historical perspectives and useful comments and references on digital analytical lines and hyperplanes. An important step in bringing different theoretical approaches together, was to establish a link between the thickness of digital hyperplanes and topology [1]: let us assume, w.l.o.g. that $0 \leq a_1 \leq \ldots \leq a_n$, the digital hyperplane $0 \leq a_0 + \sum_{i=1}^{n} a_i x_i < \omega$ is k-separating iff $\omega \geq \sum_{k+1}^{n} a_i$. With $\omega = \sum_{k+1}^{n} a_i$ the digital hyperplane is strictly k-separating, without simple points. Papers have been devoted to the study of different classes of digital hyperplanes such as naive hyperplanes [1], supercover hyperplanes [3,4,7], Graceful lines and planes [15,16], etc. An interesting sequence of papers has focused on the connectivity of digital analytical hyperplanes [8,17,39]. The problem proved to be quite difficult when it comes to digital analytical (hyper)planes with irrational coefficients. Several papers have dealt with topology especially in order to define a notion of digital surface [33,34].

Digital Analytical Hyperplanes have been defined as purely analytical digital objects. It is however quite easy to associate a digitization transform to digital analytical hyperplanes. The most obvious way is to center a digital hyperplane on the continuous hyperplane: for $H : a_0 + \sum_{i=1}^{n} a_i x_i = 0$, we define $\Delta(H) = \{p \in \mathbb{Z}^n : \frac{\omega}{2} \leq a_0 + \sum_{i=1}^{n} a_i x_i < \frac{\omega}{2}\}$. Note that the Bresenham line [12] is such a centered Reveilles line [52]. There is the question of orientation of the digital hyperplane: with a definition such as $0 \leq a_0 + \sum_{i=1}^{n} a_i x_i < \omega$, on which side do we put the "'$\leq$'" and the "$<$". One can easily switch side and obtain $0 < \omega - a_0 + \sum_{i=1}^{n} (-a_i) x_i \leq \omega$, so a choice has to be made. This question is somewhat difficult if we want coherent digitization models, so let us focus a moment on so called closed analytical digital hyperplanes $0 \leq a_0 + \sum_{i=1}^{n} a_i x_i \leq \omega$ (with two "$\leq$"). Let us suppose that we have a digitization transform Δ that is defined for hyperplanes such that, for a continuous hyperplane $H : a_0 + \sum_{i=1}^{n} a_i x_i = 0$, we have $\Delta(H) = \{p \in \mathbb{Z}^n : \frac{\omega}{2} \leq a_0 + \sum_{i=1}^{n} a_i x_i \leq \frac{\omega}{2}\}$. Under some conditions, it is possible to take this as a starting point for the construction of a general, coherent morphological digitization transform:

Definition 2. *For some classes of digitization transforms Δ defined for hyperplanes, one can extend Δ as a general and coherent morphological digitization with a structuring element $\Delta(O)$ that is defined by:*

$$For \ x \in \mathbb{R}^n, \Delta(O) = \bigcap_{\forall H \supset O} \Delta(H).$$

The idea behind this definition is basically the following: For a digitization transform to be coherent, it has to verify the condition $E \subset F \Rightarrow \Delta(E) \subset \Delta(F)$. $\Delta(O)$ has to belong to the digitization of all the hyperplanes that pass through the coordinate center O. If we consider the equality, we basically define the digitization of a point which in this case can serve as structuring element for the morphological digitization transform. The difficulty lies in the choice of ω

for the digitization transform: for a hyperplane H, we want $\Delta(H)$ to be equal to $\bigcup_{x \in H} \Delta(x)$ and that is of course not true for any random choice of ω. There are classes of digital hyperplane thickness that work, namely those that correspond to the optimal hyperplane thickness for it to be k-separating: ω is equal to the sum of the absolute values of the $n - k$ biggest coefficients of H. These thicknesses correspond to the adjacency norm $[.]_k$ based digitization transforms. It is interesting to note that, for these digitizations, the structuring element is a polytope and therefore all the linear objects, at least, can be described analytically as linear digital objects (with linear inequalities). The best known of such digitization transforms is the *Supercover* model [3,4,7,20,22,24,43,46, 57,59]. One other thickness that works is $\omega = \sqrt{\sum_{i=1}^{n} a_i^2}$. The corresponding structuring element $\Delta(O)$ is the unit hypersphere. The associated norm is the Euclidean norm. What other thicknesses work is an interesting open question.

Andres Hypersphere: The second class of digital objects that have been defined directly as digital objects are the so called Andres hyperpsheres [2,63]: $S = \left\{ x \in \mathbb{Z}^n : \omega_1 \leq \sum_{i=1}^{n} (x_i - c_i)^2 < \omega_2 \right\}$ where c is the center of the digital hypersphere and $\sqrt{\omega_2} - \sqrt{\omega_1}$ its (Euclidean) thickness. The same method (as for the hyperplanes) of centering the spherical shell can be used to associate a digitization transform. The Andres hypersphere has been proposed to overcome the limitation of the Bresenham circle [13] in particular that is only defined for integer radius, integer coordinate center and that, at the time, did not have an analytical characterization. There is one now [9,63]. An interesting property of such Andres hyperspheres is that concentric Andres hyperspheres pave digital space. This is quite useful for applications such as simulation of wave propagation [50].

nD Straight Lines: Flats in general have not been studied that much with the notable exception of straight lines: 2D analytical lines [52], 3D analytical lines [28,31], graceful lines [16], analytical nD lines [30]. The study of Digital Analytical Lines has gained a lot of traction in the arithmetical community [11] for its link to word theory. It is interesting to note that I. Debled-Rennesson's 3D line is defined as the intersection of two orthotropic naive 3D planes (thinnest planes without 6-connected holes) and thus is an analytically defined 26-connected object. However, contrary to what one could think, the 3D line one would obtain by considering naive planes and intersecting them to define a morphological digitization is usually not 26-connected. The choice of the two planes among three possible orthotropic planes depends on the orientation of the 3D line. I am not quite sure that there exists a corresponding 3D plane thickness (and thus a corresponding general digitization transform) that would define such digital 3D lines. It is an interesting question and it shows that direct analytical definitions for digital objects may lead to interesting topological properties.

Other Purely Analytically Defined Digital Objects: There are other analytically defined objects that could be considered as purely analytically defined digital objects. Let us just mention some approaches that are particularly interesting: The team around I. Debled-Rennesson proposed the notion of *Blurred* analytical objects [29] with applications in noisy digital object recognition. E. Andres, M. Rodriguez et al. proposed a notion of analytically characterized digital perpendicular bisector [8] which allowed to tackle the problem of the computation of a circumcenter of several pixels and the recognition of fuzzy circles. One could add Y. Gerard and L. Provost that proposed a notion of analytically defined curves and surfaces, named Digital Level Layers [36]. Although based on a morphological digitization, the objects are purely analytically defined.

3.2 Digitized Analytical Objects

In this section, we are going to take a look at digitized objects that have been analytically characterized. An immediate example is the Bresenham Straight line Segment [12] that has been shown to be a Reveilles straight line segment [52]. In the same way, in [9], most notions of digital circles that have been introduced have been analytically characterized [13,47]. An extension to higher dimensions has been proposed in [63] with an explicit mention of Morphological Digitizations. Let us start with morphological digitization transforms.

Supercover Digitization: One of the first analytically characterized digitization model that has been proposed is the *supercover digitization* (also called outer *Jordan digitization* [22,43]) based on the Chebychev distance d_∞ [3,4,7,20,22,24,43,46,57,59]. The supercover digitization is well-known for a long time because it has a natural geometric interpretation. The unit ball for the distance $\mathcal{B}_{d_\infty}\left(\frac{1}{2}\right)$ is a hypercube of side one. If we denote $\mathcal{V}(p)$ the voxel centered on p, Formula (2) for the Chebychev distance is the same as $\{p \in \mathbb{Z}^n : \mathcal{V}(p) \cap E \neq \varnothing\}$: a point belongs to the supercover of a continuous object E iff the corresponding voxel is cut by E. The union of all the voxels of the supercover of a continuous object covers the continuous object, thus the name *supercover*. This geometric interpretation is so natural that it has been considered long (actually as early as the 19th century [22]) before the link to the Chebychev distance has been made. We will not recall all the details on the supercover model: see [24] for general properties of the digitization transform. In [3,4,7] for the analytical characterization of the supercover digitization of m-simplice and m-flats in dimension n. In [63], the reader will find an analytical characterization of supercover 2D circles and 3D spheres.

Standard Digitization: The supercover digitization transform has many interesting topological properties. In particular, a supercover digitization of a connected object is always $(n-1)$-connected and tunnel-free but not strictly separating. When E crosses and edge or a vertex of a grid voxel then all the grid points whose voxel share this edge or vertice belong to the digitization.

This is called a bubble [3,4,7]. The supercover of a hyperplane, for instance, is $(n-1)$-connected but with possibly simple points. For theoretical [33–35] as well as practical reasons, it is interesting to have a model without bubble. Various methods have been proposed to solve this problem such as modifying the definition of a voxel [24] but that does not work [5,6]. There is however a way to solve this problem [5,6]. The idea is the following: the supercover $\mathbb{S}(H)$ of a hyperplane $H : a_0 + \sum_{i=1}^{n} a_i x_i = 0$ is given by $\mathbb{S}(H) : -\frac{\sum_{i=1}^{n} |a_i|}{2} \leq a_0 + \sum_{i=1}^{n} a_i x_i \leq \frac{\sum_{i=1}^{n} |a_i|}{2}$. It is $(n-1)$-connected, tunnel-free but it might have simple points (bubbles). The analytical hyperplane $-\frac{\sum_{i=1}^{n} |a_i|}{2} \leq a_0 + \sum_{i=1}^{n} a_i x_i < \frac{\sum_{i=1}^{n} |a_i|}{2}$ is $(n-1)$-connected, tunnel-free and strictly separating (without bubbles). The only difference comes from the "$\leq$" for the hyperplane supercover that is replaced by a "$<$" for the analytical hyperplane. So transforming one into the other comes down to choosing a side on which we change a "$\leq$" into a "$<$". We define therefore an *orientation convention*: A halfspace $H : a_0 + \sum_{i=1}^{n} a_i x_i \leq 0$ is said to have a *standard orientation* iff $a_1 > 0$ or $a_1 = 0$ and $a_2 > 0$ or ... if $a_1 = ... = a_{n-1} = 0$ and $a_n > 0$. Otherwise the halfspace is said to have a *supercover orientation*.

Since the defining structuring element for the supercover digitization transform is a unit hypercube, it is easy to see that the offset zone for a supercover linear object is a polytope defined as intersection of a finite sequence of digital half-spaces $\mathbb{S}(E) = \left\{ p \in \left(\bigcap_{i=1}^{k} H_i \right) \cap \mathbb{Z}^n; H_i : a_{i,0} + \sum_{j=1}^{n} a_{i,j} x_j \leq 0 \right\}$ where k is the cardinality of the set of halfspaces $\{H_i\}$ defining the supercover of E. For such a set of halfspaces, we replace each halfspace $H_i : a_{i,0} + \sum_{j=1}^{n} a_{i,j} x_j \leq 0$ that has a standard orientation by $H_i' : a_{i,0} + \sum_{j=1}^{n} a_{i,j} x_j < 0$ in the analytical characterization of the digital object. If the halfspace has a supercover orientation, it is not modified. This defines the *standard digitization* transform $\mathbb{S}t(E)$ of a linear Euclidean object E. It has been shown in [14] that the standard digitization produces $(n-1)$-connected, tunnel-free and strictly separating objects. See Figure 3 for examples of the standard digitization of points and a 3D triangle. The standard model keeps most of the properties of the supercover model and as such is a coherent digitization. It is not general however as it is defined only for linear objects. There is however a caution. Contrary to the supercover digitization, in general, $\mathbb{S}t(E) \neq \bigcup_{x \in E} \mathbb{S}t(x)$. The standard digitization is defined as a finite rewriting of the inequalities defining the supercover of a linear object. It does not hold for an infinite sequence of inequalities.

Grid Intersection Digitization: A popular digitization scheme is called *grid intersection digitization* [57]. For a continuous object E, the intersection points of E and the grid lines (all the straight lines $x_i = k, k \in \mathbb{Z}$) are considered and the closest grid point to these intersection points forms the digital object. This is the same as considering a structuring element corresponding to the set of polygons with vertices $\left(0, \ldots, 0, \pm\frac{1}{2}, 0, \ldots, 0, \pm\frac{1}{2}, 0, \ldots, 0 \right)$. It is very similar to the digitization with the Manhattan distance d_1. While the unit ball for this distance is a diamond shaped polytope with all the above mentioned points as vertices. The digitization is defined for all k-dimensional objects, $k > 0$.

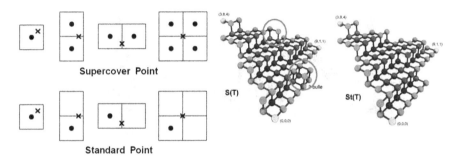

Fig. 3. Standard and Supercover digitization of points on the left and digitization of a 3D triangle on the right.

Analytical characterization can be obtained by computing the intersection of the object with one of the orthotropic faces of the structuring element or by determining the analytical charcterization of the d_1-distance digitization. The Bresenham line [12] is such an object and its characterization has been given in [52] by JP. Reveilles. In [9,63] there is the analytical characterization of d_1 digital circles and spheres.

Flake Digitization [58,62]: The analytical characterization of the supercover of a sphere S is quite complicated [63]. Most (in the geometric sense) of the offset region corresponds however simply to a translation of the continuous sphere S. Indeed, the outer and inner boundary of $\mathcal{B}_{d_\infty} \oplus S$ is in great part determined by the vertices of the ball. Let us call V_∞ the set of vertices of $\mathcal{B}_{d_\infty}$ then $V_\infty \oplus S$ corresponds largely to the same surface than the boundary of $\mathcal{B}_{d_\infty} \oplus S$. If we consider a structuring element F composed of straight line segments that join the vertices v of $\mathcal{B}_{d_\infty}$ to its reverse $\check{v}$ then $F \oplus S$ is $(n-1)$-connected and tunnel-free if S is big enough (details of S need to be bigger than a voxel [58,62]). This is true, not only for the supercover model but for all structuring elements that are polytopes, especially those corresponding to adjacency norms. The distinctive advantage is that this digitization transform is very simple to characterize analytically if the surface S is defined by an implicit equation $f(x) = 0$ such that there is a side of the surface where $f(x) < 0$ and a side where $f(x) > 0$. Let us suppose we have a surface S defined by such an implicit equation $f(x) = 0$, $x \in \mathbb{R}^n$. Let us suppose that we have a structuring element F which is a polytope, with central symmetry (for the sake of simplicity here). The vertices of F form the set v_i. Let us define the *Flake* F' formed by the straight lines joining the vertices v_i to its symmetric $\check{v}_i$ (See Fig. 1). Then $(F' \oplus S) \cap \mathbb{Z}^n$ is analytically characterized by:

$$\left\{ p \in \mathbb{Z}^n : \min_{i=1}^{n} \left(f(v_i) \right) \leq 0 \wedge \max_{i=1}^{n} \left(f(v_i) \right) \geq 0 \right\} \tag{5}$$

The idea is actually very simple: as morphological digitization, the surface S cuts a structuring element $F' \oplus p$ iff there are vertices on each side of the surface

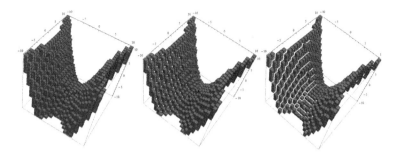

Fig. 4. Flake digitizations of the quadric $9x^2 - 4y^2 - 36z - 180 = 0$.

defined by the implicit equation. The so-defined *Flake digitization* transform $(F' \oplus S) \cap \mathbb{Z}^n$ is similar to $(F \oplus S) \cap \mathbb{Z}^n$ except may be on places where S does not fit some regularity properties [62]. The flake digital object keeps the topological properties of the original object. This is a way of defining implicit digital objects is straightforward way with the limitation that it is defined only for $(n-1)$-dimensional surfaces that are *regular enough*. See Fig. 4 for an example of a implicitly defined quadric digitized with all three 3D adjacency flakes.

4   Conclusion and Perspectives

In this paper we propose a short survey on digital analytical geometry and show what the ideas are behind the analytical characterization of digital objects. There are two key points in *digital analytical geometry* that we have not addressed in this paper due to space: transforms and object recognition. Both profit greatly of the analytical characterizations of digital objects. For the transforms, let us just cite the Quasi-Affine Transforms [23] among many others. For Object Recognition, having mathematical definitions of objects changes many things. Much has not been said and many papers have been omitted in this short survey. We have proposed several open questions along the pages of this article and many others still remain. As concluding words, let us not forget that beyond digital analytical geometry, there are many other forms of digital geometry that still need to be invented or explored: parametric digital geometry, non-Euclidean digital geometry, multiscale digital geometry, etc.

References

1. Andres, E., Acharya, R., Sibata, C.: Discrete analytical hyperplanes. GMIP **59**(5), 302–309 (1997)
2. Andres, E., Jacob, M.A.: The discrete analytical hyperspheres. IEEE Trans. Vis. Comp. Graphics **3**(1), 75–86 (1997)
3. Andres, E., Nehlig, P., Françon, J.: Supercover of straight lines, planes and triangles. In: Ahronovitz, E., Fiorio, C. (eds.) DGCI 1997. LNCS, vol. 1347. Springer, Heidelberg (1997)

4. Andres, E., Nehlig, P., Francon, J.: Tunnel-free supercover 3D polygons and poly-hedra. In: Eurographics 1997. Computer Graphics Forum, vol. 16, pp. C3–C13 (1997)
5. Andrès, É.: Defining discrete objects for polygonalization: the standard model. In: Braquelaire, A., Lachaud, J.-O., Vialard, A. (eds.) DGCI 2002. LNCS, vol. 2301, pp. 313–325. Springer, Heidelberg (2002)
6. Andres, E.: Discrete linear objects in dimension n: the standard model. Graph. Models **65**(1–3), 92–111 (2003)
7. Andres, E.: The supercover of an m-flat is a discrete analytical object. Theor. Comput. Sci. **406**(1–2), 8–14 (2008)
8. Andres, E., Largeteau-Skapin, G., Rodríguez, M.: Generalized perpendicular bisec-tor and exhaustive discrete circle recognition. Graph. Models **73**(6), 354–364 (2011)
9. Andres, E., Roussillon, T.: Analytical description of digital circles. In: Debled-Rennesson, I., Domenjoud, E., Kerautret, B., Even, P. (eds.) DGCI 2011. LNCS, vol. 6607, pp. 235–246. Springer, Heidelberg (2011)
10. Berthé, V., Jamet, D., Jolivet, T., Provençal, X.: Critical connectedness of thin arithmetical discrete planes. In: Gonzalez-Diaz, R., Jimenez, M.-J., Medrano, B. (eds.) DGCI 2013. LNCS, vol. 7749, pp. 107–118. Springer, Heidelberg (2013)
11. Berthé, V., Labbé, S.: An arithmetic and combinatorial approach to three-dimensional discrete lines. In: Debled-Rennesson, I., Domenjoud, E., Kerautret, B., Even, P. (eds.) DGCI 2011. LNCS, vol. 6607, pp. 47–58. Springer, Heidelberg (2011)
12. Bresenham, J.: Algorithm for computer control of a digital plotter. IBM Syst. J. **4**(1), 25–30 (1965)
13. Bresenham, J.: A linear algorithm for incremental digital display of circular arcs. Commun. ACM **20**(2), 100–106 (1977)
14. Brimkov, V.E., Andres, E., Barneva, R.P.: Object discretizations in higher dimen-sions. Pattern Recogn. Lett. **23**(6), 623–636 (2002)
15. Brimkov, V.E., Barneva, R.P.: Graceful planes and thin tunnel-free meshes. In: Bertrand, G., Couprie, M., Perroton, L. (eds.) DGCI 1999. LNCS, vol. 1568, pp. 53–64. Springer, Heidelberg (1999)
16. Brimkov, V.E., Barneva, R.P.: Graceful planes and lines. Theor. Comput. Sci. **283**(1), 151–170 (2002)
17. Brimkov, V.E., Barneva, R.P.: Connectivity of discrete planes. Theor. Comput. Sci. **319**(1–3), 203–227 (2004)
18. Brimkov, V.E., Coeurjolly, D., Klette, R.: Digital planarity - a review. Discrete Appl. Math. **155**(4), 468–495 (2007)
19. Brons, R.: Linguistic methods for the description of a straight line on a grid. CGIP **3**(1), 48–62 (1974)
20. Chassery, J.M., Montanvert, A.: Géométrie discrète en imagerie. Ed. Hermès, Paris (1987)
21. Chollet, A., Wallet, G., Fuchs, L., Largeteau-Skapin, G., Andres, E.: Insight in discrete geometry and computational content of a discrete model of the continuum. Pattern Recogn. **42**(10), 2220–2228 (2009)
22. Jordan, C.: Remarques sur les intégrales définies. Journal de Mathématiques, 4ème série, T.8, pp. 69–99 (1892)
23. Coeurjolly, D., Blot, V., Jacob-Da Col, M.-A.: Quasi-Affine transformation in 3-D: theory and algorithms. In: Wiederhold, P., Barneva, R.P. (eds.) IWCIA 2009. LNCS, vol. 5852, pp. 68–81. Springer, Heidelberg (2009)
24. Cohen-Or, D., Kaufman, A.E.: Fundamentals of surface voxelization. CVGIP **57**(6), 453–461 (1995)

25. Coven, E.M., Hedlund, G.: Sequences with minimal block growth. Math. Syst. Theory **7**(2), 138–153 (1973)
26. Dachille, F., Kaufman, A.E.: Incremental triangle voxelization. In: Proceeding Graphics Interface, pp. 205–212. Canadian Human-Computer Communications Society, Montréal (2000)
27. Debled-Renesson, I., Reveillès, J.P.: A new approach to digital planes. In: SPIE Vision Geometry III, vol. 2356, Boston (1994)
28. Debled-Rennesson, I.: Etude et reconnaissance des droites et plans discrets, PhD Thesis. Ph.D. thesis, Université Louis Pasteur, Strasbourg, France (1995)
29. Debled-Rennesson, I., Remy, J., Rouyer-Degli, J.: Segmentation of discrete curves into fuzzy segments. Elect. Notes Discrete Math. **12**, 372–383 (2003)
30. Feschet, F., Reveillès, J.-P.: A generic approach for n-dimensional digital lines. In: Kuba, A., Nyúl, L.G., Palágyi, K. (eds.) DGCI 2006. LNCS, vol. 4245, pp. 29–40. Springer, Heidelberg (2006)
31. Figueiredo, O., Reveillès, J.: A contribution to 3D digital lines. In: 5th DGCI, pp. 187–198, Clermont-Ferrand (1995)
32. Fiorio, C., Jamet, D., Toutant, J.L.: Discrete circles: an arithmetical approach with non-constant thickness. In: Proceeding SPIE Vision Geometry XIV, vol. 6066, pp. 1–12 (2006)
33. Francon, J.: Arithmetic planes and combinatorial manifolds. In: 5th DGCI, pp. 209–217, Clermont-Ferrand (1995)
34. Francon, J.: Discrete combinatorial surfaces. CVGIP **57**(1), 20–26 (1995)
35. Francon, J.: Sur la topologie d'un plan arithmétique. Theor. Comput. Sci. **156**(1&2), 159–176 (1996)
36. Gérard, Y., Provot, L., Feschet, F.: Introduction to digital level layers. In: Debled-Rennesson, I., Domenjoud, E., Kerautret, B., Even, P. (eds.) DGCI 2011. LNCS, vol. 6607, pp. 83–94. Springer, Heidelberg (2011)
37. Heijmans, H.J.A.M.: Morphological image operators. Academy Press, Boston (1994)
38. Herman, G.T.: Discrete multidimensional jordan surfaces. CVGIP **54**(6), 507–515 (1992)
39. Jamet, D., Toutant, J.: Minimal arithmetic thickness connecting discrete planes. Discrete Appl. Math. **157**(3), 500–509 (2009)
40. Kaufman, A.E.: Efficient algorithms for 3D scan-conversion of parametric curves, surfaces, and volumes. In: Proceeding 14th SIGGRAPH, pp. 171–179 (1987)
41. Kaufman, A.E.: Efficient algorithms for scan-converting 3D polygons. Comput. Graph. **12**(2), 213–219 (1988)
42. Kim, C.E.: Three-dimensional digital line segments. IEEE Trans. PAMI **5**(2), 231–234 (1983)
43. Klette, R., Rosenfeld, A.: Digital straightness - a review. Discrete Appl. Math. **139**(1–3), 197–230 (2004)
44. Kong, T.Y., Rosenfeld, A.: Digital topology: introduction and survey. CVGIP **48**(3), 357–393 (1989)
45. Kovalesky, V.: Finite topology and image analysis. Adv. Electron. Electron Phys. **84**, 197–259 (1992)
46. Lincke, C., Wüthrich, C.A.: Surface digitizations by dilations which are tunnel-free. Discrete Appl. Math. **125**(1), 81–91 (2003)
47. McIlroy, M.D.: Best approximate circles on integer grids. ACM Trans. Graph. **2**(4), 237–263 (1983)
48. McIlroy, M.D.: Getting raster ellipses right. ACM Trans. Graph. **11**(3), 259–275 (1992)

49. Montanari, U.: On limit properties in digitization schemes. J. ACM **17**(2), 348–360 (1970)
50. Mora, F., Ruillet, G., Andres, E., Vauzelle, R.: Pedagogic discrete visualization of electromagnetic waves. In: Eurographics 2003, Interactive Demos and Posters, pp. 123–126 (2003)
51. Morgenthaler, D.G., Rosenfeld, A.: Surfaces in three-dimensional digital images. Inf. Control **51**(3), 227–247 (1981)
52. Reveillès, J.P.: Calcul en Nombres Entiers et Algorithmique. Ph.D. thesis, Université Louis Pasteur, Strasbourg, France (1991)
53. Reveillès, J., Richard, D.: Back and forth between continuous and discrete for the working computer scientist. Ann. Math. Artif. Intell. **16**, 89–152 (1996)
54. Ronse, C., Tajine, M.: Hausdorff discretization for cellular distances and its relation to cover and supercover discretizations. J. Vis. Commun. Image Represent. **12**(2), 169–200 (2001)
55. Rosenfeld, A.: Digital topology. Amer. Math. Monthly **86**, 621–630 (1979)
56. Rosenfeld, A., Kong, T.Y., Wu, A.Y.: Digital surfaces. GMIP **53**(4), 305–312 (1991)
57. Sankar, P.: Grid intersect quantization schemes for solid object digitization. Comput. Graphics Image Process. **8**(1), 25–42 (1978)
58. Sekiya, F., Sugimoto, A.: On connectivity of discretized 2D explicit curve. In: Mathematical Progress in Expressive Image Synthesis, Symposium MEIS 2014, pp. 16–25, Japan (2014)
59. Stelldinger, P., Terzic, K.: Digitization of non-regular shapes in arbitrary dimensions. Image Vision Comput. **26**(10), 1338–1346 (2008)
60. Tajine, M., Ronse, C.: Topological properties of hausdorff discretization, and comparison to other discretization schemes. Theor. Comput. Sci. **283**(1), 243–268 (2002)
61. Taubin, G.: Rasterizing algebraic curves and surfaces. IEEE Comput. Graphics **14**(2), 14–22 (1994)
62. Toutant, J.-L., Andres, E., Largeteau-Skapin, G., Zrour, R.: Implicit digital surfaces in arbitrary dimensions. In: Barcucci, E., Frosini, A., Rinaldi, S. (eds.) DGCI 2014. LNCS, vol. 8668, pp. 332–343. Springer, Heidelberg (2014)
63. Toutant, J., Andres, E., Roussillon, T.: Digital circles, spheres and hyperspheres: from morphological models to analytical characterizations and topological properties. Discrete Appl. Math. **161**(16–17), 2662–2677 (2013)

Theoretical Foundations of Combinatorial Image Analysis – Digital Geometry and Topology

Fuzzy Connectedness Segmentation: A Brief Presentation of the Literature

Gabor T. Herman[1](✉), T. Yung Kong[1,2], and Krzysztof Chris Ciesielski[3]

[1] Computer Science PhD Program, The Graduate Center,
City University of New York, 365 Fifth Avenue, New York, NY 10016, USA
gabortherman@yahoo.com
[2] Computer Science Department, Queens College, City University of New York,
65-30 Kissena Boulevard, Flushing, NY 11367, USA
ykong@cs.qc.cuny.edu
[3] Department of Mathematics, West Virginia University,
Morgantown, WV 26506, USA
kcies@math.wvu.edu

Abstract. For any positive integer M, M-object *fuzzy connectedness* (*FC*) *segmentation* is a methodology for finding M objects in a digital image based on user-specified *seed points* and user-specified functions, called (*fuzzy*) *affinities*, which map each pair of image points to a value in the real interval $[0, 1]$. FC segmentation has been used with considerable success on biomedical and other images. We provide a brief presentation of the literature on the topic of FC segmentation.

Keywords: Segmentation · Digital image · Fuzzy connectedness · Fuzzy affinity · Seed points

1 Introduction

Image segmentation is an important and challenging task for which a multitude of different techniques have been developed; see, e.g., Sect. 1.6 of [19] and the survey articles in Part IV of that book. Our paper deals with the segmentation methodology known as *fuzzy connectedness* (or *FC*) *segmentation*, which has been used with considerable success—see, e.g., Fig. 1—on biomedical and many other kinds of images [1–11, 13–18, 21–23, 25–33]. One typical example is [1], in which FC segmentation is used to delineate nodules in computerized tomography (CT) images of the lungs of patients. In medical imaging, FC segmentation was first developed by Udupa and Samarasekera [32]. Earlier uses of FC segmentation in an entirely different context (geophysical data processing) are reported in [8–11].

Much of the theory of FC segmentation has developed along two different tracks. In one of the tracks [1, 2, 13–18, 21, 27, 28, 31, 33] two kinds of segmentation are used: *relative fuzzy connectedness* (*RFC*) segmentation and *iterative relative fuzzy connectedness* (*IRFC*) segmentation. The other track [3–7, 22, 25]

© Springer International Publishing Switzerland 2015
R.P. Barneva et al. (Eds.): IWCIA 2015, LNCS 9448, pp. 21–30, 2015.
DOI: 10.1007/978-3-319-26145-4_2

uses a third kind of segmentation that is called *multi object fuzzy segmentation* (*MOFS*). In [12] we present a general theory of FC segmentation that encompasses both tracks and unifies them.

2   Basic Definitions

Let V be the set of all points of a digital image (so that V is finite and nonempty), let M be a positive integer, and let $S_1, \ldots, S_M$ be pairwise disjoint nonempty subsets of V. Then FC (short for *fuzzy connectedness*) *segmentation* can be understood as one method of identifying M subsets $O_1, \ldots, O_M$ of V such that $S_i \subseteq O_i \subseteq S_i \cup (V \setminus \bigcup_j S_j)$ for $1 \leq i \leq M$. Each of the sets $O_1, \ldots, O_M$ that is identified is called an *object*, and (for $1 \leq i \leq M$) each point in the originally specified set S_i is called a *seed point* or simply a *seed* for the ith object O_i. In many applications one of the M objects is called the *background*.

In a practical application, the sets $S_1, \ldots, S_M$ may be specified in any way appropriate for that application. For example, to specify the seed sets that resulted in the 4-object segmentation shown at the bottom of Fig. 1, the creator of that segmentation displayed on the computer screen the image shown at the top of Fig. 1 and, for $1 \leq i \leq 4$, used mouse clicks to indicate the locations of all the points in S_i. But this is not the only way to select seed points; for example, [1] describes procedures for selecting seed points automatically in a manner that is different for the background object and the other objects (that are all nodules in [1]). Automatic seed selection has the potential of reducing the time that the user needs to spend on providing input to the segmentation process.

In addition to using the term *FC segmentation* to refer to the process by which the objects $O_1, \ldots, O_M$ are found, we will also call the sequence of objects $O_1, \ldots, O_M$ an *FC segmentation* or an *M-object FC segmentation* of the set V of image points.

An FC segmentation is not necessarily a segmentation in the most typical sense because it is not necessarily a partition of the set V of image points: It is not required that the O_i be pairwise disjoint nor that their union be the whole of V. However, FC segmentation is the only kind of segmentation we discuss here, and we will often refer to FC segmentations as "segmentations."

The objects O_i that are found by FC segmentation depend on user-specified mappings called fuzzy affinities or just affinities. An *affinity* (on V) is a mapping $\psi : V \times V \to [0,1]$ such that $\psi(v,v) = 1$ for all $v \in V$. For all $u, v \in V$, we call the value $\psi(u,v) \in [0,1]$ the ψ-*affinity value* of (u,v).

An affinity on V may be regarded as an edge-weight function of the complete digraph (with loops) on V. Affinity values are described in [32] and elsewhere as (user-specified) measures of the "hanging togetherness" of pairs of image points.

One of the ways that RFC segmentation and IRFC segmentation differ from MOFS is that, for any seed sets $S_1, \ldots, S_M$, the RFC and IRFC segmentations of V are determined by a single affinity $\psi : V \times V \to [0,1]$, whereas the MOFS segmentation of V depends on M affinities $\psi_1, \ldots, \psi_M$ (i.e., one affinity for each of the M objects).

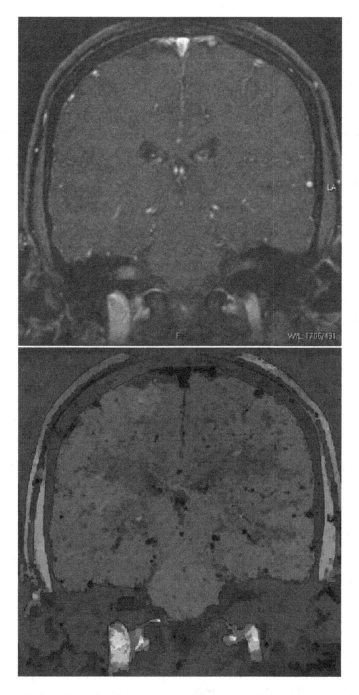

Fig. 1. Top: A slice of a patient's head obtained by magnetic resonance imaging (MRI). Bottom: A 4-object MOFS of the same slice. The number 4 reflects the fact that this FC segmentation aims at dividing up the image according to four tissue types (shown in red, green, blue, and yellow. (Reproduced from [6]) (Color figure online)

FC segmentation is unlikely to identify useful objects unless the affinity or affinities we use are appropriate for our application. The important problem of how to define appropriate affinities is discussed, e.g., in [3,7,14,15,24,25]. In MOFS each affinity ψ_i is quite frequently defined based on some statistical analysis of the image values assigned to points in neighborhoods of the seed points in S_i; see, for example, [5,6,25].

Given an affinity $\psi : V \times V \rightarrow [0,1]$ on V and $A, B, W \subseteq V$, a W-path from A to B of length l is any sequence $p = \langle w_0, \ldots, w_l \rangle$ of points in W such that $w_0 \in A$ and $w_l \in B$; the ψ-strength of $p = \langle v_0, \ldots, v_l \rangle$, denoted by $\psi(p)$, is defined by $\psi(p) = \min_{1 \leq k \leq l} \psi(v_{k-1}, v_k)$ if $l > 0$ and $\psi(p) = 1$ if $l = 0$; the ψ-strength of connectedness of $A \neq \emptyset$ to $B \neq \emptyset$ via W is defined as

$$\psi^W(A, B) = \max \{\psi(p) \mid p \text{ is a } (W \cup A \cup B)\text{-path from } A \text{ to } B\}. \quad (1)$$

For $a, b \in V$, a W-path from $\{a\}$ to $\{b\}$ will also be called a W-path from a to b. Similarly, we write $\psi^X(a, B)$, $\psi^X(A, b)$, and $\psi^X(a, b)$ for $\psi^X(\{a\}, B)$, $\psi^X(A, \{b\})$, and $\psi^X(\{a\}, \{b\})$, respectively. Note that $\psi(a, b) = \psi^\emptyset(a, b) \leq \psi^X(a, b) \leq \psi^V(a, b)$, that $\psi^X(A, B) = 1$ if $A \cap B \neq \emptyset$, and that $\psi^\emptyset(A, B) = \max_{a \in A, b \in B} \psi(a, b)$.

We say that the seed sets $S_1, \ldots, S_M$ are consistent with the affinities $\psi_1, \ldots, \psi_M$ if $\psi_i^V(S_i, S_j) < 1$, for all distinct i and j in $\{1, \ldots, M\}$. In other words, $S_1, \ldots, S_M$ are consistent with the affinities $\psi_1, \ldots, \psi_M$ if, and only if, for all distinct i and j in $\{1, \ldots, M\}$ and for every V-path $\langle v_0, \ldots, v_l \rangle$ from S_i to S_j, there exists a k, $1 \leq k \leq l$, such that $\psi_i(v_{k-1}, v_k) < 1$.

3   A Simple Multi Object Fuzzy Segmentation (MOFS) Algorithm

The following algorithm for computing the M-object MOFS of the set V for pairwise disjoint nonempty seed sets $S_1, \ldots, S_M \subset V$ and affinities $\psi_1, \ldots, \psi_M$ on V is not intended to be efficient. Rather, it is intended to be simple and concise, so as to give readers who are new to the subject a quick (yet completely accurate) understanding of the nature of the objects that are found by MOFS. A much more efficient (but less easily understood) method for computing MOFS is Algorithm 5 of [12], which is akin to Dijkstra's shortest-path algorithm [20], computes all of the objects simultaneously, and may be regarded as a simplified version of the MOFS algorithm of [6, Sect. 3].

Let $\Psi = \langle \psi_1, \ldots, \psi_M \rangle$ be any sequence of affinities on V and let $\mathcal{S} = \langle S_1, \ldots, S_M \rangle$ be any sequence of M pairwise disjoint nonempty subsets of V. Then we denote the MOFS segmentation $\langle O_1^{\text{MOFS}}, \ldots, O_M^{\text{MOFS}} \rangle$ that is produced by Algorithm 1 for affinities $\psi_1, \ldots, \psi_M$ and seed sets $S_1, \ldots, S_M$ by $\langle O_1^{\text{MOFS}}(\Psi, \mathcal{S}), \ldots, O_M^{\text{MOFS}}(\Psi, \mathcal{S}) \rangle$. We now proceed to give an alternative, non-algorithmic, characterization of this segmentation (in statements 1(a) and 1(b) below) and to state some related facts about the segmentation and Algorithm 1.

Algorithm 1. MOFS Segmentation of a Nonempty Finite Set V into M Objects

---

Data: M pairwise disjoint nonempty seed sets $S_1, \ldots, S_M \subset V$; M affinities $\psi_1, \ldots, \psi_M$ on V

Result: The MOFS segmentation $\langle O_1^{\mathrm{MOFS}}, \ldots, O_M^{\mathrm{MOFS}} \rangle$ of V

1 **for** $i \leftarrow 1$ **to** M **do** $T_i \leftarrow S_i$

2 **sort** $A = \bigcup_j \psi_j[V \times V] \setminus \{0\}$ into $1 = \alpha_1 > \ \ldots \ > \alpha_{|A|}$

3 **for** $n \leftarrow 1$ **to** $|A|$ **do**   /* the main loop */

4 $\quad$ **for** $i \leftarrow 1$ **to** M **do** $newT_i \leftarrow T_i \cup \{v \in V \setminus \bigcup_j T_j \mid \psi_i^{V \setminus \bigcup_j T_j}(T_i, v) \geq \alpha_n\}$

5 $\quad$ **for** $i \leftarrow 1$ **to** M **do** $T_i \leftarrow newT_i$

6 **for** $i \leftarrow 1$ **to** M **do** $O_i^{\mathrm{MOFS}} \leftarrow T_i$

Let $A = \bigcup_j \psi_j[V \times V] \setminus \{0\}$ and let $1 = \alpha_1 > \ \ldots \ > \alpha_{|A|}$ be the sequence obtained by sorting A into decreasing order. For $1 \leq i \leq M$ and $0 \leq n < |A|$, let T_i^n be the value of the variable T_i at the beginning of the $n + 1$st iteration of the main loop when Algorithm 1 is executed, and let $T_i^{|A|}$ be the value of T_i at the end of the $|A|$th iteration of the main loop (which is the value of T_i when Algorithm 1 terminates). Then

$$S_i \subseteq O_i^{\mathrm{MOFS}}(\Psi, \mathcal{S}) \subseteq S_i \cup (V \setminus \bigcup_j S_j)$$

for $1 \leq i \leq M$. Moreover:

1. For $1 \leq i \leq M$ we have that:
   (a) $T_i^0 = S_i$, and $T_i^n = T_i^{n-1} \cup \{v \in V \setminus \bigcup_j T_j^{n-1} \mid \psi_i^{T_i^{n-1} \cup (V \setminus \bigcup_j T_j^{n-1})}(S_i, v) = \alpha_n\}$ for $1 \leq n \leq |A|$.
   (b) $O_i^{\mathrm{MOFS}}(\Psi, \mathcal{S}) = T_i^{|A|}$.
2. $\{v \in V \setminus \bigcup_j T_j^{n-1} \mid \psi_i^{T_i^{n-1} \cup (V \setminus \bigcup_j T_j^{n-1})}(S_i, v) > \alpha_n\} = \emptyset$ for $1 \leq i \leq M$ and $1 \leq n \leq |A|$.
3. $T_i^n = T_i^{n-1} \cup \{v \in V \setminus \bigcup_j T_j^{n-1} \mid \psi_i^{T_i^k}(S_i, v) = \alpha_n\}$ for $1 \leq i \leq M$ and $1 \leq n \leq k \leq |A|$.

Now let us assume the seed sets $\mathcal{S} = \langle S_1, \ldots, S_M \rangle$ are consistent with the affinities $\Psi = \langle \psi_1, \ldots, \psi_M \rangle$. Then there is an arguably even more easily comprehended characterization of the MOFS segmentation

$$\langle O_1^{\mathrm{MOFS}}(\Psi, \mathcal{S}), \ldots, O_M^{\mathrm{MOFS}}(\Psi, \mathcal{S}) \rangle$$

than the characterization that is given by statements 1(a) and 1(b) above: The segmentation is the unique sequence of sets $\langle O_1, \ldots, O_M \rangle$ such that

$$O_i = \{v \in V \mid \max_{j \neq i} \psi_j^{O_j}(S_j, v) \leq \psi_i^{O_i}(S_i, v) \neq 0\} \text{ for } 1 \leq i \leq M. \quad (2)$$

That this is the case is stated (and proved) as part of Theorem 3.10 in [12]. The proof makes use of the concept (introduced in [12]) of a *recursively optimal* path.

Let us expand this characterization. Suppose that a sequence of sets

$$\langle O_1, \ldots, O_M \rangle$$

satisfies (2). Then, for $1 \leq i \leq M$, $S_i \subseteq O_i$ and, for any $v \in V$, $v \in O_i$ if, and only if:

1. There is an $(O_i \cup \{v\})$-path $\langle v_0, \ldots, v_l \rangle$ from S_i to v such that $\psi_i(v_{k-1}, v_k) > 0$ for $1 \leq k \leq l$. (This implies that $\psi_i^{O_i}(S_i, v) > 0$.)
2. For $1 \leq j \leq M$, the ψ_j-strength of any $(O_j \cup \{v\})$-path from S_j to v is not greater than $\psi_i^{O_i}(S_i, v)$.

Furthermore, since the characterization uniquely determines the sequence

$$\langle O_1, \ldots, O_M \rangle$$

it follows that the definition

$$\sigma(v) = \max_{1 \leq i \leq M} \psi_i^{O_i}(S_i, v) \tag{3}$$

assigns a value from $[0, 1]$ to every $v \in V$. If $\sigma(v) = 0$, then $\psi_i^{O_i}(S_i, v) = 0$ and $v \notin O_i$ for $1 \leq i \leq M$. On the other hand, if $\sigma(v) \neq 0$, then there must be at least one i such that $\max_{j \neq i} \psi_j^{O_j}(S_j, v) \leq \psi_i^{O_i}(S_i, v) = \sigma(v)$, and $v \in O_i$ for all such i. (This is a good place to point out a second difference between MOFS and either RFC segmentation or IRFC segmentation: While in MOFS the O_i may overlap, objects in any RFC or IRFC segmentation are pairwise disjoint.) When representing the outcome of such a segmentation by a color image (as in the bottom image of Fig. 1), we can associate a different pure hue with each of the *i*s (in the case of the bottom image of Fig. 1, there are four pure hues used: red, green, blue, and yellow): If $v \in O_i$, then the hue associated with i is assigned to the pixel v. In the bottom image of Fig. 1, the brightness that is assigned to each $v \in O_i$ is $\sigma(v)$, which is often regarded (see, for example, [6]) as the grade of membership of v in O_i.

4   Robustness of Fuzzy Connectedness Segmentations

In practice the seed points may be selected by the user clicking on images (as was done to produce the 4-object MOFS shown at the bottom of Fig. 1) or in some more automatic manner (as described, for instance, in [1]). In the first case, the choice of the seed points is likely to be different for different users and even for the same user at different times. Even in the second case, we have variability; the noise in the image of an object is not deterministic and an automated process for determining the location of the seeds may depend on the noise in the image (for example, if it involves finding local minima or maxima). It would make the practical usefulness of FC segmentation questionable if the outcome were highly dependent on the exact selections of the seed points. Fortunately, this has not been found to be the case: FC segmentations are generally robust with respect to small changes in seed sets.

When the affinities do not depend on the choice of seeds, we can establish mathematical results that explain this robustness. In the cases of RFC and IRFC segmentation, such results are given in [18, Sect. 2.4]. In this section we state one such result for MOFS, which shows that it is possible to introduce a large number of additional seed points without changing the resulting MOFS.

Let $\Psi = \langle \psi_1, \ldots, \psi_M \rangle$ be any sequence of affinities on V and $\mathcal{S} = \langle S_1, \ldots, S_M \rangle$ any sequence of M pairwise disjoint nonempty subsets of V that are consistent with the affinities. Then, for $1 \leq i \leq M$, we define $\mathcal{P}_i(\Psi, \mathcal{S})$ to be the collection of all subsets P of V that satisfy both of the following conditions:

1. $P \subseteq O_i^{\mathrm{MOFS}}(\Psi, \mathcal{S}) \setminus \bigcup_{j \neq i} O_j^{\mathrm{MOFS}}(\Psi, \mathcal{S})$.
2. $\psi_i^{O_i^{\mathrm{MOFS}}(\Psi, \mathcal{S})}(S_i, v) \geq \psi_i^{\emptyset}(v, V \setminus P)$ for every $v \in P$.

The *core* of $O_i^{\mathrm{MOFS}}(\Psi, \mathcal{S})$, denoted by $\mathcal{P}i$, is defined as the union of all the sets in $\mathcal{P}_i(\Psi, \mathcal{S})$. We observe that $\mathcal{P}i \subseteq O_i^{\mathrm{MOFS}}(\Psi, \mathcal{S}) \setminus \bigcup_{j \neq i} O_j^{\mathrm{MOFS}}(\Psi, \mathcal{S})$ for $1 \leq i \leq M$. This implies that the cores of distinct MOFS objects are always disjoint: $\mathcal{P}i \cap \mathcal{P}j = \emptyset$ whenever $i \neq j$.

Since $S_i \subseteq O_i^{\mathrm{MOFS}}(\Psi, \mathcal{S}) \setminus \bigcup_{j \neq i} O_j^{\mathrm{MOFS}}(\Psi, \mathcal{S})$ (which is clear when we recall from the previous section that $S_i \subseteq O_i^{\mathrm{MOFS}} \subseteq S_i \cup (V \setminus \bigcup_j S_j)$ for $1 \leq i \leq M$), we see that $S_i \in \mathcal{P}_i(\Psi, \mathcal{S})$ and therefore $S_i \subseteq \mathcal{P}i$ for $1 \leq i \leq M$. A little reflection will show that it is quite possibly the case that $\mathcal{P}i$ is a much larger set than S_i. Nevertheless, as the following theorem states, using the $\mathcal{P}i$s instead of the S_is as the seed sets for the objects does not change the resulting MOFS.

MOFS Robustness Theorem [12]: Let $\Psi = \langle \psi_1, \ldots, \psi_M \rangle$ be a sequence of affinities on V and $\mathcal{S} = \langle S_1, \ldots, S_M \rangle$ a sequence of pairwise disjoint nonempty seed sets consistent with the affinities. Let $\mathcal{R} = \langle R_1, \ldots, R_M \rangle$ be such that $S_i \subseteq R_i \subseteq \mathcal{P}i$ for $1 \leq i \leq M$. Then $O_i^{\mathrm{MOFS}}(\Psi, \mathcal{R}) = O_i^{\mathrm{MOFS}}(\Psi, \mathcal{S})$ for $1 \leq i \leq M$.

Other results regarding the robustness of MOFS with respect to changes in seed sets (e.g., [12, Corollary 5.6]) can be deduced from this theorem.

All mathematical results the authors are aware of regarding robustness of FC segmentations assume that affinities remain unchanged when seed sets change. For example, there appear to be no results in the literature regarding robustness of FC segmentations when affinities depend on statistical properties of the image values assigned to points in neighborhoods of the seeds.

5   Unified Theory of FC Segmentations

While we have repeatedly mentioned RFC and IRFC segmentations, until now we have discussed details only of MOFS. A unified theory that covers all three types of segmentations is offered in [12]: In that paper it is shown that a generally common mathematical approach is applicable in all three cases. Moreover, the methods and results stated for MOFS above (and some other methods and results for MOFS) have close analogs for RFC and IRFC segmentations.

One significant fact that emerges from the unified theory of [12] is that the IRFC segmentation for an affinity ψ and seed sets $S_1, \ldots, S_M$ consistent with ψ can be found by executing the very efficient Algorithm 5 of [12] for MOFS with $\psi_1 = \cdots = \psi_M = \psi$: Each object O_i^{IRFC} of the IRFC segmentation consists just of those points in the corresponding MOFS object O_i^{MOFS} that do not lie in any of the other $M - 1$ MOFS objects:

$$O_i^{\mathrm{IRFC}} = O_i^{\mathrm{MOFS}} \setminus \bigcup_{j \neq i} O_j^{\mathrm{MOFS}}. \tag{4}$$

For segmentation into more than two objects, this approach (which allows the M IRFC objects to be computed simultaneously) can compute IRFC segmentations more quickly than commonly-used algorithms that compute these segmentations one object at a time.

6   Conclusion

Fuzzy connectedness (FC) image segmentation, which finds objects based on user-specified seed sets and fuzzy affinity functions, is one of the most computationally efficient segmentation methodologies and is commonly used in practical image segmentation tasks (especially in biomedical imaging). As can be seen from the references cited in this brief presentation, the methodology has a growing literature which covers mathematical properties of the segmentations as well as users' practice and experience.

References

1. Badura, P., Pietka, E.: Soft computing approach to 3D lung nodule segmentation in CT. Comput. Biol. Med. **53**, 230–243 (2014)
2. Bejar, H.H.C., Miranda, P.A.V.: Oriented relative fuzzy connectedness: theory, algorithms, and applications in image segmentation. In: 27th SIBGRAPI Conference on Graphics, Patterns and Images, pp. 304–311. IEEE Computer Society, Washington, DC (2014)
3. Carvalho, B.M., Garduño, E., Santos, I.O.: Skew divergence-based fuzzy segmentation of rock samples. J. Phys. Conf. Ser. **490**, 012010 (2014)
4. Carvalho, B.M., Gau, C.J., Herman, G.T., Kong, T.Y.: Algorithms for fuzzy segmentation. Pattern Anal. Appl. **2**, 73–81 (1999)
5. Carvalho, B.M., Herman, G.T., Kong, T.Y.: Simultaneous fuzzy segmentation of multiple objects. Electron. Notes Discrete Math. **12**, 3–22 (2003)
6. Carvalho, B.M., Herman, G.T., Kong, T.Y.: Simultaneous fuzzy segmentation of multiple objects. Discrete Appl. Math. **151**, 55–77 (2005)
7. Carvalho, B.M., Souza, T.S., Garduño, E.: Texture fuzzy segmentation using adaptive affinity functions. In: 27th ACM Symposium on Applied Computing, pp. 51–53. ACM, New York (2012)

8. Chen, L.: 3-D fuzzy digital topology and its application. Geophys. Prospect. Petrol. **24**(2), 86–89 (1985). [In Chinese]
9. Chen, L.: The λ-connected segmentation algorithm and the optimal algorithm for split-and-merge segmentation. Chin. J. Comput. **14**, 321–331 (1991). [In Chinese]
10. Chen, L., Berkey, F.T., Johnson, S.A.: Application of a fuzzy object search technique to geophysical data processing. In: SPIE 2180, Nonlinear Image Processing V, pp. 300–309. SPIE, Bellingham, WA (1994)
11. Chen, L., Cheng, H., Zhang, J.: Fuzzy subfiber and its application to seismic lithology classification. Inform. Sci. Appl. **1**, 77–95 (1994)
12. Ciesielski, K.C., Herman, G.T., Kong, T.Y.: General theory of fuzzy connectedness segmentations. http://math.wvu.edu/~kcies/publications.html (Submitted)
13. Ciesielski, K.C., Miranda, P.A.V., Falcão, A.X., Udupa, J.K.: Joint graph cut and relative fuzzy connectedness image segmentation algorithm. Med. Image Anal. **17**, 1046–1057 (2013)
14. Ciesielski, K.C., Udupa, J.K.: Affinity functions in fuzzy connectedness based image segmentation I: equivalence of affinities. Comput. Vis. Image Und. **114**, 146–154 (2010)
15. Ciesielski, K.C., Udupa, J.K.: Affinity functions in fuzzy connectedness based image segmentation II: defining and recognizing truly novel affinities. Comput. Vis. Image Und. **114**, 155–166 (2010)
16. Ciesielski, K.C., Udupa, J.K.: Region-based segmentation: fuzzy connectedness, graph cut, and other related algorithms. In: Deserno, T.M. (ed.) Biomedical Image Processing, pp. 251–278. Springer, Berlin (2011)
17. Ciesielski, K.C., Udupa, J.K., Falcão, A.X., Miranda, P.A.V.: Fuzzy connectedness image segmentation in graph cut formulation: a linear-time algorithm and a comparative analysis. J. Math. Imaging Vis. **44**, 375–398 (2012)
18. Ciesielski, K.C., Udupa, J.K., Saha, P.K., Zhuge, Y.: Iterative relative fuzzy connectedness for multiple objects, allowing multiple seeds. Comput. Vis. Image Und. **107**, 160–182 (2007)
19. Deserno, T.M. (ed.): Biomedical Image Processing. Springer, Berlin (2011)
20. Dijkstra, E.W.: A note on two problems in connexion with graphs. Numer. Math. **1**, 269–271 (1959)
21. Falcão, A.X., Stolfi, J., Lotufo, R.A.: The image foresting transform: theory, algorithms, and applications. IEEE Trans. Pattern Anal. **26**, 19–29 (2004)
22. Garduño, E., Herman, G.T.: Parallel fuzzy segmentation of multiple objects. Int. J. Imag. Syst. Tech. **18**, 336–344 (2009)
23. Garduño, E., Wong-Barnum, M., Volkmann, N., Ellisman, M.H.: Segmentation of electron tomographic data sets using fuzzy set theory principles. J. Struct. Biol. **162**, 368–379 (2008)
24. Gulyás, G., Dombi, J.: Computing equivalent affinity classes in a fuzzy connectedness framework. Acta Cybernetica **21**, 609–628 (2014)
25. Herman, G.T., Carvalho, B.M.: Multiseeded segmentation using fuzzy connectedness. IEEE Trans. Pattern Anal. **23**, 460–474 (2001)
26. Palágyi, K., Tschirren, J., Hoffman, E.A., Sonka, M.: Quantitative analysis of pulmonary airway tree structures. Comput. Biol. Med. **36**, 974–996 (2006)
27. Saha, P.K., Udupa, J.K.: Iterative relative fuzzy connectedness and object definition: theory, algorithms, and applications in image segmentation. In: IEEE Workshop on Mathematical Methods in Biomedical Image Analysis (MMBIA 2000), pp. 28–35. IEEE Computer Society, Washington, DC (2000)

28. Saha, P.K., Udupa, J.K.: Relative fuzzy connectedness among multiple objects: theory, algorithms, and applications in image segmentation. Comput. Vis. Image Und. **82**, 42–56 (2001)
29. Siebra, H., Carvalho, B.M., Garduño, E.: Fuzzy clustering of color textures using skew divergence and compact histograms: segmenting thin rock sections. J. Phys. Conf. Ser. **574**, 012116 (2015)
30. Tschirren, J., Hoffman, E.A., Mclennan, G., Sonka, M.: Intrathoracic airway trees: segmentation and airway morphology analysis from low-dose CT scans. IEEE Trans. Med. Imaging **24**, 1529–1539 (2005)
31. Udupa, J.K., Saha, P.K., Lotufo, R.A.: Relative fuzzy connectedness and object definition: theory, algorithms, and applications in image segmentation. IEEE Trans. Pattern Anal. **24**, 1485–1500 (2002)
32. Udupa, J.K., Samarasekera, S.: Fuzzy connectedness and object definition: theory, algorithms, and applications in image segmentation. Graph. Model. Im. Proc. **58**, 246–261 (1996)
33. Zhuge, Y., Udupa, J.K., Saha, P.K.: Vectorial scale-based fuzzy connected image segmentation. Comput. Vis. Image Und. **101**, 177–193 (2006)

Equivalent Sequential and Parallel Subiteration-Based Surface-Thinning Algorithms

Kálmán Palágyi$^{(\boxtimes)}$, Gábor Németh, and Péter Kardos

Department of Image Processing and Computer Graphics,
University of Szeged, Szeged, Hungary
{palagyi,gnemeth,pkardos}@inf.u-szeged.hu

Abstract. Thinning is a frequently applied technique for extracting skeletons or medial surfaces from volumetric binary objects. It is an iterative object reduction: border points that satisfy certain topological and geometric constraints are deleted in a thinning phase. Sequential thinning algorithms may alter just one point at a time, while parallel algorithms can delete a set of border points simultaneously. Two thinning algorithms are said to be equivalent if they can produce the same result for each input binary picture. This work shows that it is possible to construct subiteration-based equivalent sequential and parallel surface-thinning algorithms. The proposed four pairs of algorithms can be implemented directly on a conventional sequential computer or on a parallel computing device. All of them preserve topology for $(26, 6)$ pictures.

Keywords: Discrete geometry · Discrete topology · Skeletons · Subiteration-based thinning · Equivalent thinning algorithms

1 Introduction

A *digital binary picture* on the digital space $\mathbb{Z}^3$ is a mapping that assigns a color of black or white to each point [6]. *Thinning* is an iterative object reduction until only some skeleton-like shape features (i.e., *medial curves*, *medial surfaces*, or *topological kernels*) are left [4,6,19,25]. Thinning algorithms use *reduction operators* that transform binary pictures only by changing some black points to white ones, which is referred to as deletion. Parallel thinning algorithms are comprised of *reductions* that can delete a set of border points simultaneously [4,8,26], while sequential thinning algorithms traverse the boundary of objects and may remove just one point at a time [8,26].

Two reductions are said to be *equivalent* if they produce the same result for each input picture. Similarly, a deletion rule is called *equivalent* if it yields equivalent parallel and sequential reductions. One of the authors established some sufficient conditions for equivalent deletion rules [21]. This concept can be extended to complex algorithms composed of reductions: two surface-thinning algorithms are called *equivalent* if they produce the same medial surface for each input picture.

© Springer International Publishing Switzerland 2015
R.P. Barneva et al. (Eds.): IWCIA 2015, LNCS 9448, pp. 31–45, 2015.
DOI: 10.1007/978-3-319-26145-4_3

Sequential reductions (with the same deletion rule) suffer from the drawback that different visiting orders (raster scans) of border points may yield various results. A deletion rule is called *order-independent* if it yields equivalent sequential reductions for all the possible visiting orders. In [21] one of the authors gave necessary and sufficient conditions for order-independent deletion rules. By extending this concept: a sequential reduction is said to be *order-independent* if its deletion rule is order-independent, and a sequential surface-thinning algorithm is called *order-independent* [22] if it is composed of order-independent sequential reductions.

There are three major strategies for parallel thinning [4]: a *fully parallel* algorithm applies the same parallel reduction in each iteration step; a *subiteration-based* algorithm decomposes an iteration step into $k \geq 2$ successive parallel reductions according to k deletion directions, and a subset of border points associated with the actual direction can be deleted by a parallel reduction; and in *subfield-based* algorithms the digital space is partitioned into $s \geq 2$ subfields which are alternatively activated, at a given iteration step s successive parallel reductions assigned to these subfields are performed, and some black points in the active subfield can be designated for deletion. In this paper, our attention is focussed on the subiteration approach. Since there are six kinds of major directions in 3D, 6-subiteration 3D thinning algorithms are generally proposed [3,9,11,15,19,23,27,28]. Note that there are some 8-subiteration [16] and 12-subiteration [10,17] 3D thinning algorithms as well.

In [20] one of the authors proved that the deletion rule of the known 2D fully parallel thinning algorithm proposed by Manzanera et al. [14] is equivalent. Furthermore, he also showed in [21] that a fully parallel thinning algorithm with an equivalent deletion rule yields various subfield-based algorithms that are equivalent to the original fully parallel one. As far as we know, no one showed that there exists a pair of equivalent parallel and sequential 3D subiteration-based thinning algorithms. In this paper we propose four pairs of equivalent parallel and sequential 6-subiteration surface-thinning algorithms that use the same deletion rules, but they apply diverse geometric constraints.

The rest of this paper is organized as follows. Section 2 gives an outline from basic notions and results from digital topology, topology preservation, and equivalent reductions. Then in Sect. 3 the proposed algorithms are presented. In Sect. 4 we show that our parallel algorithms are equivalent to topology-preserving sequential surface-thinning algorithms. Finally, we round off the paper with some concluding remarks.

2   Basic Notions and Results

We use the fundamental concepts of digital topology as reviewed by Kong and Rosenfeld [5,6]. Note that there are other approaches that are based on cellular/cubical complexes [7]. The most important of them uses critical kernels [2] which constitute a generalization of minimal non-simple sets. Since the topological correctness of the thinning algorithms presented in this paper is based only on the concept of simple points, we can consider the traditional paradigm of digital topology.

Consider the digital space $\mathbb{Z}^3$, the three frequently applied adjacency relations on $\mathbb{Z}^3$ (see Fig. 1), and a point $p \in \mathbb{Z}^3$. Then, we denote by $N_j(p) \subset \mathbb{Z}^3$ the set of points that are j-adjacent to point p and let $N_j^*(p) = N_j(p) \setminus \{p\}$ ($j = 6, 18, 26$).

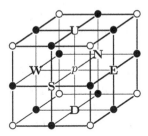

Fig. 1. The considered three adjacency relations on $\mathbb{Z}^3$. The set $N_6(p)$ contains p and the six points marked 'U', 'D', 'N', 'E', 'S', and 'W'; the set $N_{18}(p)$ contains $N_6(p)$ and the twelve points marked '•'; the set $N_{26}(p)$ contains $N_{18}(p)$ and the eight points marked '○'.

The sequence of distinct points $\langle x_0, x_1, \ldots, x_n \rangle$ is called a j-path of length n from x_0 to x_n in a non-empty set of points X if each point of the sequence is in X and x_i is j-adjacent to x_{i-1} for each $i = 1, \ldots, n$. Note that a single point is a j-path of length 0. Two points are said to be j-connected in the set X if there is a j-path in X between them. A set of points X is j-connected in the set of points $Y \supseteq X$ if any two points in X are j-connected in Y. A j-component of a set of points X is a maximal j-connected subset of X.

A $(26, 6)$ *binary digital picture* on $\mathbb{Z}^3$ is a quadruple $(\mathbb{Z}^3, 26, 6, B)$, where each point in $B \subseteq \mathbb{Z}^3$ is called a *black point*; each point in $\mathbb{Z}^3 \setminus B$ is said to be a *white point*. A *black component* (or *object*) is a 26-component of B, while a *white component* is a 6-component of $\mathbb{Z}^3 \setminus B$. A black point p is an *interior point* if all points in $N_6^*(p)$ are black. A black point is said to be a *border point* if it is not an interior point (i.e., it is 6-adjacent to at least one white point).

Topology preservation [5,6,12] is a major concern of thinning algorithms. A black point is called *simple point* if its deletion is a topology-preserving reduction. A sequential reduction and a sequential thinning algorithm is topology-preserving if their deletion rules delete only simple points. Various authors gave characterizations of simple points in $(26, 6)$ pictures [5,6,13,24]. We make use of the following one:

Theorem 1. [13,24] *A black point p is simple in picture $(\mathbb{Z}^3, 26, 6, B)$ if and only if all of the following conditions hold:*

1. The set $N_{26}^*(p) \cap B$ contains exactly one 26-component.

2. $N_6^*(p) \setminus B \neq \emptyset$.
3. *Any two points in $N_6^*(p) \setminus B$ are 6-connected in the set $N_{18}^*(p) \setminus B$.*

Theorem 1 states that the simplicity of a point p in $(26, 6)$ pictures is a local property (i.e., it can be decided by examining $N_{26}^*(p)$).

Existing sufficient conditions for topology-preserving parallel reductions are generally based on (or derived from) the notion of minimal non-simple sets [5, 12]. One of the authors established a new sufficient condition for arbitrary pictures with the help of equivalent deletion rules [21]. Let us recall and rephrase his results:

Theorem 2. *Let R be a deletion rule, let $(\mathbb{Z}^3, 26, 6, B)$ be a picture, and let $p \in B$ be a black point that can be deleted by R from that picture.*
If deletion rule R deletes only simple points and the deletability of any black point in $q \in B \setminus \{p\}$ by R does not depend on the 'color' of p, then the followings hold:

1. *The parallel reduction with deletion rule R is topology-preserving.*
2. *The sequential reduction with deletion rule R is order-independent and topology-preserving.*
3. *The parallel and the sequential reductions with deletion rule R are equivalent.*

Thinning algorithms generally classify the set of black points of the input picture into two (disjoint) subsets: the set of *interesting points* (i.e., potentially deletable points) for which the deletion rule associated with a thinning phase is evaluated and the *constraint set* whose black points are not taken into consideration (i.e., safe points that cannot be deleted). Since a phase of a subiteration-based algorithm cannot delete interior points and border points that do not fall into the actual type, these points are certainly in the constraint set.

Conventional thinning algorithms preserve some simple (border) points called *endpoints* that provide relevant geometrical information with respect to the shape of the object. Here we consider the following characterizations of surface-endpoints.

Definition 1. [19] *A black point p in picture $(\mathbb{Z}^3, 26, 6, B)$ is a* surface-endpoint *of type 1 if there is no interior point in $N_6^*(p) \cap B$ (i.e., p is not 6-adjacent to any interior point).*

Definition 2. [17] *A black point p in picture $(\mathbb{Z}^3, 26, 6, B)$ is a* surface-endpoint *of type 2 if at least one of the three pairs of points in $N_6^*(p) \cap B$ (**U**,**D**), (**N**,**S**), and (**E**,**W**) is formed by two white points.*

Definition 3. [3] *A black point p in picture $(\mathbb{Z}^3, 26, 6, B)$ is* **not** *a* surface-endpoint of type 3 *if $\|N_{26}^*(p) \cap B\| \geq 8$, or $4 \leq \|N_{26}^*(p) \cap B\| \leq 7$ and $N_6^*(p) \cap B$ contains three mutually 26-adjacent points. ($\|S\|$ stands for the count of elements in set S).*

Definitions 1–3 make us possible to specify three pairs of endpoint-based surface-thinning algorithms (**P-6-SI-**i, **S-6-SI-**i) ($i = 1, 2, 3$), see Algorithms 1 and 2 in Sect. 3.

Some advanced thinning algorithms preserve accumulated *isthmuses* [1] (i.e., generalization of curve interior points):

Definition 4. [1] *A black point p in a* $(26, 6)$ *picture is a* surface-isthmus of type 4 *if p is a non-simple border point (i.e., Condition 1 of Theorem 1 or Condition 3 of Theorem 1 is violated).*

Definition 4 helps us to give a pair of isthmus-based surface-thinning algorithms (**P-6-SI-**4, **S-6-SI-**4), see Algorithms 1 and 2 in Sect. 3.

Note that surface-endpoints of type 4 and surface-isthmuses of type i ($i = 1, 2, 3$) are not defined (i.e., there is neither surface-endpoint of type 4 nor surface-isthmuses of type i ($i = 1, 2, 3$)). Definitions 1–4 make us possible to give a unified description of the proposed endpoint-based and isthmus-based surface-thinning algorithms (see Algorithms 1 and 2 in Sect. 3).

3   Parallel and Sequential 6-Subiteration Surface-Thinning Algorithms

In this section, four pairs of 3D parallel and sequential 6-subiteration surface-thinning algorithms (**P-6-SI-**i, **S-6-SI-**i) are presented for $(26, 6)$ pictures ($i = 1, 2, 3, 4$). All of these algorithms use the same deletion rule, but diverse pairs of them apply different constraint sets. The proposed parallel thinning algorithms **P-6-SI-**i and the sequential algorithms **S-6-SI-**i are given by Algorithms 1 and 2, respectively.

It is easy to see that the first three pairs of algorithms (**P-6-SI-**i, **S-6-SI-**i) are *endpoint-based* ($i = 1, 2, 3$), and the fourth pair of algorithms (**P-6-SI-**4, **S-6-SI-**4) is *isthmus-based* (see Definitions 1–4).

By comparing the parallel algorithm **P-6-SI-**i (see Algorithm 1) and sequential algorithm **S-6-SI-**i (see Algorithm 2), we can state that in the parallel case the initial set of black points P is considered when the deletability of all the interesting points are investigated. On the contrary, the set of black points S is dynamically altered when a sequential reduction/subiteration is performed; the deletability of the actual point is evaluated in a modified picture (in which some previously visited interesting points are white).

The applied deletion rules that specify d-DELETABLE points ($d = $ **U**, **D**, **N**, **E**, **S**, **W**) are given by $3 \times 3 \times 3$ matching templates depicted in Fig. 2. Note that the six deletion rules were originally proposed by Gong and Bertrand [3] in their endpoint-based 6-subiteration surface-thinning algorithm with respect to surface-endpoints of type 3 (see Definition 3).

A period of six subiterations/reductions (i.e., the kernel of the **repeat** cycle in Algorithms 1 and 2) is decomposed into six successive subiterations according to the six main directions in 3D, and this period is repeated until stability is reached (i.e., no point is deleted within the last six subiterations/reductions).

Algorithm 1. Parallel thinning algorithm **P-6-SI-**i $(i = 1, 2, 3, 4)$

Input: set of black points B
Output: set of black points P
$P = B$
$I = \emptyset$
repeat
    // a period of six subiterations/reductions
    for each $d \in \{\mathbf{U}, \mathbf{D}, \mathbf{N}, \mathbf{E}, \mathbf{S}, \mathbf{W}\}$ **do**
        // accumulating isthmuses
        $I = I \cup \{\, p \mid p \in P \text{ is a surface-isthmus of type } i \,\}$
        // forming the constraint set
        $C = I \cup \{\, p \mid p \in P \text{ is not a } d\text{-border point} \,\}$
        $C = C \cup \{\, p \mid p \in P \text{ is a surface-endpoint of type } i \,\}$
        // forming the set of interesting points
        $X = P \setminus C$
        // deletion
        $D_d = \{\, p \mid p \in X \text{ is a } d\text{-DELETABLE point in } P \,\}$
        $P = P \setminus D_d$
until $D_\mathbf{U} \cup D_\mathbf{D} \cup D_\mathbf{N} \cup D_\mathbf{E} \cup D_\mathbf{S} \cup D_\mathbf{W} = \emptyset$

Algorithm 2. Sequential thinning algorithm **S-6-SI-**i $(i = 1, 2, 3, 4)$

Input: set of black points B
Output: set of black points S
$S = B$
$I = \emptyset$
repeat
    // a period of six subiterations/reductions
    for each $d \in \{\mathbf{U}, \mathbf{D}, \mathbf{N}, \mathbf{E}, \mathbf{S}, \mathbf{W}\}$ **do**
        // accumulating isthmuses
        $I = I \cup \{\, p \mid p \in P \text{ is a surface-isthmus of type } i \,\}$
        // forming the constraint set
        $C = I \cup \{\, p \mid p \in S \text{ is not a } d\text{-border point} \,\}$
        $C = C \cup \{\, p \mid p \in P \text{ is a surface-endpoint of type } i \,\}$
        // forming the set of interesting points
        $X = S \setminus C$
        // traversal of the elements in X
        $del(d) = 0$
        for each $p \in X$ **do**
            if p is a d-DELETABLE point in S **then**
                // deletion
                $S = S \setminus \{p\}$
                $del(d) = del(d) + 1$
until $del(\mathbf{U}) + del(\mathbf{D}) + del(\mathbf{N}) + del(\mathbf{E}) + del(\mathbf{S}) + del(\mathbf{W}) = 0$

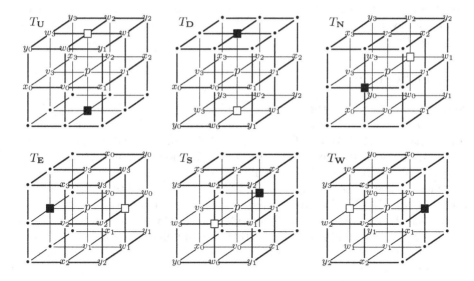

Fig. 2. Matching template T_d associated with d-DELETABLE points (d = U, D, N, E, S, W). Notations: the central position marked p matches an interesting (black) point; the position marked '■' matches a (black) point in the constraint set; the position marked '□' matches a white point; if the position marked 'v_k' coincides with a white point, then the position marked 'w_k' coincides with a white point ($k = 0, 1, 2, 3$); if all the three positions marked 'v_k', '$x_{(k+1) \bmod 4}$', and '$v_{(k+1) \bmod 4}$' coincide with white points, then the position marked '$y_{(k+1) \bmod 4}$' coincides with a white point ($k = 0, 1, 2, 3$); each '·' (don't care) matches either a black or a white point.

We propose the following ordered list of the deletion directions: ⟨U, D, N, E, S, W⟩. Note that a subiteration-based thinning algorithm is sensitive to the order of directions. Hence choosing another order of the deletion directions yields another algorithm.

An interesting black point ($p \in X$) is d-DELETABLE if template T_d matches it ($d =$ U, D, N, E, S, W). Note that the templates assigned to the deletion direction d give the condition to delete certain d-border points, and templates associated with the last five deletion directions can be obtained by proper rotations of the templates that give U-DELETABLE points.

In experiments the proposed pairs of equivalent algorithms (**P-6-SI-i**, **S-6-SI-i**) were tested on objects of different shapes. Here we have room to present four illustrative examples, see Figs. 3, 4, 5 and 6. The results of our four pairs of algorithms can be compared to the medial surfaces produced by two existing 3D parallel surface thinning algorithms proposed by Manzanera et al. [14] and Palágyi [18]. The numbers refer to the count of black points in the pictures. We can state that the isthmus-based pair of algorithms (**P-6-SI-4**, **S-6-SI-4**) result less object points than the other five methods. The authors note that, unfortunately, there is no known method for quantitative comparison of surface-thinning algorithms.

4   Verification

Now we will show that the 6-subiteration parallel surface-thinning algorithm **P-6-SI-**i and the sequential algorithm **S-6-SI-**i are equivalent and topology-preserving ($i = 1, 2, 3, 4$). It will also be proved that the sequential algorithms are order-independent. It is sufficient to show that deletion rules of these algorithms (see Sect. 3) satisfy all conditions of Theorem 2.

Since the five templates $T_\mathbf{D}$, $T_\mathbf{N}$, $T_\mathbf{E}$, $T_\mathbf{S}$, and $T_\mathbf{W}$ assigned to d-DELETABLE points ($d = \mathbf{D}, \mathbf{N}, \mathbf{E}, \mathbf{S}, \mathbf{W}$) are rotated versions of the template $T_\mathbf{U}$ (see Fig. 2), it is sufficient to prove that the deletion rule associated with the first subiteration deletes only simple points, and the deletability of a point does not depend on the 'color' of a deletable point. It can be carried out for the remaining five deletion rules in the same way.

Let us state some important properties of the **U**-DELETABLE points:

Lemma 1. *All* **U**-DELETABLE *points are simple.*

It is obvious by a careful examination of the matching template $T_\mathbf{U}$ that all conditions of Theorem 1 hold.

Lemma 2. *The deletability of a point by template* $T_\mathbf{U}$ *does not depend on the 'color' of a* **U**-DELETABLE *point.*

Proof. Let us assume that the (interesting) black point p is **U**-DELETABLE. Since **U**-DELETABLE points are given by a $3 \times 3 \times 3$ matching template, it is sufficient to investigate the deletability of interesting black points in $N_{26}^*(p)$.

Due to the symmetries that are present in template $T_\mathbf{U}$ (see Fig. 2), it is sufficient to check the eight template positions marked '★' in Fig. 7a. Hence it is assumed that the **U**-DELETABLE point p coincides with these eight positions. Consider the deletability of an interesting black point $q \in N_{26}^*(p)$ with the help of the eight corresponding configurations depicted in Figs. 7b–i.

Let us investigate all that eight cases:

- If the **U**-DELETABLE point p coincides with a position marked '★' in Fig. 7b, d, e, f, g, and i, then the deletability of q does not depend on the 'color' of p.
- If point p coincides with a position marked '★' in Fig. 7c (after its deletion), then the interesting point q is in the constraint set. Hence we arrived at a contradiction.
- If point p coincides with a position marked '★' in Fig. 7h (before its deletion), then the interesting point p is in the constraint set. Hence we arrived at a contradiction.   □

We are now ready to state our main theorem — as an easy consequence of Theorem 2, Lemmas 1 and 2.

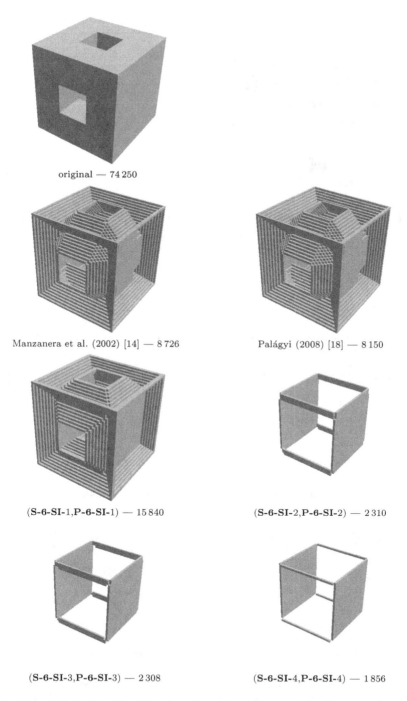

original — 74 250

Manzanera et al. (2002) [14] — 8 726

Palágyi (2008) [18] — 8 150

(**S-6-SI**-1,**P-6-SI**-1) — 15 840

(**S-6-SI**-2,**P-6-SI**-2) — 2 310

(**S-6-SI**-3,**P-6-SI**-3) — 2 308

(**S-6-SI**-4,**P-6-SI**-4) — 1 856

Fig. 3. The original $45 \times 45 \times 45$ image of a cube with two tunnels and its six medial surfaces produced by two existing algorithms and the proposed four pairs of equivalent 6-subiteration surface-thinning algorithms.

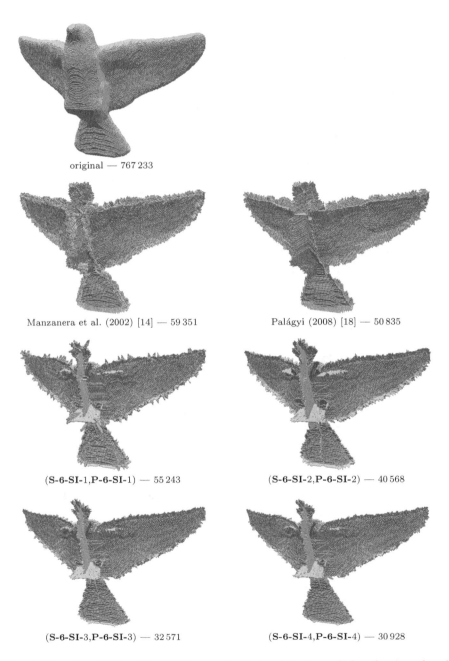

original — 767 233

Manzanera et al. (2002) [14] — 59 351

Palágyi (2008) [18] — 50 835

(**S-6-SI**-1,**P-6-SI**-1) — 55 243

(**S-6-SI**-2,**P-6-SI**-2) — 40 568

(**S-6-SI**-3,**P-6-SI**-3) — 32 571

(**S-6-SI**-4,**P-6-SI**-4) — 30 928

Fig. 4. The original $321 \times 153 \times 227$ image of a bird and its six medial surfaces produced by two existing algorithms and the proposed four pairs of equivalent 6-subiteration surface-thinning algorithms.

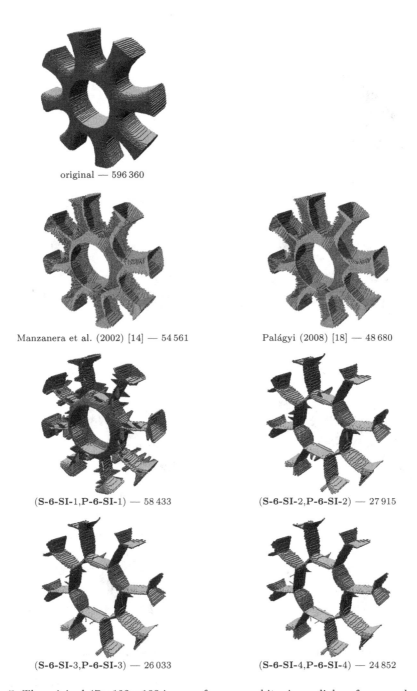

Fig. 5. The original $47 \times 193 \times 193$ image of a gear and its six medial surfaces produced by two existing algorithms and the proposed four pairs of equivalent 6-subiteration surface-thinning algorithms.

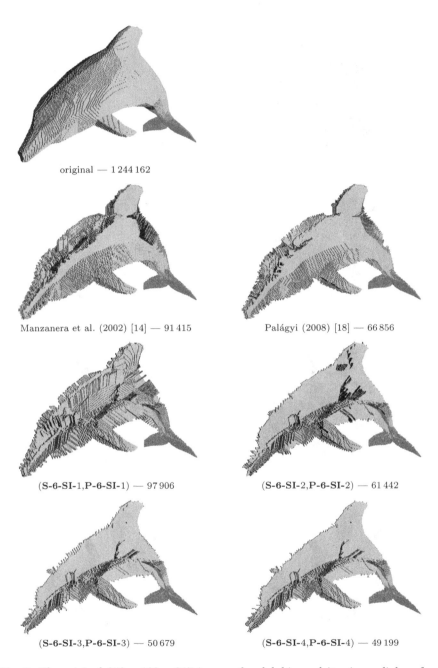

original — 1 244 162

Manzanera et al. (2002) [14] — 91 415

Palágyi (2008) [18] — 66 856

(**S-6-SI**-1,**P-6-SI**-1) — 97 906

(**S-6-SI**-2,**P-6-SI**-2) — 61 442

(**S-6-SI**-3,**P-6-SI**-3) — 50 679

(**S-6-SI**-4,**P-6-SI**-4) — 49 199

Fig. 6. The original $350 \times 132 \times 217$ image of a dolphin and its six medial surfaces produced by two existing algorithms and the proposed four pairs of equivalent 6-subiteration surface-thinning algorithms.

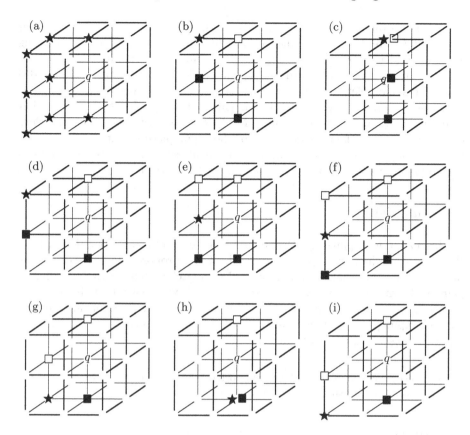

Fig. 7. The eight positions marked '★' are checked by Lemma 2 (a), and the eight configurations (b)–(i) are associated with these positions. Points marked '■' are in the constraint set, and points marked '□' are white.

Theorem 3. *The followings hold for the proposed algorithms:*

1. *The sequential surface-thinning algorithm* **S-6-SI-***i is order-independent* ($i = 1, 2, 3, 4$).
2. *The sequential surface-thinning algorithm* **S-6-SI-***i is topology-preserving for* $(26, 6)$ *pictures* ($i = 1, 2, 3, 4$).
3. *The parallel surface-thinning algorithm* **P-6-SI-***i is topology-preserving for* $(26, 6)$ *pictures* ($i = 1, 2, 3, 4$).
4. *Algorithms* **S-6-SI-***i and* **P-6-SI-***i are equivalent, i.e., they produce the same result for each input picture* ($i = 1, 2, 3, 4$).

Note that Lemma 2 is valid for arbitrary constraint sets — not only for the four kinds of sets that are used by algorithms (**S-6-SI-***i*, **P-6-SI-***i*) ($i = 1, 2, 3, 4$). Other constraint sets coupled with the set of templates depicted in Fig. 2 yield additional pairs of equivalent 6-subiteration parallel and sequential thinning algorithms.

It is important to emphasize that the parallel algorithm **P-6-SI-**3 coincides with the 6-subiteration 3D parallel surface-thinning algorithm proposed by Gong and Bertrand in 1990 [3]. According to Theorem 3, that existing parallel algorithm is equivalent to the (order-independent) sequential algorithm **S-6-SI-**3 (with the same deletion rules and constraint set). In addition, the topological correctness of an existing parallel thinning algorithm is also confirmed. Note that Gong and Bertrand sketched a proof in [3] to show that their algorithm does not change the topological properties of the input pictures. At that time (i.e., in 1990) they could not apply the very first sufficient conditions for topology-preserving 3D parallel reductions reported by Ma in 1994 [12].

5   Conclusions

In this paper four pairs of 3D 6-subiteration sequential and parallel surface-thinning algorithms were presented. Each of the proposed algorithm uses the same deletion rules that are given by $3 \times 3 \times 3$ matching templates, but different pairs of algorithms apply diverse constraint sets. It was shown that the proposed pairs of algorithms are equivalent (i.e., they produce the same medial surfaces for each input picture). It was also proved that all the reported algorithms are topology-preserving for $(26, 6)$ pictures.

Acknowledgements. This work was supported by the grant OTKA K112998 of the National Scientific Research Fund.

References

1. Bertrand, G., Couprie, M.: Transformations topologiques discrètes. In: Coeurjolly, D., Montanvert, A., Chassery, J. (eds.) Géométrie Discrète et Images Numériques, pp. 187–209. Hermès Science Publications, Paris (2007)
2. Bertrand, G., Couprie, M.: New 2D parallel thinning algorithms based on critical kernels. In: Reulke, R., Eckardt, U., Flach, B., Knauer, U., Polthier, K. (eds.) IWCIA 2006. LNCS, vol. 4040, pp. 45–59. Springer, Heidelberg (2006)
3. Gong, W.X., Bertrand, G.: A simple parallel 3D thinning algorithm. In: Proceedings of the 10th IEEE International Conference Pattern Recognition, ICPR 1990, pp. 188–190 (1990)
4. Hall, R.W.: Parallel connectivity-preserving thinning algorithms. In: Kong, T.Y., Rosenfeld, A. (eds.) Topological Algorithms for Digital Image Processing, pp. 145–179. Elsevier Science B.V., Amsterdam (1996)
5. Kong, T.Y.: On topology preservation in 2D and 3D thinning. Int. J. Pattern Recogn. Artif Intell. **9**, 813–844 (1995)
6. Kong, T.Y., Rosenfeld, A.: Digital topology: introduction and survey. Comput. Vis. Graph. Image Process. **48**, 357–393 (1989)
7. Kovalevsky, V.A.: Geometry of Locally Finite Spaces. Publishing House, Berlin (2008)
8. Lam, L., Lee, S.-W., Suen, S.-W.: Thinning methodologies - a comprehensive survey. IEEE Trans. Pattern Anal. Mach. Intell. **14**, 869–885 (1992)

9. Lee, T., Kashyap, R.L., Chu, C.: Building skeleton models via 3D medial surface/axis thinning algorithms. CVGIP: Graph. Models Image Process. **56**, 462–478 (1994)
10. Lohou, C., Bertrand, G.: A 3D 12-subiteration thinning based on P-simple points. Discrete Appl. Math. **139**, 171–195 (2004)
11. Lohou, C., Bertrand, G.: A 3D 6-subiteration curve thinning algorithm based on P-simple points. Discrete Appl. Math. **151**, 198–228 (2005)
12. Ma, C.M.: On topology preservation in 3D thinning. CVGIP: Image Underst. **59**, 328–339 (1994)
13. Malandain, G., Bertrand, G.: Fast characterization of 3D simple points. In: International Conference on Pattern Recognition, ICPR 1992, pp. 232–235 (1992)
14. Manzanera, A., Bernard, T.M., Pretêux, F., Longuet, B.: n-dimensional skeletonization: a unified mathematical framework. J. Electron. Imaging **11**, 25–37 (2002)
15. Palágyi, K., Kuba, A.: A 3D 6-subiteration thinning algorithm for extracting medial lines. Pattern Recogn. Lett. **19**, 613–627 (1998)
16. Palágyi, K., Kuba, A.: Directional 3D thinning using 8 subiterations. In: Bertrand, G., Couprie, M., Perroton, L. (eds.) DGCI 1999. LNCS, vol. 1568, pp. 325–336. Springer, Heidelberg (1999)
17. Palágyi, K., Kuba, A.: A parallel 3D 12-subiteration thinning algorithm. Graph. Models Image Process. **61**, 199–221 (1999)
18. Palágyi, K.: A 3D fully parallel surface-thinning algorithm. Theoret. Comput. Sci. **406**, 119–135 (2008)
19. Palágyi, K., Németh, G., Kardos, P.: Topology preserving parallel 3D thinning algorithms. In: Brimkov, V.E., Barneva, R.P. (eds.) Digital Geometry Algorithms. LNCVB, pp. 165–188. Springer, Heidelberg (2012)
20. Palágyi, K.: Equivalent 2D sequential and parallel thinning algorithms. In: Barneva, R.P., Brimkov, V.E., Šlapal, J. (eds.) IWCIA 2014. LNCS, vol. 8466, pp. 91–100. Springer, Heidelberg (2014)
21. Palágyi, K.: Equivalent sequential and parallel reductions in arbitrary binary pictures. Int. J. Pattern Recogn. Artif. Intell. **28**, 1460009-1–1460009-16 (2014)
22. Ranwez, V., Soille, P.: Order independent homotopic thinning for binary and grey tone anchored skeletons. Pattern Recogn. Lett. **23**, 687–702 (2002)
23. Raynal, B., Couprie, M.: Isthmus-based 6-directional parallel thinning algorithms. In: Debled-Rennesson, I., Domenjoud, E., Kerautret, B., Even, P. (eds.) DGCI 2011. LNCS, vol. 6607, pp. 175–186. Springer, Heidelberg (2011)
24. Saha, P.K., Chaudhuri, B.B.: Detection of 3D simple points for topology preserving transformations with application to thinning. IEEE Trans. Pattern Anal. Mach. Intell. **16**, 1028–1032 (1994)
25. Siddiqi, K., Pizer, S. (eds.): Medial Representations - Mathematics, Algorithms and Applications. Computational Imaging and Vision, vol. 37. Springer, New York (2008)
26. Suen, C.Y., Wang, P.S.P. (eds.): Thinning Methodologies for Pattern Recognition. Series in Machine Perception and Artificial Intelligence, vol. 8. World Scientific, Singapore (1994)
27. Tsao, Y.F., Fu, K.S.: A parallel thinning algorithm for 3-D pictures. Comput. Graph. Image Process. **17**, 315–331 (1981)
28. Xie, W., Thompson, P., Perucchio, R.: A topology-preserving parallel 3D thinning algorithm for extracting the curve skeleton. Pattern Recogn. **36**, 1529–1544 (2003)

Relative Convex Hull Determination from Convex Hulls in the Plane

Petra Wiederhold$^{(\boxtimes)}$ and Hugo Reyes

Department of Automatic Control, Centro de Investigación y de Estudios
Avanzados (CINVESTAV-IPN), Av. I.P.N. 2508,
Col. San Pedro Zacatenco, 07000 Mexico, D.F., Mexico
pwiederhold@gmail.com, hrb87@hotmail.com

Abstract. A new algorithm for the determination of the relative convex hull in the plane of a simple polygon A with respect to another simple polygon B which contains A, is proposed. The relative convex hull is also known as geodesic convex hull, and the problem of its determination in the plane is equivalent to find the shortest curve among all Jordan curves lying in the difference set of B and A and encircling A. Algorithms solving this problem known from Computational Geometry are based on the triangulation or similar decomposition of that difference set. The algorithm presented here does not use such decomposition, but it supposes that A and B are given as ordered sequences of vertices. The algorithm is based on convex hull calculations of A and B and of smaller polygons and polylines, it produces the output list of vertices of the relative convex hull from the sequence of vertices of the convex hull of A.

Keywords: Relative convex hull · Geodesic convex hull · Shortest Jordan curve · Shortest path · Minimal length polygon · Minimal perimeter polygon

1 Introduction

The relative convex hull (RCH), also called geodesic convex hull, recently has received increasing attention in Computational Geometry [25], in particular related to shortest path problems which appear in a variety of applications as in robotics, industrial manufacturing, networking, or processing of geographical data [13,26]. It was earlier defined in the context of Digital Geometry and Topology and their applications in Digital Image Analysis, where the RCH and related structures based on geodesic metrics have been proposed as approximations of digital curves and surfaces and for multi-grid convergent estimations of curve length or surface area [1,2,7–11,17,20,22,23,27,28].

The convex hull of a set S in the Euclidean space is obtained by filling up S with all points lying on straight line segments having end points in S. If S is contained in another set T, to construct the RCH of S with respect to T, points

© Springer International Publishing Switzerland 2015
R.P. Barneva et al. (Eds.): IWCIA 2015, LNCS 9448, pp. 46–60, 2015.
DOI: 10.1007/978-3-319-26145-4_4

lying on straight line segments with end points in S are added whenever these segments already belong to B.

In the Euclidean plane and for sets $S \subset T$, the RCH of S with respect to T, denoted by $CH_T(S)$, is obtained by allocating a tight thread around A but within B, see Fig. 1(a). In this paper we study the RCH for simple polygons S, T. In [2], the RCH was considered for the more general situation where S is a finite point set and T is a polygonal domain. A distinct definition of RCH applies to disjoint simple polygons S, T, then $CH_T(S)$ is the weakly simple polygon formed by the shortest closed polygonal path without self-crossings which circumscribes S but excludes T [26], see Fig. 1(b).

Under special conditions for the polygons S, T, $S \subset T$, the RCH coincides with the Minimum Perimeter Polygon (MPP) of S with respect to T, also called the Minimum Length Polygon (MLP), whose frontier is the shortest Jordan curve among all Jordan curves which circumscribe S but are contained in T [8,22,23]. The MPP was first defined for polygons S, T which are point set unions of cell complexes within plane mosaics modelling the digital plane where the pixels are identified with convex not necessarily uniform tiles [18–21], see Fig. 1(c). These polygons S, T are constructed as the Inner and Outer Jordan digitization of a subset of the Euclidean plane which is the interior of a given Jordan curve γ. For the digital plane modeled by the standard quadratic complex where all pixels are grid squares of the same size, S, T are isothetic simple polygons and $(T \setminus S)$ is a union of grid squares called grid continuum, see Fig. 1(d). In this case, the length of the frontier of the RCH is a multi-grid convergent estimator of the length of the Jordan curve γ [8,22,23]. Several efficient MLP algorithms are known, for example the corrected version of [6] in [11,15], but these can be applied only to digital continua or polyominoes.

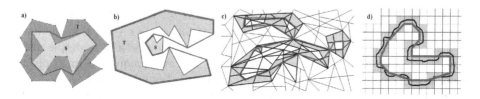

Fig. 1. (a) RCH of a set with respect to a superset, (b) RCH for two disjoint sets, (c) MPP of a subcomplex of a mosaic, (d) MLP of a grid continuum being a digital model of a Jordan curve.

In this paper we propose a novel algorithm for the determination of the ordered list of all vertices of the RCH, for the general situation of given simple plane polygons A, B such that $A \subset B$. The algorithm does not use previous triangulation or similar decompositions. Each input polygon is given as ordered set of its vertices. Our algorithm adopts some basic ideas of the algorithm published in [4] but presents essential corrections and improvements. A preliminary version of our algorithm was developed in [16].

2   Preliminaries

Recall that a non-empty set $S \subset \mathbb{R}^2$ is *convex* if for any $p, q \in S$, the straight line segment $\overline{pq}$ is contained in S, where $\overline{pq}$ is the set of all points $r = \lambda_1 p + \lambda_2 q$ such that $\lambda_1, \lambda_2 \in \mathbb{R}$, $\lambda_1, \lambda_2 \geq 0$, $\lambda_1 + \lambda_2 = 1$. The *convex hull* of S denoted by $CH(S)$, is the intersection of all convex sets which contain S. Equivalently, $CH(S)$ is the set of all points which belong to straight line segments with end points in S. For basic topological notions we refer to [14], we will denote the (topological) interior of S by $int(S)$ and its frontier by $fr(S)$. A non-convex set is distinct from its convex hull via the presence of holes or cavities: Any bounded connected component of $(\mathbb{R}^2 \setminus S)$ is a *hole* of S. The closure of any connected component of $(CH(S) \setminus S)$ which is not a hole of S, is a *cavity* of S.

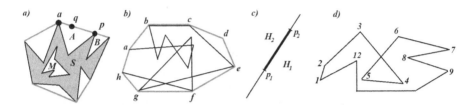

Fig. 2. (a) M is a hole of S, the cavities A, B are distinct although they share the point p. The straight line segment $\overline{aq}$ is not a cover of the cavity A although it belongs to $(fr(A) \setminus fr(S))$, $\overline{ap}$ is the cover of A. (b) A polyline and its convex hull given by the vertex sequence $\langle a, b, c, d, e, f, g, h \rangle$. (c) Right and left halfplanes determined by $\overrightarrow{p_1 p_2}$. (d) Points 1,2,5,6,7,9 are examples of convex vertices (right turns), points 8 and 12 are concave vertices (left turns) of the closed polyline traced in clockwise sense.

A *curve* $\gamma = \{f(s) = (x(s), y(s)) \in \mathbb{R}^2 : s \in [0,1]\}$ ($f : [0,1] \rightarrow \mathbb{R}^2$ continuous), is closed if $f(0) = f(1)$, simple if for any $s, t \in [0,1]$ such that $0 \leq s < t < 1$ it follows $f(s) \neq f(t)$; γ is a *Jordan curve* if it is simple and closed. A Jordan curve γ separates the plane into two uniquely defined open disjoint regions: the *interior of the Jordan curve* is bounded and encircled by γ, and the exterior of the Jordan curve is not bounded [14]. A curve is named *polyline* if there exists a finite sequence of points $\{s_0, s_1, s_2, \cdots, s_k\}$, with $0 = s_0 < s_1 < s_2 < \cdots < s_k = 1$ such that all curve segments $\{f(s) : s_i \leq s \leq s_{i+1}\}$ $(i = 0, 1, \cdots, k-1)$ are straight line segments. The points $\{s_0, s_1, s_2, \cdots, s_k\}$ are named vertices whenever no three consecutive points are collinear. A polyline is uniquely determined by the sequence of its vertices. A closed polyline corresponds to a closed curve, a simple polyline is a simple curve. A vertex p of a polyline γ is called *extreme vertex* if its x-coordinate is extreme (that is, maximal or minimal) among the x-coordinates of all vertices of γ or, if its y-coordinate is extreme among all y-coordinates of vertices of γ. Any extreme vertex of a polyline γ is a vertex of the convex hull $CH(\gamma)$. A *simple polygon* is defined as any non-empty bounded closed set $P \subset \mathbb{R}^2$ whose frontier forms a simple closed polyline. Hence the frontier of a simple polygon is a Jordan curve and can be represented by the finite cyclic sequence of its vertices. The convex hull of a simple polygon coincides with the convex hull of the finite set of its vertices. A simple polygon does not have holes, therefore it is non-convex if and only if it has at least one cavity. For any non-convex simple polygon S in the plane and any cavity M of S, define the **cover** of M as straight line segment of maximal length belonging to $fr(M) \setminus S$. The requirement of maximal length guarantees that the cover for each cavity M is unique, see Fig. 2(a). For any ordered triple of points $p_1 = (x_1, y_1)$, $p_2 = (x_2, y_2)$, $p_3 = (x_3, y_3)$ in the plane, its orientation is characterized by the sign of the determinant $D(p_1, p_2, p_3) = x_1 y_2 + y_1 x_3 + x_2 y_3 - (x_3 y_2 + x_2 y_1 + x_1 y_3)$. The oriented line segment $\overrightarrow{p_1 p_2}$ defines an oriented line which separates $\mathbb{R}^2$ into a right halfplane H_1 and a left halfplane H_2, see Fig. 2(c). (p_1, p_2, p_3) forms a *right turn* if $p_3 \in H_1$, (p_1, p_2, p_3) forms a *left turn* if $p_3 \in H_2$. Using the standard cartesian coordinate system in the plane, for a closed (simple) polyline L traced in clockwise sense, see Fig. 2(d), for any three consecutive vertices p_1, p_2, p_3 of L we have the following: (p_1, p_2, p_3) forms a right turn if and only if $D(p_1, p_2, p_3) < 0$; then p_2 is called a *convex vertex*; (p_1, p_2, p_3) forms a left turn if and only if $D(p_1, p_2, p_3) > 0$, then p_2 is called a *concave vertex*. p_1, p_2, p_3 are collinear points if and only if $D(p_1, p_2, p_3) = 0$.

3   Definition and Properties of the Relative Convex Hull

Definition 1. *Let $A, B \subset \mathbb{R}^n$ be non-empty sets such that $A \subseteq B$. Then A is called B-**convex** if any straight line segment lying in B whose both end points belong to A, is contained in A. The **relative convex hull of A with respect to B**, denoted by $CH_B(A)$, is defined as the intersection of all B-convex sets which contain A.*

It is evident that each set A is A-convex, and that if A is convex and $A \subset B$ then A also is B-convex. The following properties can be derived from the definitions of $CH(A)$ and $CH_B(A)$:

Lemma 1.   *(i) $A \subset CH_B(A) \subset B$, B is the largest B-convex set which contains A whereas $CH_B(A)$ is the smallest such set.*

(ii) $CH_B(A) \subset CH(A)$.
(iii) A is convex if and only if $CH_B(A) = CH(A) = A$.
(iv) $CH(A) \subset B$ if and only if $CH_B(A) = CH(A)$.
(v) If B is convex then $CH_B(A) = CH(A)$.

Proof. The definitions and constructions of $CH(A)$ and $CH_B(A)$ imply (i) and (ii); (iii) follows from (ii) and since A is convex if and only if $CH(A) = A$. (iv) Suppose $CH(A) \subset B$. Because of (ii), only $CH(A) \subset CH_B(A)$ remains to be proved. Let $p \in CH(A)$ and $M \subset B$ be any B-convex set containing A. We have to prove that $p \in M$. For $p \in A$ this is trivial, so assume $p \notin A$. Since $CH(A)$ is the set of all straight line segments having end points in A, p belongs to some straight line segment with end points $a, b \in A$. But then a, b belong also to $M \subset B$. The segment $\overline{ab}$ is contained in $CH(A)$ and hence, by the hypothesis, to B. Since M es B-convex, $p \in \overline{ab} \subset M$ which completes the proof of $CH_B(A) = CH(A)$. On the other hand, $CH_B(A) = CH(A)$ means in particular that $CH(A)$ is contained in each B-convex set which contains A, but B is such a set, implying $CH(A) \subset B$. (v) $A \subset B$ with B convex implies $CH(A) \subset CH(B) = B$, then (iii) gives the result. □

As a corollary, it can be proved that a necessary condition for $CH_B(A) \neq CH(A)$ is that some concave vertex of B lies in the interior of a cavity of A. In this paper we study the RCH only for simple polygons A and B in the plane, $A \subset B \subset \mathbb{R}^2$. The following properties are important for the determination of the RCH:

Theorem 1. *Let A, B be simple polygons such that $A \subset int(B)$.*

(i) *$CH_B(A)$ exists and is a uniquely defined simple polygon.*
(ii) *The frontier of the polygon $CH_B(A)$ is the Jordan curve which among all Jordan curves circumscribing A and lying in B, has the shortest length.*
(iii) *Each convex vertex of $CH_B(A)$ is a convex vertex of A, and each concave vertex of $CH_B(A)$ is a concave vertex of B.*

This was given by Theorem 3 from [22] and Theorem 4.6 from [23]. When the condition is weakened to $A \subset B$ then the polygon $CH_B(A)$ is simple or weakly simple, that means, its frontier can touch itself but does not cross itself, and the other properties are still valid [25].

Theorem 2. *For simple polygons A, B such that $A \subset B$, all vertices of $CH(A)$ are vertices of $CH_B(A)$.*

Proof. Any vertex of $CH(A)$ belongs to $A \subset CH_B(A)$. To prove that any vertex of $CH(A)$ is a vertex of $CH_B(A)$, we apply Lemma 1(i) and the well-known fact that any convex simple polygon is a finite intersection of halfplanes which are determined by the straight lines generated by the polygon edges. The convex simple polygon $CH(A)$ has $k \geq 3$ vertices $a_1, a_2, \cdots, a_k$, where no three consecutive points are collinear, and k edges $\overline{a_1 a_2}, \overline{a_2 a_3}, \cdots, \overline{a_k a_1}$. Supposing a clockwise tracing of the Jordan curve $fr(CH(A))$, let H_i be the right halfplane of the oriented straight line generated by the line segment $\overrightarrow{a_i a_{i+1}}$ for $i = 1, 2, \cdots k-1$, and H_k be the right halfplane of $\overrightarrow{a_k a_1}$. Then $CH(A) = H_1 \cap H_2 \cap \cdots \cap H_k$, and all these halfplanes are pairwise distinct. For any vertex a_i of $CH(A)$, a_{i-1}, a_i, a_{i+1} belong to $A \subset CH_B(A) \subset CH(A) \subset H_{i-1} \cap H_i \cap H_{i+1}$. This implies that $a_{i-1}, a_i, a_{i+1} \in fr(CH_B(A))$ and that $CH_B(A)$ cannot contain elements of the straight line generated by the segment $\overline{a_i a_{i+1}}$ but lying outside this segment. In consequence, in particular a_i is a vertex of $CH_B(A)$. Note that the argument of our proof is independent of a discussion weather $\overline{a_i a_{i+1}}$ belongs to B or not. □

The last theorem was briefly mentioned on p. 126 of [23] without proof, and it was stated in [4] with a wrong proof.

4   Previous Algorithms of Determining the Relative Convex Hull for Simple Polygons in the Plane

G. Toussaint proposed in [24,25] to transform the problem of determining $CH_B(A)$ into the problem of finding the shortest path between two vertices of a new simple polygon which first is triangulated. That algorithm has linear time complexity in terms of the total number k of vertices of A and B, but it makes essential use of the triangulation of M which can be achieved by a complicated process in $\mathcal{O}(k\,log(log(k)))$ time. In [22,23] several ideas for algorithms to determine the RCH were suggested, which are based on decompositions of the polygons such as trapezoidation or partition into pseudomonotone polygons. In the context of digital curve analysis, some algorithms not based on triangulations for calculating the MPP or MLP are known, for example [6,8,11,15,20,21], but these algorithms solve the RCH problem only for special difference sets $(B \setminus int(A))$ such as grid continua or polyominoes or special cell complexes.

The algorithm published in [3–5,11] starts with calculating the convex hulls of A and B. The list of vertices of $CH(A)$ is completed by inserting vertices from cavities of B until the output list of all vertices of $CH_B(A)$ is obtained. The construction of the output list follows a recursive process which searches for intersections of cavities of A and B. Whenever such intersection is detected, a new outer polygon O and a new inner polygon I are formed, and the problem of finding $CH_O(I)$ is treated to obtain missing RCH vertices of $CH_B(A)$. Subsequently, the recursive process works in each step with smaller newly generated outer and inner polygons and calculates their convex hulls. The author affirms that after sufficiently many recursion steps, the base case of the recursion is achieved where the new inner polygon is a triangle. The idea of such a recursive process was first suggested by two theorems on the shortest path between two vertices of a polygon and a series of drawings on pp. 122–124 in [23] where the explanation was not detailed at all. In certain situations, the algorithm from [4] does not produce the correct result of all vertices of $CH_B(A)$. The reason for this lies in the geometric nature of the RCH problem for general simple polygons A, B which was oversimplified in [4]; its recursion is theoretically not justified. The new polygon I sometimes is not contained in O or is not a simple polygon. For finding the missing vertices, additional regions have to be investigated in each step. It is also possible that the process stops when I becomes convex but is not a triangle.

5   A New Algorithm of Determining the Relative Convex Hull for Simple Polygons in the Plane

5.1   Vertex Lists, Convex Hull Determination and Cavity Detection

The new algorithm will be explained with the help of the example shown in Fig. 3. The input data consist of two simple polygons A, B satisfying $A \subset B$, given as ordered sequence of vertices: $A = \langle p_1, p_2, \ldots, p_n \rangle$, $B = \langle q_1, q_2, \ldots, q_m \rangle$

representing the frontier of each polygon due to the clockwise tracing. We suppose p_1 as an extreme vertex of A, q_1 extreme for B which can be achieved by a simple pre-processing of both lists. Hence p_1 is a vertex of $CH(A)$ and hence of $CH_B(A)$, by Theorem 2. The algorithm produces an ordered list of all vertices of $CH_B(A)$ as output data, starting with p_1 and corresponding to a clockwise tracing of the frontier of $CH_B(A)$.

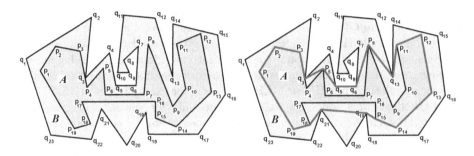

Fig. 3. Left: example of input data given by an inner polygon A and an outer polygon B. Right: the sides of $CH_B(A)$ are marked by heavy red lines.

Our algorithm starts with determining all vertices of the convex hulls of both A, B which are stored in the vertex lists $CH(A), CH(B)$, respecting the clockwise tracing. This can be done for example by the efficient Melkman-Algorithm [12]. As a particularity of this algorithm, the last vertex which was confirmed as vertex of the convex hull and hence appears at the end of the output list, is repeated in that list as first point, we eliminate this first point from the list. So we obtain the vertex list $CH(A)$ starting with p_1 and containing a selection of points from the list A whose original ordering and internal indices are preserved, similarly for $CH(B)$ starting with q_1. The vertex list $CH(A)$ is considered as initial output list of the vertices of the RCH. By subsequent steps of our algorithm, all other RCH vertices are found and inserted into this list $CH(A)$ at appropriate positions. Therefore, the format of a double ended queue owned by the vertex list $CH(A)$ as output of the Melkman-Algorithm, cannot be preserved during subsequent steps of our method. We apply later again the Melkman-Algorithm [12] which produces the vertex list of the convex hull for any input vertex list of a polyline not necessarily closed or forming a simple polygon, and it always respects the order in the input vertex list.

In each vertex list A, B, $CH(A)$, $CH(B)$, we copy its first point as added at the list end but having a new index. This permits to study all sides of each polygon, including the line segment connecting the last vertex with the first one, without producing errors in the indices when performing our algorithm. For our example, this produces $CH(A) = (p_1, p_2, p_{12}, p_{13}, p_{14}, p_{19}, p_{20} = p_1)$, $CH(B) = (q_1, q_2, q_{11}, q_{12}, q_{14}, q_{15}, q_{16}, q_{17}, q_{20}, q_{23}, q_{24} = q_1)$.

Since each point of the vertex list $CH(A)$, besides having an $CH(A)$-index i, also preserves its original index from the vertex list A, a cavity of the polygon

A is easily detected during tracing the list $CH(A)$: When consecutive vertices have a difference strictly mayor than 1 between their own indices, $CH(A)_i = p_k$, $CH(A)_{i+1} = p_l$, and $|k - l| \geq 2$, then A has a cavity whose cover is given by the line segment $\overline{p_k p_l}$. Cavities of B can be detected in the same manner from the list $CH(B)$. This idea was adopted from [4]. In our example, $i = 2$ indicates that $\overline{p_2 p_{12}}$ is the cover of a first cavity of A.

5.2   Processing of One Cavity

As in [4], whenever a cavity of A is found, it is considered as a *new polygon* O determined by its vertices $\langle CH(A)_i = p_k, p_{k+1}, p_{k+2}, \cdots, p_{k+r} = p_l = CH(A)_{i+1} \rangle$ for some $r \geq 1$ which always is a simple closed polyline in counterclockwise order. For our example, $i = 2$, $O = \langle p_2, p_3, p_4, \cdots, p_{11}, p_{12} \rangle$.

The next step is to construct a *new polyline* I whose convex hull, if it has at least three vertices, provides vertices of B which are vertices of the RCH and should be inserted in the list $CH(A)$ between $CH(A)_i$ and $CH(A)_{i+1}$. Let I be the sequence starting with $CH(A)_{i+1}$, $CH(A)_i$ and then containing all vertices from B, in the same order as in B, which belong to the set $(O \setminus \overline{CH(A)_i CH(A)_{i+1}})$ which is the polygon O with exception of its cover $\overline{CH(A)_i CH(A)_{i+1}}$. Only in the case that all those vertices selected from B are vertices of the same cavity of B, our definition of I coincides with that of [4]. For our example, $I = \langle p_{12}, p_2, q_3, q_4, q_5, q_6, q_7, q_8, q_9, q_{10}, q_{13} \rangle$ represents a closed polyline in counter-clockwise sense, but it does not form a simple polygon, and the curve is not completely contained in O. All points q_k of I with exception of q_{13} belong to the same cavity of B.

The Melkman-Algorithm [12] is applied to determine the convex hull of I. In our example, this produces the output $CH(I) = \langle q_{13}, p_{12}, p_2, q_3, q_5, q_6, q_{13} \rangle$. After eliminating the first point which is repeated and the end points of the cover which already belong to $CH(A)$, we obtain the following new vertices which will be inserted into the list $CH(A)$: q_3, q_5, q_6, q_{13}. The updated list $CH(A)$ then contains vertices both from A, B: $CH(A) = (p_1, p_2, q_3, q_5, q_6, q_{13}, p_{12}, p_{13}, p_{14}, p_{19}, p_{20})$. This current list $CH(A)$ represents two special line segments, each one connecting a vertex from A with a vertex from B: $\overline{p_2 q_3}$ and $\overline{q_{13} p_{12}}$. We will use these segments to form polylines whose convex hulls will provide eventually missing vertices of the RCH. These polylines were not defined or used in the algorithm of [4].

Definition 2. *Let $b_1, b_2, \cdots, b_k$ be the vertices of $CH(I)$ which were inserted into $CH(A)$ at the index i due to the procedure described above in order to generate the current list*

$$CH(A) = (CH(A)_1, CH(A)_2, \cdots, CH(A)_i, b_1, b_2, \cdots, b_k, CH(A)_{i+k+1}, \cdots).$$

*Define a **starting O-polygon** O_S by the vertex sequence starting with $CH(A)_i$, $CH(A)_{i+1}$ and then containing all vertices which in the vertex list B are previous to $CH(A)_{i+1} = b_1$, copying them in reversed order, until the first vertex which* *lies outside O. Let I_S be the polyline starting with $CH(A)_{i+1}$, $CH(A)_i$ and then containing all vertices from the vertex list A, copying their ordering, which belong to $(O \setminus \overline{CH(A)_i b_1})$.*

*Similarly, define an **ending O-polygon** O_E by the vertex sequence starting with $CH(A)_{i+k+1}$, $CH(A)_{i+k} = b_k$ and then containing all vertices which in the vertex list B are subsequent to $CH(A)_{i+k} = b_k$, copying their ordering, until the first vertex which lies outside O. Let I_E be the polyline starting with $CH(A)_{i+k+1}$, $CH(A)_{i+k} = b_k$ and then containing all vertices from the vertex list A, copying their ordering, which belong to $(O \setminus \overline{b_k CH(A)_{i+k+1}})$.*

By this definition, O_S is generated in counter-clockwise sense whereas I_S, O_E and I_E are polylines traced in clockwise sense. I_S, I_E are used for our algorithm whereas O_S, O_E are polygons only needed in its correctedness proof.

Lemma 2. *All vertices of $CH(I)$, $CH(I_S)$, $CH(I_E)$ are vertices of $CH_B(A)$.*

Idea of Proof: Let O be a cavity of A with cover $\overline{pq}$ and at least one vertex of B inside $O \setminus \overline{pq}$. O is a simple polygon. Due to Theorem 1(ii), all vertices of $CH_B(A)$ belonging to $R(O)$ are vertices of the shortest polygonal Jordan path which circumscribes A but lies in B. As consequence, the polygonal subpath from p to q is the shortest path between p, q as vertices of the weakly simply polygon $O \cap B$. By Theorem 4.4 of [23] (whose validity has to be generalized from a simple to a weakly simple polygon), this subpath is contained in $CH(I)$. Together with the fact that all vertices and edges of $CH_B(A)$ cannot intersect $int(A)$, it can be proved that each vertex of $CH(I)$ is a vertex of $CH_B(A)$. The polygons O_S, O_E, I_S, I_E are simple and $I_S \subset O_S$, $I_E \subset O_E$. The subpath of $fr(CH_B(A))$ from $CH(A)_i$ to b_1 passing through certain vertices of A (if any), is the shortest path between these vertices of the simple polygon $O_S \cap B$, it also belongs to $fr(CH_{O_S}(I_S))$. By Theorem 2, all vertices of $CH(I_S)$ are vertices of $CH_B(A)$; similarly for I_E. *(End of Idea of Proof)*

The Melkman-Algorithm [12] is applied for calculating the lists $CH(I_S)$, $CH(I_E)$, which after eliminating the points which are repeated or already belonging to the list $CH(A)$, have to be inserted into the list $CH(A)$: new points provided by $CH(I_S)$ are inserted between $CH(A)_i$ and $CH(A)_{i+1} = b_1$, new points from $CH(I_E)$ are inserted between $CH(A)_{i+k} = b_k$ and $CH(A)_{i+k+1}$. In our example, $i = 2$, $CH(A)_i = p_2$, $CH(A)_{i+1} = b_1 = q_3$, $k = 4$, $CH(A)_{i+k} = b_k = q_{13}$, $CH(A)_{i+k+1} = p_{12}$, $O_S = \langle p_2, q_3, q_2 \rangle$, $I_S = \langle q_3, p_2, p_3 \rangle$ is convex and provides the new point p_3 to be inserted between p_2 and q_3. $O_E = \langle p_{12}, q_{13}, q_{14} \rangle$, $I_E = \langle p_{12}, q_{13}, p_{11} \rangle$ is convex, so that only p_{11} has to be inserted between q_{13} and p_{12}. The new list is $CH(A) = \langle p_1, p_2, p_3, q_3, q_5, q_6, q_{13}, p_{11}, p_{12}, p_{13}, p_{14}, p_{19}, p_{20} \rangle$. This completes to process the cavity of A starting at the vertex with $CH(A)$-index i. Note that during the whole procedure just described, this starting index i is not changed and points are inserted only after that index. Comparing the current list $CH(A)$ with Fig. 3 we see that within the actual cavity, more RCH vertices have to been detected, but the list $CH(A)$ will guide us naturally to discover these missing points.

5.3   Detection and Processing of Subsequent Cavities

The algorithm continues tracing the vertex list $CH(A)$ which has been updated by processing the cavity previously detected, increasing the $CH(A)$-index i and looking for consecutive vertices whose own indices have a difference more than 1. This test is done only for consecutive vertices which both are from A, or both from B. When two points are consecutive in $CH(A)$ but one is from A and the other from B, then the point from B was inserted as result of the treatment of the special polygons O_S or O_E, and no more vertices of the RCH are missing between these two points.

Whenever in the list $CH(A)$ two consecutive points of A, $CH(A)_i = p_k$ and $CH(A)_{i+1} = p_l$, such that $|k - l| \geq 2$, are found, then $p_k p_l$ covers some kind of "cavity" of A and the whole "Processing of one cavity" described in the previous section, is performed. This includes the analysis of the polygons and polylines O, I, O_S, I_S, O_E, I_E, resulting in an updated vertex list $CH(A)$. The same is done when such two consecutive points of B, $CH(A)_i = q_k$ and $CH(A)_{i+1} = q_l$, are detected, but then the "Processing of one cavity" is applied with the roles of A and B interchanged (points q_j instead of p_j and vice versa).

In our example, the next such situation is found for $i = 4$ and points of B: $CH(A)_4 = q_3, CH(A) = q_5$. Following faithfully the procedure with roles of A and B interchanged, we obtain $O = \langle q_3, q_4, q_5 \rangle$ which is a cavity of a cavity of B with one vertex of A inside, giving $I = \langle q_5, q_3, p_5 \rangle$. I is convex and provides only the new point p_5. The special segments $\overline{q_3 p_5}$ and $\overline{p_5 q_5}$ generate $O_S = \langle q_3, p_5, p_4 \rangle$ and $O_E = \langle q_5, p_5, p_6 \rangle$ which both do not contain vertices of B,

Fig. 4. The relative convex hull determined by the new algorithm implemented in Matlab, and by the algorithm of [4], for an example developed in [16].

Input: Simple polygons A, B with $A \subset B$ given by vertex lists $A = \langle p_1, p_2, \ldots, p_n \rangle$,
$\quad B = \langle q_1, q_2, \ldots, q_m \rangle$ (clockwise traced).
Output: List of vertices of the relative hull $CH_B(A)$ stored in the actualized list
$\quad CH(A)$ (clockwise traced).

1: Initialize $CH(A) = \emptyset$, $i = 1$.
2: Determine the vertices of convex hull of A by the Melkman algorithm stored in the
   list $CH(A)$ which has s elements.
3: Eliminate the first element of $CH(A)$.
4: Extend the lists A, B y $CH(A)$ adding at the end a copy of the first element with
   new index.
5: **while** $i < s$ **do**
6: **if** a cavity is detected between $CH(A)_i$ and $CH(A)_{i+1}$, **then**
7:   **if** The cavity is between points of A, **then**
8:    CAV$(CH(A), i, s, A, B)$
9:   **else**
10:    CAV$(CH(A), i, s, B, A)$
11:   **end if**
12: **end if**
13: i=i+1
14: **end while**
15: The actualized list $CH(A)$ contains all vertices of the relative convex hull $CH_B(A)$.

Fig. 5. Pseudocode of the new RCH algorithm (Main program).

hence $I_S = \langle p_5, q_3 \rangle$ and $I_E = \langle q_5, p_5 \rangle$ are degenerated to line segments and do not provide more points to be inserted into the vertex list. We obtain as current list $CH(A) = (p_1, p_2, p_3, q_3, p_5, q_5, CH(A)_7 = q_6, q_{13}, p_{11}, p_{12}, p_{13}, p_{14}, p_{19}, p_{20})$.

The next jump in the indices is found at $i = 7$ again with points from B: $CH(A)_7 = q_6, CH(A)_7 = q_{13}$. We should be careful using geometrical concepts, the segment $\overline{q_6 q_{13}}$ covers some part of B which is neither a cavity nor a cavity of a cavity of B. We obtain $O = \langle q_6, q_7, q_8, q_9, q_{10}, q_{11}, q_{12}, q_{13} \rangle$, $I = \langle q_{13}, q_6, p_8 \rangle$ which is convex and provides only the new point p_8. O_S, O_E are not interesting since I_S, I_E degenerate to line segments and do not provide more points: $CH(A) = (p_1, p_2, p_3, q_3, p_5, q_5, q_6, p_8, q_{13}, p_{11}, p_{12}, p_{13}, p_{14}, p_{19}, p_{20})$. The next pair of points to be treated is found as p_{14}, p_{19}, where the polylines O, I provide the new RCH vertices q_{19} and q_{21}, and then we need O_S, I_S to discover p_{15} and also O_E, I_E to detect the last RCH vertex p_{18} which completes the correct determination of the RCH shown in Fig. 3.

5.4 Pseudocode, Implementation, and Complexity

Figure 4 shows an example where the RCH was calculated by our algorithm implemented in Matlab R2012a. The example was designed in [16] to contain several interesting situations, such as a convex cavity of A with vertices of B inside, a non-convex cavity of A with vertices of B inside, vertices of A inside

1: **procedure** CAV($CH(A)$, i, s, P1, P2)
2:     Initialize local variables $u = 0, v = 0, w = 0$.
3:     Generate the polygon O by all vertices of P1, from $CH(A)_i$ up to $CH(A)_{i+1}$.
4:     Form the polyline I by $CH(A)_{i+1}$, $CH(A)_i$, and all vertices of $P2$ inside O or collinear with its frontier with exception of the cover $\overline{CH(A)_{i+1}CH(A)_i}$. I has N vertices.
5:     **if** $N > 2$ **then**
6:         Determine the list $CH(I)$ which has S vertices.
7:         Insert between $CH(A)_i$ and $CH(A)_{i+1}$ the vertices of $CH(I)$, with exception of the first point and $CH(A)_i$, $CH(A)_{i+1}$.
8:         u=S-3
9:         s=s+u
10:         Generate the polygon O_S by $CH(A)_i$, $CH(A)_{i+1}$ and all vertices of $P2$ previous to $CH(A)_{i+1}$ up to a first point found outside O.
11:         Form the polygon I_S by $CH(A)_{i+u+1}$, $CH(A)_{i+u}$ and all vertices of $P1$ inside O_S or collinear with its frontier (with exception of the line segment $\overline{CH(A)_{i+u+1}CH(A)_{i+u}}$). I_S has N_S vertices.
12:         **if** $N_S > 2$ **then**
13:             Determine the list $CH(I_S)$ which has S elements.
14:             Insert between $CH(A)_i$ and $CH(A)_{i+1}$ the elements of $CH(I_S)$ with exception of the first one and $CH(A)_{i+u+1}$ and $CH(A)_{i+u}$.
15:             v=S-3
16:             s=s+v
17:             w=u+v
18:         **end if**
19:         Generate the polygon O_E by $CH(A)_{i+w+1}$, $CH(A)_{i+w}$ and all vertices of $P2$ subsequent to $CH(A)_{i+w}$ up to the first point found outside O.
20:         Form the polygon I_E by $CH(A)_{i+w+1}$, $CH(A)_{i+w}$ and all vertices of $P1$ inside O_E or collinear with its frontier (with exception of the line segment $\overline{CH(A)_{i+w+1}CH(A)_{i+w}}$). I_E has N_E vertices.
21:         **if** $N_E > 2$ **then**
22:             Determine the list $CH(I_E)$ which has S elements.
23:             Insert between $CH(A)_{i+w}$ and $CH(A)_{i+w+1}$ the elements of the list $CH(I_E)$ with exception of the first one and $CH(A)_{i+w}$ and $CH(A)_{i+w+1}$.
24:             x=S-3
25:             s=s+x
26:         **end if**
27:         i=i-1
28:     **end if**
29:     **return** $CH(A), i, s$
30: **end procedure**

Fig. 6. Pseudocode of the new RCH algorithm (Cavity processing procedure).

interesting parts of B, a part of $fr(B)$ collinear with the frontier of a cavity of A. In the left part of each figure, both polygons A, B are isothetic and the difference set $(B \setminus A)$ looks like a grid continuum, such that in this part we apply our algorithm to solve the MLP problem. The figure shows that the RCH problem, even for the MLP case, cannot be solved by the recursion of [4].

Figures 5 and 6 present a pseudocode of our algorithm which is not yet optimized. To estimate the time complexity of our method, suppose that the input polygons A and B have n and m vertices, respectively. Not only the Melkman-Algorithm is applied and computes the convex hull of any polyline given as ordered sequence of k vertices in linear time $\mathcal{O}(k)$. In several steps, our method needs to decide whether a point belongs to the right or left halfplane of a straight line segment, where the determinant described in Sect. 2 is used. Also it has to be determined whether a point lies inside or outside a simple polygon given by its vertex list. When this list corresponds to a clockwise order tracing, then a point is inside the polygon if it belongs to the right halfplanes of all polygon edges. Such verifications are needed in our algorithm for polygons given by small subsets of vertices of A, B, so that their time complexity can be considered as linear in dependance of $m + n$.

Up to three distinct convex hulls have to be computed for each "cavity" intersection of A and B. A has a maximum number of $\lfloor n/2 \rfloor$ cavities. Each such cavity of A could have vertices of B inside. These vertices belong to the set of concave vertices of B which could have almost m elements. This gives a quadratic time complexity in the worst case. Another problem is the possible existence of interleaved and interlaced cavities within other cavities. Although our algorithm is not recursive but iterative, each cavity lying inside another cavity, when not treated immediately, is detected later when tracing the updated vertex list $CH(A)$ and then treated. So, as also observed in [4], only in cases when the "deepness" of such "stacked cavities" is bounded by some constant and the cavities in general are "well distributed" then our algorithm can present a nearly linear time complexity behaviour.

6   Conclusion and Future Work

This paper proposes an algorithm for the determination of the list of all vertices of the relative convex hull, for the general situation of given simple plane polygons A, B such that $A \subset B$. This algorithm does not use triangulation or similar decompositions of the difference set between B and A as preprocessing. The ordered input vertex sequences of A and B are processed going forward to generate the output list of vertices of $CH_B(A)$ by inserting points iteratively into the list of vertices of the convex hull of A.

Near future work previews to complete the formal proof of correctness of our algorithm and the solution of some pendent details such as the insertion of the new vertices found from the convex hull of the polyline I into the current vertex list $CH(A)$ in the "correct" order, or the treatment of the presence of collinear (non-consecutive) vertices of A or B, a situation which interestingly is forbidden for algorithms based on triangulation [25].

Acknowledgement. The first author gratefully acknowledges support for this research from SEP and CONACYT Mexico, grant No. CB-2011-01-166223. The authors would like to thank very much to the reviewers for their careful study of the work, and for their constructive criticism and helpful comments which were important to improve the presentation of the paper.

References

1. Biswas, A., Bhowmick, P., Sarkar, M., Bhattacharya, B.B.: A linear-time combinatorial algorithm to find the orthogonal hull of an object on the digital plane. Inf. Sci. **216**, 176–195 (2012)
2. Ishaque, M., Toth, C.D.: Relative convex hulls in semi-dynamic arrangements. Algorithmica **68**(2), 448–482 (2014)
3. Klette, G.: A recursive algorithm for calculating the relative convex hull. In: Proceedings of 25th International Conference on Image and Vision Computing, New Zealand, pp. 1-7. IEEE Computer Society (2010). doi:10.1109/IVCNZ.2010. 6148857, 978-1-4244-9631-0/10
4. Klette, G.: Recursive calculation of relative convex hulls. In: Debled-Rennesson, I., Domenjoud, E., Kerautret, B., Even, P. (eds.) DGCI 2011. LNCS, vol. 6607, pp. 260–271. Springer, Heidelberg (2011)
5. Klette, G.: Recursive computation of minimum-length polygons. Comput. Vis. Image Underst. **117**, 386–392 (2012)
6. Klette, R., Kovalevsky, V., Yip, B.: On the length estimation of digital curves. In: SPIE Proceedings of Vision Geometry VIII, vol. 3811, pp. 117–129. SPIE (1999)
7. Klette, R.: Multigrid convergence of geometric features. In: Bertrand, G., Imiya, A., Klette, R. (eds.) Digital and Image Geometry. LNCS, vol. 2243, pp. 318–338. Springer, Heidelberg (2002)
8. Klette, R., Rosenfeld, A.: Digital Geometry - Geometric Methods for Digital Picture Analysis. Morgan Kaufmann Publ., Elsevier, USA (2004)
9. Lantuejoul, C., Beucher, S.: On the use of the geodesic metric in image analysis. J. Microsc. **121**(1), 39–49 (1981)
10. Lantuejoul, C., Maisonneuve, F.: Geodesic methods in quantitative image analysis. Pattern Recoglnition **17**(2), 177–187 (1984)
11. Li, F., Klette, R.: Euclidean Shortest Paths, Exact or Approximate Algorithms. Springer, London (2011)
12. Melkman, A.: On-line construction of the convex hull of a simple polyline. Inf. Process. Lett. **25**, 11–12 (1987)
13. Mitchell, J.S.B.: Geometric shortest paths and network optimization. In: Sack, J.R., Urrutia, J. (eds.) Handbook of Computational Geometry, pp. 633–701. Elsevier, Amsterdam (2000)
14. Munkres, J.R.: Topology, 2nd edn. Prentice Hall, USA (2000)
15. Provençal, X., Lachaud, J.-O.: Two linear-time algorithms for computing the minimum length polygon of a digital contour. In: Brlek, S., Reutenauer, C., Provençal, X. (eds.) DGCI 2009. LNCS, vol. 5810, pp. 104–117. Springer, Heidelberg (2009)
16. Reyes Becerril, H.: Versión revisada de un algorítmo que determina la cubierta convexa relativa de polígonos simples en el plano, Master Thesis. Dept. of Automatic Control, CINVESTAV-IPN, Mexico City, September 2013
17. Robert, L., Faugeras, O.D.: Relative 3D positioning and 3D convex hull computation from a weakly calibrated stereo pair. Image Vis. Comput. **13**(3), 189–196 (1995)

18. Sklansky, J.: Recognition of convex blobs. Pattern Recognition **2**, 3–10 (1970)

19. Sklansky, J.: Measuring cavity on a rectangular mosaic. IEEE Trans. Comput. **C–21**(12), 1355–1364 (1972)

20. Sklansky, J., Kibler, D.F.: A theory of nonuniformly digitized binary pictures. IEEE Trans. Syst. Man Cybern. **6**(9), 637–647 (1976)

21. Sklansky, J., Chazin, R.L., Hansen, B.J.: Minimum perimeter polygons of digitized silhouettes. IEEE Trans. Comput. **21**(3), 260–268 (1972)

22. Sloboda, F., Stoer, J.: On piecewise linear approximation of planar Jordan curves. J. Comput. Appl. Math. **55**, 369–383 (1994)

23. Sloboda, F., Zatco, B., Stoer, J.: On approximation of planar one-dimensional continua. In: Klette, R., Rosenfeld, A., Sloboda, F. (eds.) Advances in Digital and Computational Geometry, pp. 113–160. Springer, Singapore (1998)

24. Toussaint, G.T.: An optimal algorithm for computing the relative convex hull of a set of points in a polygon. In: Proceedings of EURASIP, Signal Processing III: Theories and Applications, Part 2, pp. 853–856. North-Holland (1986)

25. Toussaint, G.T.: Computing geodesic properties inside a simple polygon. Invited paper, Special Issue on Geometric Reasoning, Revue D'Intelligence Artificielle **3**(2), 9–42 (1989)

26. Toussaint, G.T.: On separating two simple polygons by a single translation. Discrete Comput. Geom. **4**(1), 265–278 (1989)

27. Wiederhold, P., Villafuerte, M.: Triangulation of cross-sectional digital straights segments and minimum length polygons for surface area estimation. In: Wiederhold, P., Barneva, R.P. (eds.) Progress in Combinatorial Image Analysis, pp. 79–92. Research Publishing Services, Singapore (2009)

28. Yu, L., Klette, R.: An approximative calculation of relative convex hulls for surface area estimation of 3D digital objects. ICPR **1**, 131–134 (2002)

Spatiotemporal Barcodes for Image Sequence Analysis

Rocio Gonzalez-Diaz$^{(\boxtimes)}$, Maria-Jose Jimenez,
and Belen Medrano

Department of Applied Mathematics (I), University of Seville,
Av. Reina Mercedes, s/n, 41012 Seville, Spain
{rogodi,majiro,belenmg}@us.es

Abstract. Taking as input a time-varying sequence of two-dimensional (2D) binary images, we develop an algorithm for computing a *spatiotemporal* 0–barcode encoding lifetime of connected components on the image sequence over time. This information may not coincide with the one provided by the 0–barcode encoding the 0–persistent homology, since the latter does not respect the principle that it is not possible to move backwards in time. A cell complex K is computed from the given sequence, being the cells of K classified as *spatial* or *temporal* depending on whether they connect two consecutive frames or not. A *spatiotemporal path* is defined as a sequence of edges of K forming a path such that two edges of the path cannot connect the same two consecutive frames. In our algorithm, for each vertex $v \in K$, a spatiotemporal path from v to the "oldest" *spatiotemporally-connected* vertex is computed and the corresponding *spatiotemporal* 0–bar is added to the *spatiotemporal* 0–barcode.

Keywords: Persistent homology · Barcodes · Spatiotemporal data · Digital image sequence analysis

1 Introduction

Persistent homology [3,5,12] and zigzag persistence [2] provides information about lifetime of homology classes along a filtration of cell complexes. Such a filtration might be determined by time in a set of spatiotemporal data. Our general aim is to compute the "spatiotemporal" topological information of such filtration, taking into account that it is not possible to move backwards in time (which is not obvious if we use the known algorithms for computing (zigzag) persistent homology).

In the context of mobile sensor networks, [4] is devoted to a problem related with the one posed here: can a moving intruder avoid being detected by the sensors? If the answer is yes, the path that describes the intruder over time is called an *evasion path*. In the study of evasion paths in [4], the region covered

Author partially supported by IMUS, Junta de Andalucia under grant FQM-369, Spanish Ministry under grant MTM2012-32706 and ESF ACAT program.

R.P. Barneva et al. (Eds.): IWCIA 2015, LNCS 9448, pp. 61–70, 2015.
DOI: 10.1007/978-3-319-26145-4_5

by sensors at time t is encoded using Rips complex. A single cell complex SR is computed by stacking the Rips complexes $R(t)$ for all times t. Theorem 7 of [4] proves that there is no evasion path in a given mobile sensor network under a "homological" criterion. Using zig-zag persistent homology, an equivalent condition is provided in [1]. Nevertheless, no general necessary and sufficient condition for the existence of an evasion path is given. The problem is how to capture in the cell complex SR, the idea that an intruder cannot move backwards in time. In [6], the authors analyze time-varying coverage properties in dynamic sensor networks by means of zigzag persistent homology. Coverage holes are tracked in the network by using representative cycles of 1–homology classes.

In this paper, we are concerned with the treatment of time-varying sequences of 2D binary images and the tracking of connected components over time inspired by persistent homology methods.

An overview of the main tools used in this paper: basics of persistent homology and AT-models are given in Sect. 2. We state the problem of computing the "correct" topological information of spatiotemporal data encoded in a single cell complex in Sect. 3, through two simple examples. Our method to solve the problem in dimension 0 is then introduced in Sect. 4. Cell complexes encoding spatiotemporal information of time-varying sequences of 2D binary images is given in Sect. 5. We conclude in Sect. 6 and describe possible directions for future work.

2   Persistent Homology Through AT-models

Roughly speaking, a cell complex K is a general topological structure by which a space is decomposed into basic elements (cells) of different dimensions that are glued together by their boundaries (see the definition of CW-complex in [10]). If the cells in K are p–dimensional cubes (vertices, edges, square faces, cubes, ...) then K is a *cubical complex*. The dimension of a cell $\sigma \in K$ is denoted by $dim(\sigma)$. A cell $\mu \in K$ is a p–*face* of a cell $\sigma \in K$ if μ lies in the boundary of σ and $p = dim(\mu) < dim(\sigma)$.

A p–*chain* is a formal sum of p–cells in K. Since we work with coefficients that are either 0 or 1, we can think of a p–chain as a set of p–cells, namely those with coefficients equals to 1. In set notation, the sum of two p–chains is their symmetric difference. The p–chains together with the addition operation form a group denoted as $C_p(K)$. Besides, the set $\{C_p(K)\}_p$ is denoted by $C(K)$. A set of homomorphism $\{f_p : C_p(K) \to C_p(K')\}_p$ is called a *chain map* and denoted by $f : C(K) \to C(K')$. Given two p–cells $\sigma \in K$ and $\sigma' \in K'$, we say that $\sigma' \in f(\sigma)$ if σ' belongs to the p–chain $f_p(\sigma)$ (in set notation). The *boundary map* $\partial : C(K) \to C(K)$ is defined on a p–cell σ as the sum of its $(p-1)$–faces. This way, for a p–chain, $c = \sum_{i \in I} \sigma_i$, the boundary of c is the sum of the boundaries of its cells, $\partial_p c = \sum_{i \in I} \partial_p \sigma_i$.

A *filtration* of K is an increasing sequence of cell complexes: $\emptyset = K_0 \subset K_1 \subset \cdots \subset K_n = K$. The partial ordering given by such a filtration can be extended to a total ordering of the cells of K: $\{\sigma_1, \ldots, \sigma_m\}$, satisfying that for each i, $1 \leq i \leq m$, the faces of σ_i lies in the set $\{\sigma_1, \ldots, \sigma_i\}$. Then, the map $index : K \to \mathbb{Z}$ is defined by $index(\sigma_i) := i$.

Informally, the p–th persistent homology groups [3,12] can be seen as a collection of p–homology classes (representing connected components when $p = 0$, holes when $p = 1$, cavities when $p = 2$, ...) that are born at or before we go from K_{i-1} to K_i and die after we go from K_i to K_{i+1}. A p–barcode [7] is a graphical representation of the p–th persistent homology groups as a collection of horizontal line segments (bars) in a plane. Axis corresponds to the indices of the cells in K. For example, if a p–homology class was born at time i (i.e. when σ_i is added) and died at time j ($1 \leq i < j \leq m$), then a bar with endpoints (i, i) and (j, i) is added to the p–barcode.

In [8] the authors establish a correspondence between the incremental algorithm for computing AT-models [9] and the one for computing persistent homology. The first approach provides a rich algebraic information encoded by a chain homotopy operator ϕ, that "connects" any p–cell to the corresponding surviving cell.

An AT-model for a cell complex K is a quintuple (f, g, ϕ, K, H), where:

– K is the cell complex.
– $H \subseteq K$ describes the homology of K, in the sense that it contains a distinct p–cell for each p–homology class of a basis, for all p. The cells in H are called surviving cells. The set of all the surviving p–cells together with the addition operation form the group $C_p(H)$ for all p.
– $g : C(H) \to C(K)$ is a chain map that maps each p–cell h in H to one representative cycle $g_p(h)$ of the corresponding homology class $[g_p(h)]$.
– $f : C(K) \to C(H)$ is a chain map that maps each p–cell in K to a sum of surviving cells, satisfying that if $a, b \in C_p(K)$ are two homologous p–cycles then $f_p(a) = f_p(b)$.
– $\phi : C(K) \to C(K)$ is a chain homotopy (see [11]). Intuitively, for a p–cell σ, $\phi_p(\sigma)$ returns the $(p + 1)$–cells needed to be contracted to "bring" σ to a surviving p–cell contained in $f_p(\sigma)$.

In the case of a 0–cell $v \in K$, $\phi_0(v)$ will provide a path in K (a sequence of edges of K connecting a sequence of different vertices) from the vertex v to the oldest (i.e., with lowest index) vertex in the same connected component.

3   Stating the Problem

Our general goal is to compute spatiotemporal p–barcodes for a time-varying sequence of nD binary images in the sense that they can represent evolution of homology classes over time. In this paper, we focus our effort in computing spatiotemporal 0–barcodes for time-varying sequences of 2D binary images.

In order to give some intuition about the problem we want to state, let us consider the simple examples given in Fig. 1, in which two sequences of a few 4–connected pixels appearing, moving and disappearing over time, are shown.

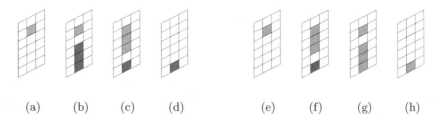

Fig. 1. Pixels appearing, moving and disappearing over time.

To encode the spatiotemporal information of the two sequences, we construct associated cell complexes by replacing each pixel by a vertex and adding an edge between two vertices if:

– The corresponding pixels are 4–connected (in the same frame).
– The vertices correspond to the same pixel at different times.

The resulting cell complexes K and K' are shown in Fig. 2.

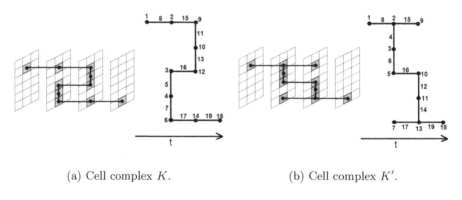

(a) Cell complex K.   (b) Cell complex K'.

Fig. 2. Cell complexes K and K' obtained, respectively, from the sequence showed in Fig. 1(a)–(d) and (e)–(h).

Now, to compute 0–persistent homology on these two cell complexes K and K', we should select an appropriate filtration. Since we want to capture the variation of homology classes over time, we first classify the cells of K and K' in *spatial* and *temporal*:

– All vertices are spatial (since vertices represent pixels).
– An edge is spatial if its endpoints represent pixels of the same frame.
– If an edge is not spatial then it is temporal.

Therefore, we have the following *spatial subcomplexes* of K: $T_1 = \{1\}$, $T_2 = \{2, 3, 4, 5, 6, 7\}$, $T_4 = \{9, 10, 11, 12, 13, 14\}$, $T_6 = \{18\}$. And the following sets of temporal cells: $T_3 = \{8\}$, $T_5 = \{15, 16, 17\}$, $T_7 = \{19\}$, where numbers correspond to the labels of the cells showed in Fig. 2(a). The filtration $\emptyset = K_0 \subset K_1 \subset \cdots \subset K_7 = K$ is obtained by interleaving the temporal cells after the correspondent spatial subcomplexes. That is, $K_i = K_{i-1} \cup T_i$, $i = 1, \ldots, 7$.

Besides, the filtration on K' coincides with the filtration on K, where numbers now correspond to the labels of the cells showed in Fig. 2(b).

If we compute 0–persistent homology of K and K' using the above filtrations, we will obtain, in both cases, that a connected component (0–homology class) is born when cell 1 is added and survives until the end. So, in both cases, a bar with endpoints $(1, 1)$ and $(19, 1)$ is added to the 0–barcode.

However, we can observe that Fig. 1(a)–(d) cannot represent a connected component that is moving from the very beginning until the end while Fig. 1(e)–(h) can. So we wonder if we could modify the 0–barcode of the first sequence (Fig. 1(a)–(d)) so that it codifies the connected components that can survive along time. The idea is to replace the bar with endpoints $(1, 1)$ and $(19, 1)$ by respective bars from $(1, 1)$ to $(13, 1)$ and from $(3, 3)$ to $(19, 3)$, what will be formally described in next section.

4   Our Method

In this section, our aim is to design an algorithm to compute the spatiotemporal 0–barcode of a cell complex K encoding spatiotemporal data.

Suppose that K is composed by a stack of (*spatial*) complexes and a set of (*temporal*) cells such that each temporal cell connects two (consecutive) spatial complexes. Hence, our starting point is a (*spatiotemporal*) filtration of K, that is, a filtration $\emptyset = K_0 \subset K_1 \subset \cdots \subset K_n = K$ such that, for all i, $1 \leq i \leq n$, the set $T_i = K_i \setminus K_{i-1}$ is:

– a set of spatial cells if $i = 1$ or i is even;
– a set of temporal cells if $i > 1$ is odd.

A *spatiotemporal path* c in K is a path in K such that $\#(c \cap T_i) \leq 1$, for any i odd, $1 < i \leq n$. That is, there are not two temporal edges connecting the same consecutive spatial complexes, which follows from the idea that it is not possible to move backwards in time. Two vertices are *spatiotemporally-connected* if there is a spatiotemporal path between them.

Algorithm 1 extends the incremental algorithm for computing AT-models given in [9]. The eleven last lines of Algorithm 1 are original in this paper.

Although Algorithm 1 follows the same idea behind the algorithm given in [9] (by which, for each cell σ, $\phi(\sigma)$ "connects" the cell σ to a surviving cell), the computation of the map ϕ' is new in this paper.

In Algorithm 1, we compute a path $\phi'(v)$ from v to a surviving cell and, if $\phi'(v)$ is not spatiotemporal, we break it in pieces that are spatiotemporal paths. Then, a spatiotemporal path $\phi'(v)$ is obtained from each vertex $v \in K$ to the

---

Algorithm 1. Spatiotemporal 0–barcode.

---

1 **Input:** An ordering of the cells of K extending the partial ordering imposed by a spatiotemporal filtration.

2 **Output:** An AT-model for K and a spatiotemporal 0–barcode $\mathcal{B}$.

3 $H := \emptyset$.

4 **for** $i = 1$ **to** m **do**

5    $f(\sigma_i) := 0,\ \phi(\sigma_i) := 0,\ \phi'(\sigma_i) := 0$.

6 **for** $i = 1$ **to** m **do**

7   **if** $f\partial(\sigma_i) = 0$ **then**

8      $f(\sigma_i) := \sigma_i,\ g(\sigma_i) := \sigma_i + \phi\partial(\sigma_i),\ H := H \cup \{\sigma_i\}$.

9   **if** $\dim(\sigma_i) = 0$ **then**

10   Add to $\mathcal{B}$ a point at (i,i).

11   **if** $f\partial(\sigma_i) \neq 0$ **then**

12   Let $\sigma_j \in f\partial(\sigma_i)$ s.t. $j = \max\{index(\mu) : \mu \in f\partial(\sigma_i)\}$

13      $H := H \setminus \{\sigma_j\}$

14   **foreach** $x \in K$ s.t. $\sigma_j \in f(x)$ **do**

15       $f(x) := f(x) + f\partial(\sigma_i),\ \phi(x) := \phi(x) + \sigma_i + \phi\partial(\sigma_i)$.

16   **if** $\dim(\sigma_i) = 1$ **then**

17   Let $v, w, v', w' \in K$ s.t. $\partial(\sigma_i) = v + w,\ v' = \partial\phi'(v) + v,$ $w' = \partial\phi'(w) + w$ and $index(v') < index(w')$.

18   Add to $\mathcal{B}$ the bar with endpoints $\{(index(v'), index(v')),$ $(i, index(v'))\}$ and the bar with endpoints $\{(index(w'), index(w')),$ $(i, index(w'))\}$.

19   **if** $v \in T_\ell$ and $w, w' \in T_{\ell'}$ for some ℓ, ℓ', s.t. $1 \leq \ell \leq \ell' \leq n$ **then**

20   **foreach** $x \in K,\ x \neq w, w'$ s.t. $\partial\phi'(x) + x = w'$ **do**

21         $\phi'(x) := \phi'(x) + \phi'(w) + \sigma_i + \phi'(v)$.

22        $\phi'(w') := \phi'(w) + \sigma_i + \phi'(v)$;

23        $\phi'(w) := \sigma_i + \phi'(v)$.

"oldest" spatiotemporally-connected vertex. Regarding the spatiotemporal 0–barcode, at time i, we elongate a bar only if $dim(\sigma_i) = 1$ and the connected component that represents the bar is spatiotemporally connected to some of the endpoints of the edge σ_i. Otherwise, we do not elongate the bar. This is different from classical barcodes in which, for example, the bar corresponding to a connected component that appear in time i and does not merge to other connected component later, is elongated until the very end.

Proposition 1. *If v is a vertex in K then, $\phi'(v)$ is a spatiotemporal path.*

Proof. Let us prove the proposition by construction. At the beginning of the algorithm, $\phi'(v) = 0$ for every vertex $v \in K$. Suppose the algorithm is running and we are in step i, $1 \leq i \leq m$. Suppose that σ_i is an edge of K. Then, $\partial(\sigma_i) = v + w$ being v and w two vertices of K. Besides, by induction, $\phi'(v)$ and

$\phi'(w)$ are spatiotemporal paths. Then, $\partial\phi'(v) + v = v'$ and $\partial\phi'(w) + w = w'$ for some vertices $v' \in T_\ell$ and $w' \in T_{\ell'}$ being $1 \leq \ell, \ell' \leq n$.

We can assume that $index(v') < index(w')$. The case $index(v') = index(w')$ can only occur when $v' = w'$, what means that $f\partial(\sigma_i) = 0$ (a 1-cycle is being closed) and neither $\mathcal{B}$ nor ϕ' are modified in this case.

Now, let $c_w = \sigma + \phi'(v)$, $c_{w'} = \phi'(w) + \sigma + \phi'(v)$ and $c_x = \phi'(x) + \phi'(w) + \sigma + \phi'(v)$, for any $x \in K$ such that $\partial\phi'(x) + x = w'$. Then, $\partial(c_w) = w + v'$, $\partial(c_{w'}) = w' + v'$ and $\partial(c_x) = x + v'$. We have to consider the following cases:

– If σ_i is spatial, then $\sigma_i, v, w \in T_j$ for some j, $1 \leq j \leq n$. We have to consider the following cases:
  • If $\ell' < j$ then ϕ' is not updated.
  • If $\ell' = j$ then $\phi'(x) \subseteq T_j$ for any $x \in K$ s.t. $\partial\phi'(x) + x = w'$ and, therefore, c_w, $c_{w'}$ and c_x are spatiotemporal paths.
– If σ_i is temporal, then $v \in T_j$ and $w \in T_{j'}$ for some $j \neq j'$, $1 \leq j, j' \leq n$.
  • If $j < j'$. We consider two cases:
    * If $\ell' = j'$ then $\phi'(x) \subseteq T_{j'}$ for any $x \in K$ s.t. $\partial\phi'(x) + x = w'$ and, therefore, c_w, $c_{w'}$ and c_x are spatiotemporal paths.
    * If $\ell' < j'$ then ϕ' is not updated.
  • If $j' < j$ then ϕ' is not updated.   □

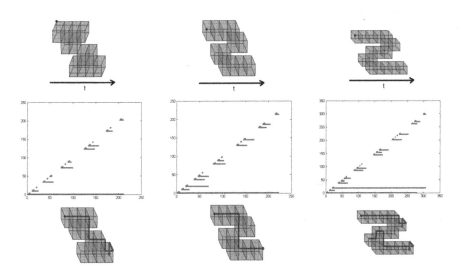

Fig. 3. Top: Three simple examples of stacked cubical complexes (t being the temporal dimension). Middle: The associated spatiotemporal barcodes obtained by applying Algorithm 1. Bottom: The spatiotemporal paths of the longest-lived 0–homology classes (in blue) (Color figure online).

5   Spatiotemporal Representation of Image Sequences

In this section, we explain how to compute a spatiotemporal filtration representing a time-varying sequence of 2D binary images, inspired by the stack complexes described in [1,4].

Consider $\mathbb{Z}^2$ as the set of points with integer coordinates in 2D space $\mathbb{R}^2$. A *2D binary image* is a set $I = (\mathbb{Z}^2, 8, 4, B)$, where $B \subset \mathbb{Z}^2$ is the *foreground*, $B^c = \mathbb{Z}^2 \backslash B$ the *background*, and $(8, 4)$ is the adjacency relation for the foreground and background, respectively. A point $p \in \mathbb{Z}^2$ can be interpreted as a unit closed square (called *pixel*) in $\mathbb{R}^2$ centered at p with edges parallel to the coordinate axes. The set of pixels centered at the points of B together with their faces (edges and vertices) constitute a cubical complex denoted by $Q(I)$. A p–cell in I can be identified by its barycentric coordinates $(x_\sigma, y_\sigma) \in \mathbb{R}^2$.

Following the construction given in [4], a cubical complex in which consecutive images are stacked to include a third, temporal dimension, is defined.

Definition 1. *Consider a sequence of 2D binary images $S = \{I_1, \ldots, I_n\}$ and the associated (2D) cubical complexes $Q(I_1), \ldots, Q(I_n)$. The stacked (3D) cubical complex $SQ[S]$ is obtained as follows. Let $Q(I_i) \times \{i\}$, $1 \leq i \leq n$, be the cubical complex obtained by adding a third coordinate i to the barycentric coordinates of the cells of $Q(I)$. Initially, $SQ[S] = \sqcup_{i=1}^{n}(Q(I_i) \times \{i\})$. Now, if a p-cell σ with barycentric coordinates (x_σ, y_σ) belongs to $Q(I_i) \cap Q(I_{i+1})$ for some i, $1 \leq i < n$, add to $SQ[S]$ the $(p+1)$–cell $\tau = \sigma \times [i, i+1]$. This way, the barycentric coordinates of τ are $(x_\sigma, y_\sigma, i + \frac{1}{2})$.*

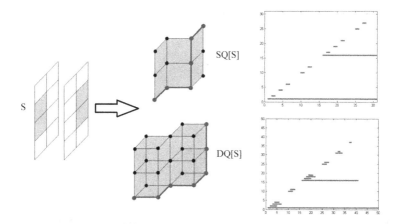

Fig. 4. A sequence S of two 2D images, the associated 3D cubical complexes $SQ[S]$ and $DQ[S]$ and the corresponding spatiotemporal 0–barcodes.

Since each cell $\sigma \in SQ[S]$ can be identified by its barycentric coordinates $(x_\sigma, y_\sigma, t_\sigma) \in \mathbb{R}^3$, then σ is spatial if, for some $i \in \mathbb{Z}$, $t_\sigma = i$; and it is temporal otherwise. For example, a cube $\tau \in SQ[S]$ is always temporal, and with respect to its faces we find: 6 spatial vertices; 8 spatial and 4 temporal edges; and 2 spatial and 4 temporal squares.

Let us denote by $Q(I_i, I_{i+1})$ the set of temporal cells with faces in $Q(I_i)$ and $Q(I_{i+1})$. The spatiotemporal filtration $\emptyset \subset SQ_0 \subset SQ_1 \subset \cdots \subset SQ_n = SQ[S]$ is given by: $SQ_i = Q(I_1)$, if $i = 1$; $SQ_i = SQ_{i-1} \cup Q(I_{j+1})$, if $i = 2j$ and $j > 0$; and $SQ_i = SQ_{i-1} \cup Q(I_j, I_{j+1})$, if $i = 2j + 1$ and $j > 0$.

Figure 3 shows three simple examples of stacked cubical complexes. The associated spatiotemporal 0–barcodes are computed using Algorithm 1. From left to right, the first and second spatiotemporal 0–barcodes have only one long bar, while third one has two. Notice that in this last case, the classical 0–barcode would produce only one long bar.

Observe that we could construct the 3D cubical complex $DQ[S]$, just considering every pixel centered at point (x, y, t) as a voxel (unit cube with faces parallel to the coordinate planes) centered at point (x, y, t). We have the following result:

Proposition 2. *Given a sequence of 2D binary images $S = \{I_1, \ldots, I_n\}$, the 3D cubical complexes $SQ[S]$ and $DQ[S]$ are homotopy equivalent.*

Proof. In our approach, to construct the 3D cubical complex $SQ[S]$, we build a cube only when two pixels in same spatial locations (i.e., with identical barycentric coordinates) belong to two consecutive frames; the other approach is to consider pixels as voxels (cubes) to directly obtain a 3D cubical complex $DQ[S]$ (see Fig. 4). To prove that $SQ[S]$ and $DQ[S]$ are homotopy equivalent, we describe how to collapse one complex, $DQ[S]$, to the other one, $SQ[S]$. For this aim, we first apply the translation $\tau(x, y, t) = (x, y, t + 1/2)$ to cells in $DQ[S]$. Consider a pixel (square cell) $\sigma \in SQ[S]$ centered at $(x_\sigma, y_\sigma, t_\sigma)$ that belongs to a cube c centered at $(x_\sigma, y_\sigma, t_\sigma + 1/2)$ in $SQ[S]$. Let c_σ be the voxel in $DQ[S]$ centered at $(x_\sigma, y_\sigma, t_\sigma)$. Then clearly $\tau(c_\sigma) = c \in SQ[S]$. Now, the idea is to successively collapse all the cells that are in $\tau(DQ[S])$ but not in $SQ[S]$. First, if σ does not belong to any cube in $SQ[S]$ centered at (x, y, t) with $t = t_\sigma + 1/2$, then collapse the square face centered at $(x_\sigma, y_\sigma, t_\sigma + 1)$ in $DQ[S]$. Similarly if an edge e of σ centered at (x_e, y_e, t_σ) does not belong to any cube in $SQ[S]$ centered at (x, y, t) with $t = t_\sigma + 1/2$, then collapse the edge centered at $(x_e, y_e, t_\sigma + 1)$. Finally, if a vertex v of σ with coordinates (x_v, y_v, t_σ) does not belong to any cube in $SQ[I]$

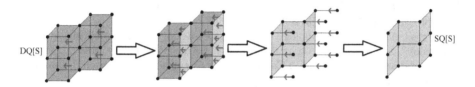

Fig. 5. A sequences of collapses starting from the complex $DQ[S]$ and ending at the complex $SQ[S]$. First, 4 square faces collapse, then 12 edges collapse and finally, 9 vertices collapse.

centered at (x, y, t) with $t = t_\sigma + 1/2$, then collapse the vertex with coordinates $(x_v, y_v, t_\sigma + 1)$. See Fig. 5. □

In this paper, we use the construction $SQ[S]$ instead of $DQ[S]$ because we considered that, in $SQ[S]$, the notion of spatial and temporal cells is more intuitive.

6   Conclusions and Future Work

In this paper, we have computed a modified 0–barcode for a temporal sequence of 2D binary images respecting the time nature of the data. This is part of an ongoing project to define and compute spatiotemporal p–barcodes for sequences of nD binary images.

Acknowledgments. We want to thank the valuable suggestions and comments made by the reviewers to improve the final version of this paper.

References

1. Adams, H., Carlsson, G.: Evasion paths in mobile sensor networks. I. J. Rob. Res. **34**(1), 90–104 (2015)
2. Carlsson, G.E., de Silva, V.: Zigzag persistence. Found. Comput. Math. **10**(4), 367–405 (2010)
3. Edelsbrunner, H., Letscher, D., Zomorodian, A.: Topological persistence and simplification. In: FOCS 2000, pp. 454–463. IEEE Computer Society (2000)
4. de Silva, V., Ghrist, R.: Coordinate-free coverage in sensor networks with controlled boundaries via homology. I. J. Rob. Res. **25**(12), 1205–1222 (2006)
5. Edelsbrunner, H., Harer, J.: Computational Topology - An Introduction. American Mathematical Society, Providence (2010)
6. Gamble, J., Chintakunta, H., Krim, H.: Coordinate-free quantification of coverage in dynamic sensor networks. Sign. Proces. **114**, 1–18 (2015)
7. Ghrist, R.: Barcodes: the persistent topology of data. Bull. Am. Math. Soc. **45**, 61–75 (2008)
8. Gonzalez-Diaz, R., Ion, A., Jimenez, M.J., Poyatos, R.: Incremental-decremental algorithm for computing AT-models and persistent homology. In: Real, P., Diaz-Pernil, D., Molina-Abril, H., Berciano, A., Kropatsch, W. (eds.) CAIP 2011, Part I. LNCS, vol. 6854, pp. 286–293. Springer, Heidelberg (2011)
9. Gonzalez-Diaz, R., Real, P.: On the cohomology of 3D digital images. Discrete Appl. Math. **147**(2–3), 245–263 (2005)
10. Hatcher, A.: Algebraic Topology. Cambridge University Press, Cambridge (2002)
11. Munkres, J.: Elements of Algebraic Topology. Addison-Wesley Co., Reading (1984)
12. Zomorodian, A., Carlsson, G.: Computing persistent homology. Discrete Comput. Geom. **33**(2), 249–274 (2005)

Characterization and Construction of Rational Circles on the Integer Plane

Papia Mahato$^{(\boxtimes)}$ and Partha Bhowmick

Department of Computer Science and Engineering,
Indian Institute of Technology, Kharagpur, India
papiamahatostar@gmail.com, bhowmick@gmail.com

Abstract. Discretization of geometric primitives in the integer space is a well-researched topic in the subject of digital geometry. In this paper, we present some novel results related to discretization of circles on the integer plane when the center and the radius are specified by arbitrary rational numbers. These results reveal elementary number-theoretic properties of rational circles on the integer plane and lead to useful characterization in terms of certain integer intervals defined by the circle parameters. We show how it finally culminates to an efficient algorithm for construction of rational circles using integer operations. Related experimental results exhibit interesting similitudes between the characteristic patterns of rational circles and those of integer circles.

Keywords: Discrete circle · Rational circle · Discrete curve · Digital geometry · Number theory

1 Introduction

Amongst different geometric primitives on the integer plane, one of the most important primitives is circle. As it is discretized on the integer plane by an optimum set of integer points or pixels, which are well-connected in a particular topological model, it is referred to as discrete or digital circle. It is found to possess interesting digital-geometric properties that are derivable by digital calculus and elementary number theory [1,4,10,15]. These properties provide analytical insights on the composition of a digital circle, which can be harnessed to make out efficient solutions to many algorithmic and application-bound problems in computer graphics, image analysis, and computer vision. There exists a multitude of work related to digital circle in these areas, some of which can be seen in [1–5,8,9,11–13,15] and the bibliographies therein.

In this paper, we present a novel study on some of the elementary number-theoretic properties of digital circle when its center and radius are specified by rational numbers. To the best of our knowledge, this is the first work related to this problem. Its motivation lies in the fact that with rational specification of digital circle, we can design an efficient integer-based algorithm, which guarantees a correct output for any input. A real specification, on the contrary, does

© Springer International Publishing Switzerland 2015
R.P. Barneva et al. (Eds.): IWCIA 2015, LNCS 9448, pp. 71–85, 2015.
DOI: 10.1007/978-3-319-26145-4_6

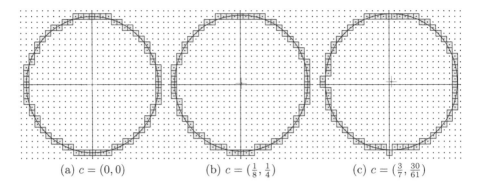

Fig. 1. Instances of three naive circles for radius $\rho = 12$. Notice that, in spite of fixed integer radius, the symmetry of the circle in (a) gets lost in (b, c), as the center c moves away from $(0,0)$.

not guarantee this due to finite precision of the computing environment. Based on the proposed number-theoretic properties, we derive a simple-yet-effective number-theoretic characterization of a digital circle in terms of certain integer intervals defined by the circle parameters. We show how it eventually champions to an efficient integer algorithm by dint of its computational adequacy with simple integer operations. Figure 1 shows a small set of results produced by our algorithm, which indicate the inherent intricacies and asymmetries in digital circles when they are specified by rational parameters.

The paper is organized as follows. In Sect. 1.1, we explain the preliminary concepts and the theoretical framework adopted in our work. In Sect. 2, we derive some of the number-theoretic properties of naive rational circle. In Sect. 3, we extend the above properties to derive integer intervals and a set of useful recurrences, which leads to an efficient algorithm for construction of the circle using integer operations. In Sect. 4, we present some test results, which show a strong similarity in the characteristic patterns between the family of naive rational circles and that of naive integer circles, although the former is found to be asymmetric opposed to the latter. We discuss few interesting open problems in that section, which can be pursued to invent further properties of naive rational circles.

1.1 Preliminaries

We fix here some basic definitions and metrics in 2D space, which are used in the sequel. For further details, we refer to [10].

An *integer point* means a point with integer coordinates. A *pixel* or *2-cell*, which is perceived as a unit square on xy-plane, is thus uniquely identified by its center, as it is an integer point. Two pixels are said to be *0-adjacent* if they share (at least) a vertex (0-cell) and *1-adjacent* if they share an edge (1-cell). A *digital curve* is either 0- or 1-connected on the digital plane. A *0-connected*

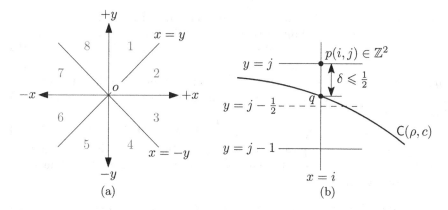

Fig. 2. (a) Eight octants defined w.r.t. $o = (0,0)$. (b) For each $p(i,j) \in C_1(\rho,c)$, there is a point $q(i, j - \delta) \in C(\rho,c)$, where $-\frac{1}{2} \leq \delta \leq \frac{1}{2}$.

curve is a sequence of pixels such that every two consecutive pixels are 0-adjacent. For a 1-*connected curve*, every two consecutive pixels in the sequence have to be 1-adjacent.

Between two points $p(i,j)$ and $p'(i',j')$, *x-distance* and *y-distance* are given by $d_x(p,p') = |i - i'|$ and $d_y(p,p') = |j - j'|$ respectively. Consequently, the isothetic distance between p and p' is taken as the Minkowski norm [10], given by $d_\infty(p,p') = \max\{d_x(p,p'), d_y(p,p')\}$. The isothetic distance of the point $p(i,j)$ from a 2D curve C is given by $d_\perp(p,C) = \min\{d_x(p,C), d_y(p,C)\}$; here, $d_x(p,C) = d_x(p,q)$ if there exists a (the nearest, if there is more than one) point $q(x,j)$ on C, and ∞ otherwise; and $d_y(p,C) = d_y(p,q)$ if there exists a point $q(i,y)$ on C, and ∞ otherwise.

We denote by $\mathbb{O}_t$ the tth octant, where $1 \leq t \leq 8$. A simple illustration is shown in Fig. 2a. So, $t(\bmod \ 4) \in \{0,1\}$ for $\mathbb{O}_1$, $\mathbb{O}_8$, $\mathbb{O}_4$, and $\mathbb{O}_5$; and $t(\bmod \ 4) \in \{2,3\}$ for $\mathbb{O}_2$, $\mathbb{O}_3$, $\mathbb{O}_6$, and $\mathbb{O}_7$. Notice that $|i| \leq |j|$ for the former case and $|j| \leq |i|$ for the latter. We define h as (absolute value of) the *major coordinate* corresponding to each octant, i.e., $h = |j|$ for the former case and $h = |i|$ for the latter. For a digital curve or a digital circle C, we denote by $h_{\max}$ the maximum value of h over all points of C in a particular octant. So, $h_{\max}$ assumes the maximum value of $|j|$ in C if $t(\bmod \ 4) \in \{0,1\}$, and the maximum value of $|i|$ in C if $t(\bmod \ 4) \in \{2,3\}$.

2 Naive Rational Circle

A *digital circle* means a set of integer points obtained by discretization of a real circle [4]. Depending on the topological model used for discretization, a digital circle (or any digital curve, in general) can be of different types [7,15]. Our work is focused on *naive circle*, which is 0-connected digital circle with minimum number of pixels.

A (naive) digital circle is said to be *integer circle* if the corresponding real circle has integer radius and integer center [14]. When the real circle has rational radius and rational center, we call its discretization as *naive rational circle*. We denote by $C(\rho, c)$ the real circle and by $\mathcal{C}(\rho, c)$ the naive rational circle having radius $\rho \in \mathbb{Q}$ and center $c = (x_c, y_c) \in \mathbb{Q}^2$, where $\mathbb{Q}$ represents the set of rational numbers. We use $C_t(\rho, c)$ and $\mathcal{C}_t(\rho, c)$ to denote the arcs of $C(\rho, c)$ and of $\mathcal{C}(\rho, c)$ lying in $\mathbb{O}_t$, where $t = 1, 2, \ldots, 8$.

We assume $\rho = r + \frac{r_1}{r_2}$, $x_c = \frac{i_1}{i_2}$, $y_c = \frac{j_1}{j_2}$, where $0 \le \frac{r_1}{r_2}, \frac{i_1}{i_2}, \frac{j_1}{j_2} < 1$, and $r, r_1, r_2, i_1, i_2, j_1, j_2$ are all positive integers. Note that if $\frac{r_1}{r_2} = \frac{i_1}{i_2} = \frac{j_1}{j_2} = 0$, then the naive rational circle just becomes a naive integer circle. We define naive rational circle as follows.

Definition 1. *A naive rational circle $\mathcal{C}(\rho, c)$ is the 0-connected set of integer points obtained by discretization of $C(\rho, c)$ such that each point has an isothetic distance of at most $\frac{1}{2}$ from $C(\rho, c)$.*

A naive integer circle has eight symmetric octants [4]. Hence, it can be generated by reflecting its integer points in the 1st octant to other seven octants about the lines of symmetry. In case of a naive rational circle, this symmetry may not be present, which makes its generation more difficult compared to generation of an integer circle.

2.1 Number-Theoretic Properties

We introduce here some general theoretical results when a circle is specified by real radius and real center. Hence, these results are applicable also for any rational specification. For discretization of a real circle $C(\rho, c)$ to the naive circle $\mathcal{C}(\rho, c)$, we first derive an equation for each of the eight octants. We use these equations to obtain the canonical representation of the complete naive circle. We first introduce the following lemma.

Lemma 1. *An integer point $p(i, j)$ belongs to $\mathcal{C}_1(\rho, c)$ if and only if*

$$\rho^2 - (j - y_c)^2 - (j - y_c) - \frac{1}{4} \le (i - x_c)^2 \le \rho^2 - (j - y_c)^2 + (j - y_c) - \frac{1}{4}. \quad (1)$$

Proof. Let $p(i, j)$ be a point of $\mathcal{C}_1(\rho, c)$ and $(i, j - \delta)$ be the point of intersection of the real circle $C(\rho, c)$ with the vertical line $x = i$ in the first octant as shown in Fig. 2b. Then by Definition 1, we get $-\frac{1}{2} \le \delta \le \frac{1}{2}$. Now, $(i, j - \delta)$ lies on $C_1(\rho, c)$

$$\Leftrightarrow (i - x_c)^2 + (j - \delta - y_c)^2 = \rho^2$$

$$\Leftrightarrow \delta = (j - y_c) - \sqrt{\rho^2 - (i - x_c)^2}$$

$$\Leftrightarrow -\frac{1}{2} \le (j - y_c) - \sqrt{\rho^2 - (i - x_c)^2} \le \frac{1}{2},$$

whence the condition follows.   □

Lemma 1 provides all the integer points of $\mathcal{C}(\rho, c)$ belonging to the first octant. A similar result is given in [9, Eq. 2] for center and radius in the real domain. For rational circle, we require Lemma 1 for its definiteness, which leads to several theoretical results presented in the forthcoming sections. We first extend this lemma to get similar equations for each of the eight octants, as stated in the following theorem.

Theorem 1. *An integer point $p(i, j)$ belongs to $\mathcal{C}_t(\rho, c)$ if and only if*

$$\rho^2 - (M + \frac{1}{2})^2 \leq m^2 \leq \rho^2 - (M - \frac{1}{2})^2 \tag{2}$$

where,

$$M = \begin{cases} |j| - \sigma_y y_c \text{ if } t \mod 4 \in \{0, 1\} \\ |i| - \sigma_x x_c \text{ otherwise,} \end{cases}$$

$$m = \begin{cases} |i| - \sigma_x x_c \text{ if } t \mod 4 \in \{0, 1\} \\ |j| - \sigma_y y_c \text{ otherwise,} \end{cases}$$

$\sigma_x = \dfrac{i}{|i|}$ *and* $\sigma_y = \dfrac{j}{|j|}$.

Proof. By Lemma 1, we get the result for first octant. For each other octant, we can show the result with appropriate substitution of σ_x and σ_y, and adding or subtracting x_c and y_c from $|i|$ and $|j|$, depending on the combination of i and j considering their relative values and signs. □

Theorem 1 can be used to decide whether a given integer point belongs to the naive circle of a given radius and center. However, it does not lead to the requisite conditions that can be used to design an algorithm for construction of a naive circle. Such conditions, in particular, are certain recurrences from where the integer points comprising the digital circle can be computed efficiently with simple integer operations only. Hence, we extend this theorem for further refinement, as stated shortly in Theorem 2. For proof of this theorem, we need the following lemmas.

Lemma 2. *For an octant $\mathbb{O}_t$,*

$$h_{\max} = \begin{cases} r - 1 \text{ if } \beta_c \in (-1, -\frac{1}{2}] \\ r \quad\;\; \text{ if } \beta_c \in (-\frac{1}{2}, \frac{1}{2}] \\ r + 1 \text{ if } \beta_c \in (\frac{1}{2}, \frac{3}{2}] \\ r + 2 \text{ if } \beta_c \in (\frac{3}{2}, 2) \end{cases}$$

where,

$$\beta_c = \begin{cases} \rho - r + \sigma_y y_c \text{ if } t \mod 4 \in \{0, 1\} \\ \rho - r + \sigma_x x_c \text{ otherwise.} \end{cases}$$

Proof. W.l.o.g., we prove it for $\mathbb{O}_1$. Observe that in $\mathbb{O}_1$, $h_{\max} = \lfloor \rho + y_c \rfloor = r + \lfloor (\rho - r) + y_c \rfloor$, which becomes r, $r + 1$, or $r + 2$, depending on whether $(\rho - r) + y_c$ lies in $(-1, -\frac{1}{2}]$, $(-\frac{1}{2}, \frac{1}{2}]$, $(\frac{1}{2}, \frac{3}{2}]$, or $(\frac{3}{2}, 2)$, respectively. □

Lemma 3. *An integer point* $(i, h_{\max} - k)$ *belongs to* $\mathcal{C}_1(\rho, c)$ *if and only if* $(i - x_c)^2$ *lies in the interval* $Q_k = [\rho^2 - r_k^2 - r_k - \frac{1}{4}, \rho^2 - r_k^2 + r_k - \frac{1}{4}]$, *where* $r_k = h_{\max} - k - y_c$ *and* k *is a non-negative integer.*

Proof. By setting $j = h_{\max} - k$ in Lemma 1, the point $(i, h_{\max} - k)$ belongs to $\mathcal{C}_1(\rho, c)$ if and only if $(i - x_c)^2$ lies in the given interval. Hence the proof.   □

By Lemma 3, we get the integer points of $\mathcal{C}(\rho, c)$ only in the first octant, on setting $k = 0, 1, 2, \ldots$. To get the points of $\mathcal{C}(\rho, c)$ in the remaining octants, we have to generalize the result of the above lemma, as stated in the following theorem.

Theorem 2. *An integer point* (i, j) *belongs to* $\mathcal{C}_t(\rho, c)$ *if and only if* m^2 *lies in the interval* $Q_k = [\rho^2 - (r_k + \frac{1}{2})^2, \rho^2 - (r_k - \frac{1}{2})^2]$, *where*

$$|j| = h_{\max} - k \text{ if } t \bmod 4 \in \{0, 1\}$$
$$|i| = h_{\max} - k \text{ otherwise},$$

$$r_k = \begin{cases} h_{\max} - k - \sigma_y y_c \text{ if } t \bmod 4 \in \{0, 1\} \\ h_{\max} - k - \sigma_x x_c \text{ otherwise}, \end{cases}$$

$k \geq 0$, *and* m *follows the equation given in Theorem 1.*

Proof. By Lemma 3, we get the result for $\mathbb{O}_1$. For each other octant, we get the result by considering an appropriate value of r_k.   □

The above theorem gives us the intervals in the domain of rationals from where we can compute the integer points that constitute a naive circle. To simplify the computation, we make further refinement of the intervals so that they are in the integer domain. This is explained in the next section.

3 Integer Intervals and the Algorithm

By Theorem 2, we get the intervals in the domain of rationals. We obtain integer intervals from these rational intervals using appropriate multiplication by the denominators of the rational terms. For this, we introduce the following six terms.

$$k_c = 2r_2 i_2 j_2 \tag{3a}$$
$$k_1 = 2r_1 i_2 j_2 \tag{3b}$$
$$k_2 = 2r_2 i_2 j_1 \tag{3c}$$
$$k_3 = 4r_2^2 i_2^2 j_1 j_2 \tag{3d}$$
$$k_4 = r_2^2 i_2^2 j_2^2 \tag{3e}$$
$$k_5 = 2i_1 r_2 j_2 \tag{3f}$$

Note that the formulas given above for k_2, k_3, k_5 are for $\mathbb{O}_1$ and they are generalized later for other octants. The formulas of all other terms are applicable for all octants. We use the above terms in the following lemma.

Lemma 4. *An integer point (i, j) belongs to $C_1(\rho, c)$ if and only if*

$$(rk_c + k_1)^2 - (jk_c - k_2)^2 - (jk_c^2 - k_3) - k_4 \le (ik_c - k_5)^2 \le$$
$$(rk_c + k_1)^2 - (jk_c - k_2)^2 + (jk_c^2 - k_3) - k_4. \tag{4}$$

Proof. By Lemma 1 and by multiplying Eq. 1 by k_c^2, we get the result. □

Lemma 4 provides the integer intervals to find all the integer points of $C(\rho, c)$ lying in the first octant. We extend this lemma to get similar equations for each of the eight octants. We introduce M_1, M_2, m_1, and generalize the formulations of k_2, k_3, k_5, and use them in the subsequent theorems.

$$k_2 = \begin{cases} 2r_2 i_2 j_1 \text{ if } t \bmod 4 \in \{0, 1\} \\ 2r_2 i_1 j_2 \text{ otherwise} \end{cases} \tag{5a}$$

$$k_3 = \begin{cases} 4r_2^2 i_2^2 j_1 j_2 \text{ if } t \bmod 4 \in \{0, 1\} \\ 4r_2^2 j_2^2 i_1 i_2 \text{ otherwise} \end{cases} \tag{5b}$$

$$k_5 = \begin{cases} 2i_1 r_2 j_2 \text{ if } t \bmod 4 \in \{0, 1\} \\ 2j_1 r_2 i_2 \text{ otherwise} \end{cases} \tag{5c}$$

$$M_1 = \begin{cases} |j|k_c - \sigma_y k_2 \text{ if } t \bmod 4 \in \{0, 1\} \\ |i|k_c - \sigma_x k_2 \text{ otherwise} \end{cases} \tag{5d}$$

$$M_2 = \begin{cases} |j|k_c^2 - \sigma_y k_3 \text{ if } t \bmod 4 \in \{0, 1\} \\ |i|k_c^2 - \sigma_x k_3 \text{ otherwise} \end{cases} \tag{5e}$$

$$m_1 = \begin{cases} |i|k_c - \sigma_x k_5 \text{ if } t \bmod 4 \in \{0, 1\} \\ |j|k_c - \sigma_y k_5 \text{ otherwise} \end{cases} \tag{5f}$$

Theorem 3. *An integer point (i, j) belongs to $C_t(\rho, c)$ if and only if*

$$(rk_c + k_1)^2 - M_1^2 - M_2 - k_4 \le m_1^2 \le (rk_c + k_1)^2 - M_1^2 + M_2 - k_4. \tag{6}$$

Proof. By Lemma 4, we get the result for 1st octant. For each other octant, we get the result with appropriate substitution of σ_x, σ_y, and the coefficients and the related terms given in (3a), (3b), (3e), (5). □

To get the integer points in the all the octants using only integer operations, we need the following theorem.

Theorem 4. *Let (i, j) be an integer point in $\mathbb{O}_t$ such that $|j| = h_{\max} - k$ if $t \bmod 4 \in \{0, 1\}$ and $|i| = h_{\max} - k$ otherwise, for $k \ge 0$. Then the point (i, j) belongs to $C_t(\rho, c)$ if and only if m_1^2 lies in the interval $I_k = [u_k, v_k := u_k + l_k - 1]$, where u_k and l_k are given as follows.*

$$u_k = \begin{cases} (rk_c + k_1)^2 - M_{1(k=0)} - M_{2(k=0)} - k_4 \ if \ k = 0 \\ u_{k-1} + l_{k-1} - 1 \qquad\qquad otherwise \end{cases}$$
$$l_k = \begin{cases} 2M_{2(k=0)} + 1 \ if \ k = 0 \\ l_{k-1} - 2k_c^2 \quad otherwise \end{cases} \tag{7}$$

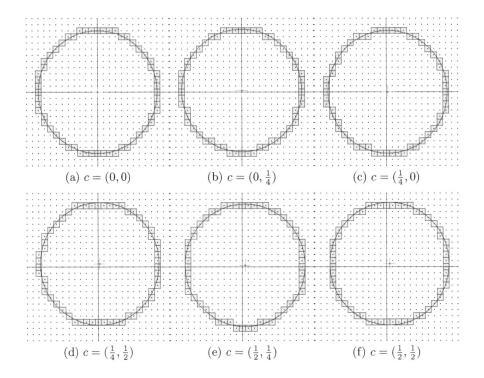

(a) $c = (0,0)$ (b) $c = (0, \frac{1}{4})$ (c) $c = (\frac{1}{4}, 0)$

(d) $c = (\frac{1}{4}, \frac{1}{2})$ (e) $c = (\frac{1}{2}, \frac{1}{4})$ (f) $c = (\frac{1}{2}, \frac{1}{2})$

Fig. 3. Naive circles for radius 10 with different centers.

Proof. We use Theorem 3 to derive u_0 by substituting $k = 0$ in (5d) and (5e). By Theorem 3 again, we get $l_k = v_k - u_k + 1 = 2M_2 + 1$, which becomes $2M_{2(k=0)} + 1$ when we substitute $k = 0$ in (5e).

To get the recurrence of u_k, observe that M_1 and M_2 depend on $|i|$ or $|j|$, and hence on k, whereas k_c, k_1, k_4 do not. Further, for every unit increment of k, it follows from (3) and (5) that $v_{k-1} - u_k = -M_{1(k-1)}^2 + M_{2(k-1)} + M_{1(k)}^2 + M_{2(k)} = 0$. Hence, $u_k = v_{k-1} = u_{k-1} + l_{k-1} - 1$.

To get the recurrence of l_k, observe that $l_k = 2M_{2k} + 1$, as explained above. Hence, $l_k - l_{k-1} = 2M_{2k} - 2M_{2k-1} = 2k_c^2$ by (5e).

All these holds for $t \in \{1, 2, \ldots, 8\}$, and hence the proof. $\square$

3.1 Algorithm for Naive Circle

Naive circles with non-zero rational values of center coordinates do not follow the property of 8-symmetry. Hence, to generate a naive circle, we need to consider its octants one by one and have to use the recurrences given in Theorem 4. The steps are shown in Algorithm 1.

In Algorithm 1, the **for** loop (Line 2) considers a particular octant in one iteration. In each iteration, the procedure `initializeParameters` sets the initial values of the necessary parameters (Line 3). Subsequently, the values of the

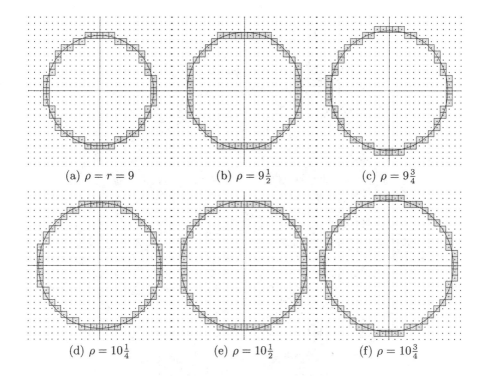

(a) $\rho = r = 9$ (b) $\rho = 9\frac{1}{2}$ (c) $\rho = 9\frac{3}{4}$

(d) $\rho = 10\frac{1}{4}$ (e) $\rho = 10\frac{1}{2}$ (f) $\rho = 10\frac{3}{4}$

Fig. 4. Naive rational circles centered at $(0,0)$ with gradually increasing radius.

parameters defining an integer interval are initialized in Line 4–Line 6. The value of the first square number in that interval is set in Line 7. After this, the algorithm works depending on whether the octant has $|i| \leq |j|$ or $|j| \leq |i|$. This is determined by the condition of the **if** statement in Line 8.

The **while** loop (Line 9 or Line 24) generates the runs of the integer points in different iterations. (A *run* means a sequence of points with same x- or y-coordinate.) Inside the **while** loop, there is **repeat-until** loop (Line 10 or Line 25) that computes the points of the $k(\geq 0)$th run of C_t using the integer interval $I_k := [u_k, v_k]$. For every perfect square s generated in succession, the condition of the **if** statement (Line 11 or Line 26) decides whether s lies in I_k and if so, the procedure drawPoint includes the corresponding point in C. After generation of all the points in kth run, the parameters are updated in Line 17–Line 22 or in Line 32–Line 37 to compute the points in $(k+1)$th run.

4   Experimental Results and Conclusion

We present here some results of naive rational circles produced by Algorithm 1. These results show the symmetry and the asymmetry of the circles when they are specified by integer parameters and non-integer parameters. In Fig. 3, all the circles have identical integer radius but different centers. When the center c is an

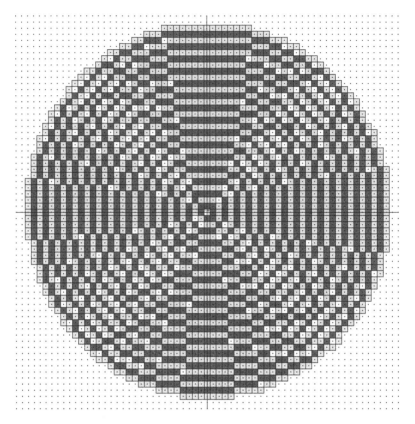

Fig. 5. Naive rational circles concentric at $(\frac{1}{5}, \frac{2}{7})$, radius $\rho = 1\frac{1}{8}, 2\frac{1}{8}, \ldots, 30\frac{1}{8}$, shown alternately in green and yellow (Color figure online).

integer point (Fig. 3a), the circle is 8-symmetric. The 8-symmetry is destroyed when c is not an integer point. However, if one between x- and y-coordinates of c is integer, then the circle has a partial symmetry. For example, in Fig. 3b, $x_c = 0, y_c = \frac{1}{4}$, and so the circle is symmetric about y-axis. Similarly in Fig. 3c, it is symmetric about x-axis, as $x_c = \frac{1}{4}, y_c = 0$. But in Fig. 3d–f, it is asymmetric about both the coordinate axes, since x_c and y_c are both non-zero.

Figure 4 shows a sequence of 6 naive rational circles for radius increasing gradually from 9 to $10\frac{3}{4}$. The 8-symmetry is present for all these circles, since the center is an integer point. It may be noticed that there are common pixels between every two consecutive circles in this sequence, since their radii differ by $\frac{1}{4}$ or $\frac{1}{2}$.

When two radii differ by unity, two concentric circles are always disjoint, as there is no common pixel between them. This is true irrespective of whether their common center is an integer point or a non-integer point. This is clearly depicted in the results shown in Fig. 5, which contains a set of circles concentric at a rational center with radius increasing by unity. It forms an interesting

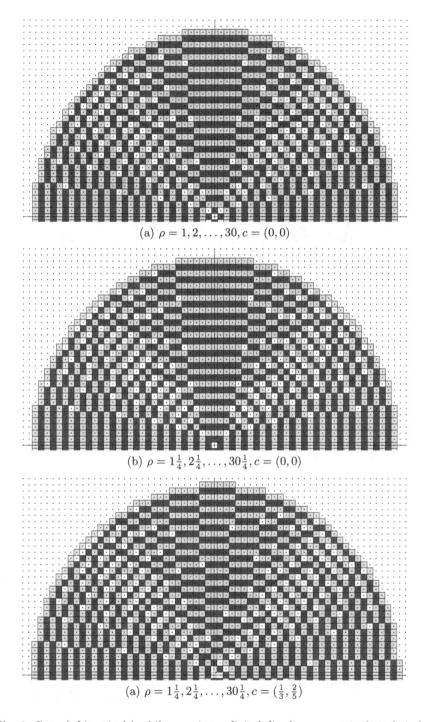

(a) $\rho = 1, 2, \ldots, 30, c = (0,0)$

(b) $\rho = 1\frac{1}{4}, 2\frac{1}{4}, \ldots, 30\frac{1}{4}, c = (0,0)$

(a) $\rho = 1\frac{1}{4}, 2\frac{1}{4}, \ldots, 30\frac{1}{4}, c = (\frac{1}{3}, \frac{2}{5})$

Fig. 6. Gaps (white pixels) while covering a digital disc by concentric digital circles.

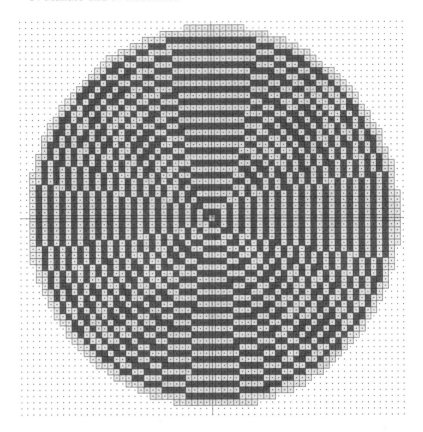

Fig. 7. An example showing no gaps for a set of naive rational circles concentric at $(\frac{1}{5}, \frac{2}{7})$, radius $\rho(= r + \frac{r_1}{r_2}) = \frac{5}{8}, 1\frac{1}{8}, 1\frac{5}{8}, 2\frac{1}{8}, 2\frac{5}{8}, \ldots, 30\frac{5}{8}$. Pixels are shown in green and in yellow for odd and even values of r, respectively (Color figure online).

similitude of naive rational circles with naive integer circles, since the latter class also exhibits the aforesaid disjoint-circle property, as shown in [3].

Another similitude between the class of naive rational circles and the class of naive integer circles is that there are *gaps* in a set of circles in both the classes when the circles have consecutive radii differing by unity[1]. This is depicted through the results in Fig. 6. For the integer class, related theoretical results on gap characterization can be seen in [3]. Figure 6a shows a result for this class, and Fig. 6b–c shows results for the rational class. Notice that the gaps are quite asymmetric as the center moves away from an integer point. Characterization of the gaps for the rational class, as we can foresee, would be an interesting work related to covering of rational discs by naive rational circles.

In Fig. 7, we have shown a set of naive rational circles, which are concentric and having radius increasing in steps of $\frac{1}{2}$. The resultant set contains no gap,

---

[1] In 2D discretization, a *gap* means a missing 2-cell [6]; it is also termed as *absentee* in [3].

Algorithm 1. DrawNaiveCircle

Input: Radius $\rho = r + \frac{r_1}{r_2}$, center $c = \left(\frac{i_1}{i_2}, \frac{j_1}{j_2}\right)$
Output: Naive circle

1   int $i, j, M_1, M_2, m_1, k_c, k_1, k_2, k_3, k_4, k_5, u_k, v_k, l_k, s$
2   **for** $t \leftarrow 1$ *to* 8 **do**
3   initializeParameters
4      $u_k \leftarrow (rk_c + k_1)^2 - M_1^2 - M_2 - k_4$ $\triangleright$ Theorem 3
5      $l_k \leftarrow 2M_2$ $\triangleright$ Theorem 4
6      $v_k \leftarrow u_k + l_k$ $\triangleright$ Theorem 4
7      $s \leftarrow m_1^2$
8   **if** $t \bmod 4 \in \{0, 1\}$ **then**
9   **while** $i \leq j$ **do**
10   **repeat**
11   **if** $u_k \leq s \leq v_k$ **then**
12   drawPoint(i, j)
13               $i \leftarrow i + 1$
14               $m_1 \leftarrow ik_c - \sigma_x k_5$ $\triangleright$ Equation 5f
15               $s \leftarrow m_1^2$
16   **until** $s > v_k$;
17            $u_k \leftarrow v_k$
18            $l_k \leftarrow l_k - 2k_c^2$ $\triangleright$ Theorem 4
19            $v_k \leftarrow v_k + l_k$ $\triangleright$ Theorem 4
20            $j \leftarrow j - 1$
21            $m_1 \leftarrow ik_c - \sigma_x k_5$ $\triangleright$ Equation 5f
22            $s \leftarrow m_1^2$
23   **else**
24   **while** $j \leq i$ **do**
25   **repeat**
26   **if** $u_k \leq s \leq v_k$ **then**
27   drawPoint(i, j)
28               $j \leftarrow j + 1$
29               $m_1 \leftarrow jk_c - \sigma_y k_5$ $\triangleright$ Equation 5f
30               $s \leftarrow m_1^2$
31   **until** $s > v_k$;
32            $u_k \leftarrow v_k$
33            $l_k \leftarrow l_k - 2k_c^2$ $\triangleright$ Theorem 4
34            $v_k \leftarrow v_k + l_k$ $\triangleright$ Theorem 4
35            $i \leftarrow i - 1$
36            $m_1 \leftarrow jk_c - \sigma_y k_5$ $\triangleright$ Equation 5f
37            $s \leftarrow m_1^2$

---

and hence successfully covers the disc. A pertinent issue is, therefore, whether a digital disc can always be covered by a set of concentric naive rational circles with radius increasing in this manner. In particular, we have the following question:

Procedure initializeParameters

---

1 int a, b
2 **if** $1 \leq t \mod 8 \leq 4$ **then**
3 $\quad \lfloor \; \sigma_x \leftarrow 1$

4 **else**
5 $\quad \lfloor \; \sigma_x \leftarrow -1$

6 **if** $0 \leq (t+1) \mod 8 \leq 3$ **then**
7 $\quad \lfloor \; \sigma_y \leftarrow 1$

8 **else**
9 $\quad \lfloor \; \sigma_y \leftarrow -1$

10 **if** $t \mod 4 \in \{0,1\}$ **then**
11 $\quad \lfloor \; a \leftarrow 2(r_1 j_2 + \sigma_y r_2 j_1), b \leftarrow r_2 j_2$

12 **else**
13 $\quad \lfloor \; a \leftarrow 2(r_1 i_2 + \sigma_x r_2 i_1), b \leftarrow r_2 i_2$

14 **if** $a \leq -b$ **then**
15 $\quad \lfloor \; h_{max} \leftarrow r - 1$

16 **else if** $-b < a \leq b$ **then**
17 $\quad \lfloor \; h_{max} \leftarrow r$

18 **else if** $b < a \leq 3b$ **then**
19 $\quad \lfloor \; h_{max} \leftarrow r + 1$

20 **else**
21 $\quad \lfloor \; h_{max} \leftarrow r + 2$

22 $k_c \leftarrow 2r_2 i_2 j_2, \; k_1 \leftarrow 2r_1 i_2 j_2, \; k_4 \leftarrow r_2^2 i_2^2 j_2^2$ $\triangleright$ Equations 3a, 3b, 3c
23 **if** $t \mod 4 \in \{0,1\}$ **then**
24 $\quad \mid \quad i \leftarrow 0, j \leftarrow h_{max}$
25 $\quad \mid \quad k_2 \leftarrow 2r_2 i_2 j_1, \; k_3 \leftarrow 4r_2^2 i_2^2 j_1 j_2, \; k_5 \leftarrow 2i_1 r_2 j_2$ $\triangleright$ Equations 5a- 5c
26 $\quad \lfloor \quad M_1 \leftarrow jk_c - \sigma_y k_2, \; M_2 \leftarrow jk_c^2 - \sigma_y k_3, \; m_1 \leftarrow ik_c - \sigma_x k_5$ $\triangleright$ Equations 5d- 5f

27 **else**
28 $\quad \mid \quad i \leftarrow h_{max}, j \leftarrow 0$
29 $\quad \mid \quad k_2 \leftarrow 2r_2 i_1 j_2, \; k_3 \leftarrow 4r_2^2 j_2^2 i_1 i_2, \; k_5 \leftarrow 2j_1 r_2 i_2$ $\triangleright$ Equations 5a- 5c
30 $\quad \lfloor \quad M_1 \leftarrow ik_c - \sigma_x k_2, \; M_2 \leftarrow ik_c^2 - \sigma_x k_3, \; m_1 \leftarrow jk_c - \sigma_y k_5$ $\triangleright$ Equations 5d- 5f

---

what should be the maximum step in radius increment so that a digital disc can always be covered by a set of concentric circles in the rational class?

References

1. Andres, E.: Discrete circles, rings and spheres. Comput. Graph. **18**(5), 695–706 (1994)
2. Andres, E., Roussillon, T.: Analytical description of digital circles. In: Debled-Rennesson, I., Domenjoud, E., Kerautret, B., Even, P. (eds.) DGCI 2011. LNCS, vol. 6607, pp. 235–246. Springer, Heidelberg (2011)

3. Bera, S., Bhowmick, P., Stelldinger, P., Bhattacharya, B.: On covering a digital disc with concentric circles in $\mathbb{Z}^2$. Theoret. Comput. Sci. **506**, 1–16 (2013)
4. Bhowmick, P., Bhattacharya, B.B.: Number-theoretic interpretation and construction of a digital circle. Discrete Appl. Math. **156**(12), 2381–2399 (2008)
5. Bhowmick, P., Pal, S.: Fast circular arc segmentation based on approximate circularity and cuboid graph. J. Math. Imaging Vis. **49**(1), 98–122 (2014)
6. Brimkov, V.E.: Formulas for the number of $(n-2)$-gaps of binary objects in arbitrary dimension. Discrete Appl. Math. **157**(3), 452–463 (2009)
7. Brimkov, V.E., Barneva, R.P., Brimkov, B.: Connected distance-based rasterization of objects in arbitrary dimension. Graph. Models **73**(6), 323–334 (2011)
8. Fiorio, C., Jamet, D., Toutant, J.-L.: Discrete circles: An arithmetical approach with non-constant thickness. In: Latecki, L.J., Mount, D.M., Wu, A.Y. (eds.), Vision Geometry XIV, Electronic Imaging, vol. 6066, pp. 60660C. SPIE, San Jose (CA), USA (2006)
9. Fiorio, C., Toutant, J.-L.: Arithmetic discrete hyperspheres and separatingness. In: Kuba, A., Nyúl, L.G., Palágyi, K. (eds.) DGCI 2006. LNCS, vol. 4245, pp. 425–436. Springer, Heidelberg (2006)
10. Klette, R., Rosenfeld, R.: A: Digital Geometry: Geometric Methods for Digital Picture Analysis. Morgan Kaufmann, San Francisco (2004)
11. Nagy, B.: An algorithm to find the number of the digitizations of discs with a fixed radius. Electron. Notes Discrete Math. **20**, 607–622 (2005)
12. Pal, S., Bhowmick, P.: Determining digital circularity using integer intervals. J. Math. Imag. Vis. **42**(1), 1–24 (2012)
13. Pham, S.: Digital circles with non-lattice point centers. Vis. Comput. **9**(1), 1–24 (1992)
14. Pitteway, M.L.V.: Integer circles, etc.–some further thoughts. Comput. Graph. Image Process. **3**, 262–265 (1974)
15. Toutant, J.-L., Andres, E., Roussillon, T.: Digital circles, spheres and hyperspheres: from morphological models to analytical characterizations and topological properties. Discrete Appl. Math. **161**(16–17), 2662–2677 (2013)

On the Connectivity and Smoothness of Discrete Spherical Circles

Ranita Biswas[1][✉], Partha Bhowmick[1], and Valentin E. Brimkov[2]

[1] Department of Computer Science and Engineering,
Indian Institute of Technology, Kharagpur, India
biswas.ranita@gmail.com
[2] Mathematics Department, SUNY Buffalo State,
1300 Elmwood Avenue, Buffalo, NY 14222, USA

Abstract. A discrete spherical circle is a topologically well-connected 3D circle in the integer space, which belongs to a discrete sphere as well as a discrete plane. It is one of the most important 3D geometric primitives, but has not possibly yet been studied up to its merit. This paper is a maiden exposition of some of its elementary properties, which indicates a sense of its profound theoretical prospects in the framework of digital geometry. We have shown how different types of discretization can lead to forbidden and admissible classes, when one attempts to define the discretization of a spherical circle in terms of intersection between a discrete sphere and a discrete plane. Several fundamental theoretical results have been presented, the algorithm for construction of discrete spherical circles has been discussed, and some test results have been furnished to demonstrate its practicality and usefulness.

Keywords: 3D discrete circle · Discrete sphere · Spherical circle · Digital geometry

1 Introduction

The literature of digital geometry as of now contains a rich collection of work related to different discrete-geometric primitives and general surfaces [24,27]. Among these, although a multitude of work can be found on characterization and modeling of planes, hyperplanes, spheres, and hyperspheres, no perceivable progress is noticed on discrete 3D circles and related problems. In this paper, we initiate the study on characterization and modeling of discrete 3D circles, extending the knowledge gathered from the recent developments in discrete spheres and discrete planes. This study unfolds the interplay of spheres and planes in the discrete domain, and indicates the scope of ample research issues related to modeling of 3D circles in the paradigm of digital geometry.

We first make it clear that in the context of our work, by discretization of a curve (3D circle in our case) or of a surface (sphere or plane in our case), we

---

V.E. Brimkov—On leave from the Institute of Mathematics and Informatics, Bulgarian Academy of Sciences, Sofia, Bulgaria.

R.P. Barneva et al. (Eds.): IWCIA 2015, LNCS 9448, pp. 86–100, 2015.
DOI: 10.1007/978-3-319-26145-4_7

mean *rasterization* or *voxelation* whereby the real object is approximated by a set of voxels (unit cubes), subject to certain topological constraints. The notion owes its origin to the early stage of computer graphics and geometric modeling [9,17,21,23].

In any model of discretization, the connectivity of a digital curve/surface is defined in an appropriate topology, and assuring the connectivity has a paramount role. Computer-aided geometric modeling of curves and surfaces is often based on the offset of a curve/surface Γ so that the locus of points lies within a given distance τ from Γ (see, e.g., [4,5,22] and the bibliography therein). Results from algebraic geometry, such as Grobner basis computation [3,18], are used as the mathematical foundation for this kind of modeling. The topological properties of the voxel set induced by the integer points lying within the enclosure defined by the offset are usually not taken into consideration. Some recent works related to digital conics [19,20] indicate about the exceptions (in 2D).

1.1 Motivation

Although a multitude of works have been carried out on generation and analysis of discrete circles and other curves in 2D, discrete surfaces in 3D, and on representation schemes for 3D curves, very little research has been done on discretization of circles or curves in 3D space. Only in recent time, the idea of offset discretization scheme for generation of discrete circles in arbitrary dimension has been proposed in [6,7]. Real circles specified by their center, radius, and containing plane can be discretized in this manner using a hypersphere as the structuring element.

In view of the above, we have picked up the problem of discrete spherical circle to study its characteristics in different topological models and to design efficient algorithms for its construction. We mention here that, contrary to the recent work in [8], which is focused to construction of only geodesic circles on a discrete sphere, in this paper we have generalized the concept to discrete spherical circles. A *discrete spherical circle* C is a/the shortest 0-path of voxels satisfying the following conditions:

1. Each voxel of C is at most τ distance away from the real circle defined by the intersection of a given pair of a real sphere and a real plane (not necessarily geodesic). Here, τ is the maximum between the thickness of the discrete sphere and that of the discrete plane.[1]
2. Each voxel of C belongs to the discretization of the real sphere and to the discretization of the real plane, considered in an appropriate topological model of the sphere-plane pair.

1.2 Definitions and Terminologies

In this section, we fix some basic notions and notations to be used in the sequel. For more details, we refer to [24]. Some other notions will be defined in the subsequent sections.

---

[1] The meaning of thickness of some discretization is discussed in Sect. 2.

A *voxel* (also called a 3-cell) is a unit cube determined by the integer grid and fully identified by its center which is a point of $\mathbb{Z}^3$. A discrete (or digital) object is a finite set of voxels. The *supercover* of a set $M \subseteq \mathbb{R}^3$ is the set $S(M)$ of all voxels that are intersected by M.

Two voxels are said to be *0-adjacent* if they share a vertex (0-cell), *1-adjacent* if they share an edge (1-cell), and *2-adjacent* if they share a face (2-cell). Thus, two distinct voxels, $p_1(i_1, j_1, k_1)$ and $p_2(i_2, j_2, k_2)$ are *1-adjacent* if and only if $|i_1 - i_2| + |j_1 - j_2| + |k_1 - k_2| \leqslant 2$ and $\max\{|i_1 - i_2|, |j_1 - j_2|, |k_1 - k_2|\} = 1$; *2-adjacent* if and only if $|i_1 - i_2| + |j_1 - j_2| + |k_1 - k_2| = 1$; and *0-adjacent* if and only if $|i_1 - i_2| = |j_1 - j_2| = |k_1 - k_2| = 1$. Clearly, 0-adjacent (1-adjacent) voxels are not considered as adjacent while considering 1-adjacency (2-adjacency). Note that the 0-, 1-, and 2-adjacencies, as adopted in this paper, correspond respectively to the classical 26-, 18-, and 6-neighborhood notations used in [16].

For $k = 0, 1$, or 2, a *k-path* in a discrete object A links elements through j-facets, where $j \geqslant k$. More precisely, it is a sequence of discrete points from A such that every two consecutive points are k-adjacent. A is called *k-connected* if there is a k-path connecting any two points of A. A *k-component* is a maximal k-connected subset of A.

Let D be a subset of a discrete object A. If $A \setminus D$ is not k-connected, then the set D is said to be *k-separating* in A. Let D be a k-separating digital object in A such that $A \setminus D$ has exactly two k-components. A *k-simple cell* (or *k-simple point*) of D (with respect to A) is a 3-cell c such that $D \setminus \{c\}$ is still k-separating in A. A k-separating digital object in A is *k-minimal* (or *k-irreducible*) if it does not contain any k-simple cell (with respect to A). If a digital object D is not 2-separating in a digital object A, then D has *tunnels*; otherwise, it is *tunnel-free*[2].

We define *x-distance* and *y-distance* between two (real or integer) points, $p(i, j)$ and $p'(i', j')$, as $d_x(p, p') = |i - i'|$ and $d_y(p, p') = |j - j'|$, respectively. In $\mathbb{R}^3$ or in $\mathbb{Z}^3$, we have also *z-distance*, given by $d_z(p, p') = |k - k'|$, for $p(i, j, k)$ and $p'(i', j', k')$. Using these inter-point distances, we define the respective x-, y-, and z-distances between a point $p(i, j, k)$ and a surface S as follows. Let $d_x(p, S)$ denote the x-distance between a point $p(i, j, k)$ and a surface S. If there exists a point $p'(x', y', z')$ in S such that $(y', z') = (j, k)$, then $d_x(p, S) = d_x(p, p')$; otherwise, $d_x(p, S) = \infty$. The other two distances, i.e., $d_y(p, S)$ and $d_z(p, S)$, are defined in a similar way; note that the metric $d_z(p, S)$ is not defined in 2D.

The above metrics are used to define the *isothetic distance* between two points, or between a point and a curve, or between a point and a surface. Between two points, $p_1(i_1, j_1)$ and $p_2(i_2, j_2)$, the isothetic distance is taken as the *Minkowski norm* [24], given by $d_\infty(p_1, p_2) = \max\{d_x(p_1, p_2), d_y(p_1, p_2)\}$. Between a 2D point $p(i, j)$ and a curve C, it is $d_\perp(p, C) = \min\{d_x(p, C), d_y(p, C)\}$, where $d_x(p, C)$ and $d_y(p, C)$ are defined similar to $d_x(p, S)$ and $d_y(p, S)$ respectively. Between a 3D point $p(i, j, k)$ and a surface S, it is given by $d_\perp(p, S) = \min\{d_x(p, S), d_y(p, S), d_z(p, S)\}$.

---

[2] For formal definitions and details on tunnels and gaps in discrete objects, we refer to [14]. In what follows, *gap-free* means *0-gap-free*, since a 0-gap-free surface is also 1- and 2-gap-free.

2   Discrete Spheres, Types, Ordering

Sphere discretization in the integer space is a well-studied research problem and several types of discretization can be found in the current literature [1, 2, 10, 15, 20, 25, 26, 28]. Considering the commonly used topological properties like tunnel-freeness, gap-freeness, tiling, and surface connectivity, we summarize here the different types of discrete spheres in this section. Their descriptions are given in increasing order of *thickness*, which is defined as the supremum of the distances of the constituent voxels of the discrete sphere from the corresponding real sphere. Figure 1 shows examples of four types of discrete hemispheres for $r = 9$.

Naive Sphere. Naive sphere is the irreducible tunnel-free sphere. In other words, it is the thinnest possible discrete sphere that separates the interior from the exterior. For computing the naive sphere with a real value of the radius r, there is no definite algorithm till date. For an integer value of r, a closed form has recently been reported in [8]. It is given by

$$S_n(r) = \left\{ \begin{array}{c} (i, j, k) \in \mathbb{Z}^3 : r^2 - \max\{|i|, |j|, |k|\} \leqslant s < r^2 + \max\{|i|, |j|, |k|\} \\ \wedge \left(\begin{array}{c} (s \neq r^2 + \max\{|i|, |j|, |k|\} - 1) \\ \vee (\mathrm{mid}\{|i|, |j|, |k|\} \neq \max\{|i|, |j|, |k|\}) \end{array} \right) \end{array} \right\} \quad (1)$$

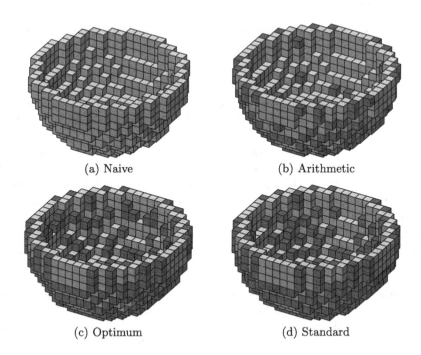

(a) Naive   (b) Arithmetic

(c) Optimum   (d) Standard

Fig. 1. Four different types of discrete hemispheres of radius 9. They are ordered here from (a) to (d) by increasing thickness. Naive voxels are shown gray, and the other voxels are shown blue. Notice that (c) optimum is not a subset of (d) standard (Color figure online).

where $s = i^2 + j^2 + k^2$. As discussed and proved in [8], Eq. 1 conforms to the fact that each point p included in $S_n(r)$ maintains an isothetic distance of (strictly) less than $\frac{1}{2}$ from the real sphere $S(r)$, which in turn implies that the Euclidean distance of p from $S(r)$ is less than $\frac{1}{2}$. The set $S_n(r)$ is also devoid of any simple voxels, and hence guarantees the topological property of 2-separableness or tunnel-freeness. There always exists a 1-path entirely contained in $S_n(r)$ between any two points on $S_n(r)$.

Arithmetic Sphere. Arithmetic sphere is thicker than naive sphere. Properties and algorithm for construction of arithmetic sphere can be seen in [2]. The arithmetic sphere of radius r is given by

$$S_a(r) = \left\{ (i,j,k) \in \mathbb{Z}^3 : \left(r - \tfrac{1}{2}\right)^2 \leqslant i^2 + j^2 + k^2 < \left(r + \tfrac{1}{2}\right)^2 \right\}. \tag{2}$$

The Euclidean distance of each voxel of $S_a(r)$ from $S(r)$ is bounded from above by $\frac{1}{2}$, but the upper bound of its isothetic distance can be greater than $\frac{1}{2}$. Therefore, $S_n(r)$ is a subset of $S_a(r)$. An arithmetic sphere maintains the property of tiling the space without any gap when the radius is increased in unit steps. Clearly, this type of sphere is not irreducible but has the property of 2-separableness due to being superset of $S_n(r)$. Similar to $S_n(r)$, between any two points on $S_a(r)$, there always exists a 1-path lying entirely on $S_a(r)$.

Standard Sphere. Standard sphere is the irreducible gap-free sphere and thicker than arithmetic sphere. For radius r, it is given by

$$S_s(r) = \left\{ (i,j,k) \in \mathbb{Z}^3 : (C_0(i,j,k) \cap S_{in}(r) \neq \emptyset) \wedge (C_0(i,j,k) \cap S_{ex}(r) \neq \emptyset) \right\} \tag{3}$$

where $C_0(i,j,k)$ denotes the set of eight 0-cells for the 3-cell corresponding to (i,j,k), and $S_{in}(r)$ and $S_{ex}(r)$ denote the interior and the exterior of $S(r)$, respectively.

It may be observed that as the radius r of the sphere is integer, none of the 0-cells can lie on the surface of $S(r)$. This directly implies that the supercover $S_c(r)$ of $S(r)$ is identical with $S_s(r)$. Each point on $S_s(r)$ maintains an Euclidean distance of strictly less than $\frac{\sqrt{3}}{2}$ from $S(r)$, since its corresponding 3-cell is intersected by $S(r)$. Further, $S_s(r)$ is the minimum set that guarantees gap-freeness of the discrete sphere surface. This is necessary in an application where any gap cannot be allowed on the surface. Between any two points on $S_s(r)$, there always exists a 2-path lying entirely on $S_s(r)$.

Offset Sphere. The impact of offset digitization on the connectivity and separability of 3- or higher-dimensional curves and surfaces has been discussed in [11,12]. In this work, we extend the concept proposed in [12] for 3D spheres. We first introduce the definition of an offset sphere; given an integer radius r, the τ-offset[3] sphere is given by

$$S_o(r,\tau) = \left\{ (i,j,k) \in \mathbb{Z}^3 : (r - \tau)^2 \leqslant i^2 + j^2 + k^2 \leqslant (r + \tau)^2 \right\}. \tag{4}$$

---

[3] The term *offset* is equivalent to *thickness* in the context of our work.

Setting τ to different values produces different types of discrete spheres. In particular, we have the following theorem.

Theorem 1 (Offset sphere). *Given a real radius r, the offset sphere $S_o(r, \tau)$ is 2-gap-free, 1-gap-free, or 0-gap-free for $\tau = \frac{1}{2}, \frac{1}{\sqrt{2}}$, or $\frac{\sqrt{3}}{2}$, respectively.*

Proof. As shown in [12], for any 3D real surface, $\tau \geqslant \frac{\sqrt{3-k}}{2}$ (where, $0 \leqslant k \leqslant 2$) produces a discrete surface that has no k-gap. In other words, $\frac{\sqrt{3-k}}{2}$ is the minimum value of τ to achieve k-gap-free discretization of any arbitrary 3D surface. For $k = 2, 1, 0$, we get the respective k-gap-free spheres with the values of τ as $\frac{1}{2}, \frac{1}{\sqrt{2}}$, or $\frac{\sqrt{3}}{2}$. □

Note that these offset values are sufficient to guarantee the required separableness property of a discrete surface, but for a specific nature of the surface, such as sphere in our case, a smaller offset value may produce the desired separableness. We briefly compare here the properties of different offset spheres with the other discrete spheres mentioned earlier. We first introduce the following two lemmas.

Lemma 1. *For any real radius r, we have*

$$S_n(r) \subseteq S_a(r) \subseteq S_o(r, \tfrac{1}{2}) \subseteq S_o(r, \tfrac{1}{\sqrt{2}}).$$

Proof. As already mentioned, both $S_n(r)$ and $S_a(r)$ maintain the topological property of tunnel-freeness. The former is thinnest, but the latter is not, wherefore $S_n(r) \subseteq S_a(r)$. By Theorem 1, $S_o(r, \frac{1}{2})$ is tunnel-free and the Euclidean distance of each of its voxels from $S(r)$ is within $\frac{1}{2}$, inclusive. But for $S_a(r)$, as evident from Eq. 2, the Euclidean distance of each voxel from $S(r)$ lies in the half-open interval $[r - \frac{1}{2}, r + \frac{1}{2})$. So, $S_a(r)$ is a subset of $S_o(r, \frac{1}{2})$. Finally, it is easy to see that $S_o(r, \frac{1}{2}) \subseteq S_o(r, \frac{1}{\sqrt{2}})$, which completes the proof. □

Lemma 2. *For any real radius r, we have*

$$S_s(r) \subseteq S_c(r) \subseteq S_o(r, \tfrac{\sqrt{3}}{2}).$$

Proof. Each of $S_s(r)$, $S_c(r)$, and $S_o(r, \frac{\sqrt{3}}{2})$ has 0-separating property. Among these, $S_s(r)$ is irreducible and hence thinnest. Between the other two, $S_c(r)$ contains all voxels intersected by $S(r)$, whereas $S_o(r, \frac{\sqrt{3}}{2})$ may contain some additional voxels (not intersected by $S(r)$). This leads to their relation. □

Note that $S_o(r, \frac{1}{\sqrt{2}})$ is a special discrete sphere, which is guaranteed to be 1-gap-free. In the context of our work, it is the sole type of sphere that 1-separates the interior and the exterior, but does not 0-separate them. However, it may not be a subset of $S_s(r)$. See, for example, the result for $r = 9$ in Fig. 1. We have the following fact.

Fact 1. *Although $S_o(r, \frac{1}{\sqrt{2}})$ is not a subset of $S_s(r)$, its thickness does not exceed that of $S_s(r)$, since the Euclidean distance of voxels in $S_s(r)$, which close its 0-gaps, would be more than $\frac{1}{\sqrt{2}}$ from* $\mathsf{S}(r)$.

Thus we have the following theorem.

Theorem 2. (Sphere ordering). *For a given real or integer radius r, the discrete spheres under different models of sphere discretization can be ordered by non-decreasing thickness as follows.*

$$\left\langle S_n(r),\ S_a(r),\ S_o(r, \tfrac{1}{2}),\ S_o(r, \tfrac{1}{\sqrt{2}}),\ S_s(r),\ S_c(r),\ S_o(r, \tfrac{\sqrt{3}}{2}) \right\rangle$$

Proof. Lemmas 1, 2, and Fact 1 indicate that the thickness of the different models follows the chain of inequalities

$$\tau(S_n(r)) \leqslant \tau(S_a(r)) \leqslant \tfrac{1}{2} \leqslant \tfrac{1}{\sqrt{2}} \leqslant \tau(S_s(r)) \leqslant \tau(S_c(r)) \leqslant \tfrac{\sqrt{3}}{2},$$

and hence the ordering.   □

The above theorem holds for radius and center in the real space. For integer specification, we have the following fact.

Fact 2. *When the center and the radius are in the integer space, $S_a(r) = S_o(r, \frac{1}{2})$ and $S_s(r) = S_c(r)$.*

We show in the forthcoming section that out of all the above seven types of discrete spheres, $S_o(r, \frac{1}{\sqrt{2}})$ is the thinnest possible sphere that guarantees the connectivity of the discretized set corresponding to any circle (or a curve, in general) lying on $\mathsf{S}(r)$. Hence, for brevity, we refer an offset discretization with $\tau = \frac{1}{\sqrt{2}}$ as *optimum discretization*, and in particular, $S_o(r, \frac{1}{\sqrt{2}})$ as *optimum sphere*, in this sequel.

3   Discrete Spherical Circles

Any 3D real circle is given by the intersection of a real sphere and a real plane. As different kinds of discrete planes have been studied in detail over the last two decades (see [13]), an effective approach towards finding the discretization of a real circle is to consider the intersection of the voxels sets obtained by discretization of its corresponding real sphere and real plane. To form this intersection set, the type of discretization has to be chosen appropriately based on the desired type of connectivity of the discrete circle and the proximity of each of its voxels from the real sphere and the real plane. In this section, we analyze the different possibilities.

As shown in [8], the set of voxels of a naive sphere intersected by a real geodesic plane always contains a geodesic cycle, which is a 1-connected closed path.

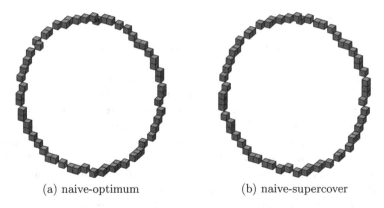

(a) naive-optimum   (b) naive-supercover

Fig. 2. Discrete spherical circles with the sphere $r = 12, c = o$ and the plane $72x + 78y - 94z = 0$. (a) Maximum distance from the real plane $= 0.662959$, maximum distance from the real sphere $= 0.424163$. (b) Maximum distance from the real plane $= 0.832225$, maximum distance from the real sphere $= 0.424163$.

The underlying idea is as follows. Given $S_n(r)$ with integer radius and integer center (which is $o(0,0,0)$, w.l.o.g.) and any two points s and t on it, we get the supercover $P_c(s,t,o)$ of the real geodesic plane, $P(s,t,o)$, that passes through s, t, o. The intersection set $I(S_n(r), P_c(s,t,o)) := S_n(r) \cap P_c(s,t,o)$ contains a 1-connected path from s to t and another 1-connected path from t to s, and the concatenation of these two paths gives the cycle. A simple mapping from the 1-connections in this cycle to 0-connections produces a 0-connected *geodesic circle* as a subset of $I(S_n(r), P_c(s,t,o))$. In this geodesic circle, although the isothetic distance of each voxel from $S(r)$ is less than $\frac{1}{2}$, the Euclidean distance from $P(s,t,o)$ can be as high as $\frac{\sqrt{3}}{2}$. The latter upper bound is relatively high and makes the path jagged w.r.t. the real geodesic plane.

In order to make the circle smoother w.r.t. the intersection plane without compromising with its distance from the real sphere, we consider a different discretization combination of the sphere and its intersection plane.[4] It is worth mentioning here at this point that it is one of the main objectives of the work proposed in this paper. Further, opposed to the work proposed in [8] where only geodesic circles have been considered, we consider here a circle resulted from the intersection of a sphere and an arbitrary plane. Figure 2 shows a comparison between an instance of a geodesic circle in $I(S_n(r), P_o(s,t,o,\frac{1}{\sqrt{2}}))$ and of another in $I(S_n(r), P_c(s,t,o))$. Here, $P_o(s,t,o,\frac{1}{\sqrt{2}})$ denotes the discrete plane obtained by offset digitization of $P(s,t,o)$ with the offset $\tau = \frac{1}{\sqrt{2}}$. Notice that the maximum distance of the constituent voxels of the circle from the real plane indicates that the former is relatively smoother than the latter.

---

[4] While the meaning of smoothness of discrete curve is perhaps intuitively clear, this will be formally defined and discussed in Sect. 3.2.

3.1 Discretization Classes

As put in Sect. 2, we have seven discretization models of a real sphere, which follow a thickness ordering given in Theorem 2. A thicker discretization admits a better connectivity of a discrete circle—or any curve, in general—lying on the discrete sphere, but at the cost of a higher deviation from the real sphere. A thinner discretization, on the contrary, can enforce a restricted deviation, but is prone to disconnectedness. Hence, to strike a balance, we address here the issue of optimum thickness. Also, while defining the discretization for a circle on a real sphere, we need to consider the discretization model of the intersection plane. Just like sphere discretization, we get seven possible models of plane discretization, which, when taken together with sphere discretization, gives us $7 \times 7 = 49$ combination pairs. Out of these 49 pairs, some do not guarantee construction of (0-connected) *discrete spherical circles* on the concerned discrete sphere, while the rest do. We thus get two unique classes of discretization pairs:

1. *Admissible Class* (AC): Guarantees the construction of discrete spherical circles from the intersection of a discrete sphere and a discrete plane.
2. *Forbidden Class* (FC): Intersection of a discrete sphere and a discrete plane may not produce a discrete spherical circle.

The following theorem explicates the members of the above two classes.

Theorem 3. (Sphere-plane class). *If a sphere and its intersection plane are both discretized to thinner than $\tau = \frac{1}{\sqrt{2}}$ (optimum discretization), then the corresponding discretization pair belongs to the forbidden class; otherwise, the pair belongs to the admissible class.*

Proof. As shown in [12, Theorem 11], an offset $\tau \geqslant \frac{1}{\sqrt{2}}$ for the discretization of any 3D curve γ guarantees its 0-connectedness in $\mathbb{Z}^3$. Here the curve γ is a circle resulted from the intersection of a sphere and a plane in $\mathbb{R}^3$. Hence, optimum discretization of this circle is always 0-connected in $\mathbb{Z}^3$. In order to ensure that this discrete circle belongs to both the discrete sphere and the discrete plane, we have to choose a discretization pair from the admissible class. For this, the intersection set (IS) between the discrete sphere (DS) and the discrete plane (DP) must be 0-connected. We show that this is ensured only if one discretization is at least 2-separating (i.e., 2-gap-free), and the other at least 1-separating (i.e., 1-gap-free).

   The ordering of the seven discretization models in Theorem 2 indicates that if both DS and DP are thinner than optimum discretization, then both are at most 2-separating in $\mathbb{Z}^3$, and so, IS is also at most 2-separating in both. However, IS may not be connected. Figure 3(a) and (b) shows such counterexamples. To circumvent this problem, if, w.l.o.g., the sphere is subject to optimum discretization, then by Theorem 1, it becomes 1-separating. As a result, its intersection with DP—whatsoever be its discretization—becomes 1-separating (i.e. 1-gap-free) in DP. There may exist 0-gaps although, which do not make IS disconnected. The rationale is as follows.

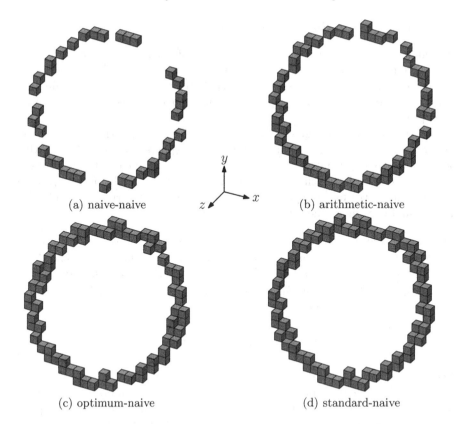

(a) naive-naive   (b) arithmetic-naive

(c) optimum-naive   (d) standard-naive

Fig. 3. Sphere-plane intersection sets in (a, b) forbidden class and (c, d) admissible class. Sphere: $r = 9, c = o$; plane: $x + 2y + 3z + 1 = 0$.

If there is a 0-gap in IS, then the two parts A and B formed by exclusion of IS from DP are 0-connected. So, there exist two voxels $a \in A$ and $b \in B$, which are 0-adjacent. As DP is at least naive, i.e., a tunnel-free discrete surface, there exist two other voxels p and q on DP, both of which are 1-adjacent to a and 2-adjacent to b, and at least 0-adjacent to each other. The voxels p and q belong to IS; for, otherwise they form 1- or 2-gaps in DP—a contradiction. Therefore, IS is 1-gap-free in at least one of DS and DP, or equivalently, IS is 0-connected.   □

By Theorems 1 and 2, optimum sphere is the thinnest model of a discrete sphere with 1-separability, and naive sphere is the thinnest with 2-separability. Similar properties hold also for a discrete plane. When we construct a discrete spherical circle, it belongs to IS, and hence to both DS and DP. To ensure its smoothness, we should make DS and DP as thin as possible. To ensure its connectedness in simultaneity, by Theorem 3, either DS or DP should be no thinner than optimum discretization. In other words, naive discretization of the sphere demands optimum discretization of the plane, and vice versa. If DS is naive, then the circle is smoothest w.r.t. the sphere; and if DP is naive, then it is smoothest w.r.t. the plane.

3.2 Smoothness of Spherical Circles

We define the *smoothness tuple* of a discrete spherical circle as $\varsigma := \langle \varsigma_s, \varsigma_p \rangle$, where ς_s and ς_p denote the respective supremum of distances of all voxels of DS and the supremum of distances of all voxels of DP from their corresponding real counterparts. Based on this tuple, we define its *smoothness* as the average of the two suprema, given by $\bar{\varsigma} = \frac{1}{2}(\varsigma_s + \varsigma_p)$.

Observe that for any given 3D real circle C, although there is a unique plane containing C, there are infinitely many real spheres on which C lies. After choosing the sphere, whatsoever discretization is used, we cannot achieve a discrete spherical circle which is better than the one we get by naive-optimum (or optimum-naive) pair of sphere-plane discretization, in terms of smoothness. More formally, we have the following theorem.

Theorem 4 (Smoothness). *Let* C *be any 3D circle defined by the intersection of a given sphere-plane combination. Let, for the discrete spherical circle obtained by any discretization* D *of* C, *the resultant smoothness tuple be* $\varsigma^{(D)} := \langle \varsigma_s^{(D)}, \varsigma_p^{(D)} \rangle$. *Let, for the discrete spherical circle obtained by the naive-optimum discretization of* C, *the tuple be* $\varsigma^{(\mathrm{opt})} := \langle \varsigma_s^{(\mathrm{opt})}, \varsigma_p^{(\mathrm{opt})} \rangle$. *Then,*

$$\bar{\varsigma}^{(\mathrm{opt})} = \min_D \left\{ \bar{\varsigma}^{(D)} \right\} < \frac{\sqrt{2}+1}{4}.$$

Proof. By Theorem 3, either the discrete sphere or the discrete plane should be no thinner than the one obtained by optimum discretization. So, with optimum discretization of the plane, in order to minimize the smoothness, naive sphere is the only option, as it is thinnest with $\varsigma_s < \frac{1}{2}$ (Sect. 2 and Theorem 2). By [12, Theorem 11], $\varsigma_p \leqslant \frac{1}{\sqrt{2}}$ for optimum discretization of plane. As a result, the smoothness is $\bar{\varsigma} = \frac{1}{2}(\varsigma_s + \varsigma_p) < \frac{\sqrt{2}+1}{4}$. □

Using this smoothness metric, for the results shown in Fig. 2, smoothness of naive-optimum result is 0.543561 ($< \frac{\sqrt{2}+1}{4} = 0.603553$) and smoothness of naive-supercover result is 0.628194 (> 0.603553).

3.3 Algorithm for Discrete Spherical Circle

Opposed to the recent work in [8], we are not restricting the spherical circles to only geodesic circles. For any real plane intersecting a given real sphere, there may exist voxels in the intersection set of the corresponding discrete plane and the discrete sphere, which are far from the concerned real circle, as shown in Fig. 4. In that case, the shortest 0-path in the intersection set may not produce the correct discrete circle. Therefore, we consider only the *circular set*, which is defined as follows. Let S and P be a real sphere and a real plane, and let C be the real circle given by their intersection. Let (D_1, D_2) be a discretization pair from the admissible class, which produces $D_1(\mathsf{S})$ and $D_2(\mathsf{P})$ as the respective discretizations of S and P. Let, w.l.o.g., D_1 produce a thicker discretization

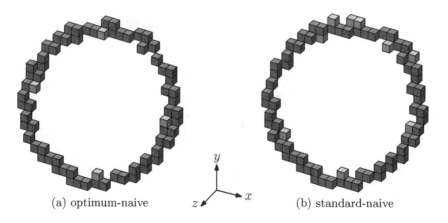

(a) optimum-naive   (b) standard-naive

Fig. 4. *Circular sets*, resulted from the last two intersection sets shown in Fig. 3. Each green voxel belongs to the circular set. Each gray voxel belongs to the intersection set but not to the circular set (Color figure online).

compared to D_2. Then the circular set comprises the voxels of $D_1(C)$ that belong to $D_1(S) \cap D_2(P)$, and hence given by $C = D_1(C) \cap D_1(S) \cap D_2(P)$.

The above definition implies that if the thicker discretization is optimum, then the circular set will be a subset of the optimum discretization of the real circle; if the thicker one is standard, then the circular set will be a subset of the standard discretization of the real circle; and so on. Examples are shown in Fig. 4.

The following theorem ensures that we always get a 0-connected discrete spherical circle from the circular set.

Theorem 5 (Circular Set). *The circular set for any discretization pair in the admissible class always has at least one 0-connected discrete spherical circle.*

Proof. By Theorem 3, D_1 is at least optimum discretization. Assume that D_2 is thinnest (i.e., naive) discretization. We have a 3-step proof as follows.

Step 1: $D_1(C)$ contains all the voxels with at most $\frac{1}{\sqrt{2}}$ distance away from C, and hence $C = D_1(C) \cap D_2(P)$.

Step 2: $D_1(C)$ is (at least 0-)connected and hence 1-separates $D_1(P)$ [12, Theorem 11]. As $D_2(\cdot) \subseteq D_1(\cdot)$ by Theorem 2, $D_1(C)$ 1-separates $D_2(P)$ too. This implies that $C := D_1(C) \cap D_2(P)$ 1-separates $D_2(P)$.

Step 3: As C 1-separates $D_2(P)$, it contains a cycle that 1-separates $D_2(P)$.

As each voxel in C satisfies the requirement needed to be a part of the discrete spherical circle, the proof follows optimum-naive combination of (D_1, D_2). When D_1 is thicker than optimum or D_2 is thicker than naive, the resultant intersection is thicker, thus completing the proof for the admissible class.   □

From the circular set, the discrete spherical circle is computed by the procedure of prioritized-BFS, as explained in [8]. This will ensure that we get the correct discretization of the real circle as a 0-path, as shown in Fig. 5. The steps are as follows.

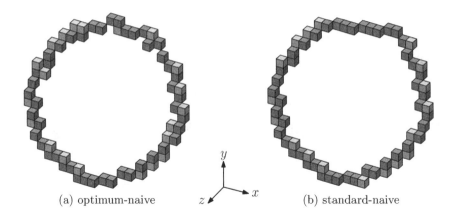

(a) optimum-naive   (b) standard-naive

Fig. 5. Discrete spherical circles (green) obtained as cycles (0-paths) from the circular sets shown in Fig. 4. Gray voxels belong to the circular set but not to the discrete spherical circle (Color figure online).

1. Generate intersection set from the specific discretization pair of sphere and plane from the admissible class.
2. From the intersection set, find the circular set consisting of those voxels that are intersected by the real circle.
3. Generate an adjacency list of the voxels in the circular set.
4. Run prioritized-BFS to get 0-connected shortest paths.

4   Results and Discussion

Figures 4 and 5 contain the step-by-step results obtained by our algorithm to finally get a discrete spherical circle for a typical sphere-plane combination. From the result shown in Fig. 6, it is evident that the intersection set of the discrete sphere and the discrete plane in the admissible class may contain many voxels that are far from the corresponding real circle if the plane is almost tangential to the sphere. Even in that case, our consideration of circular set ensures a proper discrete spherical circle which is any one among the shortest paths from the circular set, maintaining the properties of a discrete spherical circle as stated in Sect. 1.1.

The discrete spherical circle generated as a cycle from the circular set need not be symmetric in any sense. However, this may be considered as a future scope of work where the 48-symmetric property of discrete sphere as explained in [8] can be utilized to compute a better discrete spherical circle. A discrete 3D circle can alternatively be produced as a 0-connected shortest path lying in the supercover of the real circle defined by the intersection of a real plane with a real sphere. Whether this would be better or worse in terms of smoothness can be studied in line with the proposed work. The issues of other metrics, one being 'smoothness' as explained in Sect. 3.2, may also be explored to chalk out further scopes of improvement.

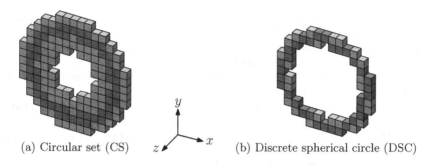

(a) Circular set (CS)   (b) Discrete spherical circle (DSC)

Fig. 6. DSC for optimum sphere ($r = 17, c = o$) and naive plane $z + 16 = 0$. (a) Green voxels comprise the CS; gray voxels belong to the intersection set but not to CS. (b) Green voxels comprise the DSC; gray voxels belong to the CS but not to the DSC (Color figure online).

References

1. Andres, E., Jacob, M.: The discrete analytical hyperspheres. IEEE Trans. Visual Comput. Graphics **3**(1), 75–86 (1997)
2. Andres, E.: Discrete circles, rings and spheres. Comput. Graphics **18**(5), 695–706 (1994)
3. Anton, F.: Voronoi diagrams of semi-algebraic sets. Ph.D. thesis, University of British Columbia, Vancouver, British Columbia, Canada (2004)
4. Anton, F., Emiris, I.Z., Mourrain, B., Teillaud, M.: The offset to an algebraic curve and an application to conics. In: Gervasi, O., Gavrilova, M.L., Kumar, V., Laganá, A., Lee, H.P., Mun, Y., Taniar, D., Tan, C.J.K. (eds.) ICCSA 2005. LNCS, vol. 3480, pp. 683–696. Springer, Heidelberg (2005)
5. Arrondo, E., Sendra, J., Sendra, J.: Genus formula for generalized offset curves. J. Pure Appl. Algebr. **136**(3), 199–209 (1999)
6. Aveneau, L., Andres, E., Mora, F.: Expressing discrete geometry using the conformal model. In: AGACSE (2012). http://hal.archives-ouvertes.fr/hal-00865103
7. Aveneau, L., Fuchs, L., Andres, E.: Digital geometry from a geometric algebra perspective. In: Barcucci, E., Frosini, A., Rinaldi, S. (eds.) DGCI 2014. LNCS, vol. 8668, pp. 358–369. Springer, Heidelberg (2014)
8. Biswas, R., Bhowmick, P.: On finding spherical geodesic paths and circles in $\mathbb{Z}^3$. In: Barcucci, E., Frosini, A., Rinaldi, S. (eds.) DGCI 2014. LNCS, vol. 8668, pp. 396–409. Springer, Heidelberg (2014)
9. Bresenham, J.E.: Algorithm for computer control of a digital plotter. IBM Syst. J. **4**(1), 25–30 (1965)
10. Brimkov, V.E., Barneva, R.P.: On the polyhedral complexity of the integer points in a hyperball. Theor. Comput. Sci. **406**(1–2), 24–30 (2008)
11. Brimkov, V.E., Barneva, R.P., Brimkov, B.: Minimal offsets that guarantee maximal or minimal connectivity of digital curves in nD. In: Brlek, S., Reutenauer, C., Provençal, X. (eds.) DGCI 2009. LNCS, vol. 5810, pp. 337–349. Springer, Heidelberg (2009)
12. Brimkov, V.E., Barneva, R.P., Brimkov, B.: Connected distance-based rasterization of objects in arbitrary dimension. Graph. Models **73**, 323–334 (2011)

13. Brimkov, V.E., Coeurjolly, D., Klette, R.: Digital planarity–a review. Discrete Appl. Math. **155**(4), 468–495 (2007)
14. Brimkov, V.: Formulas for the number of $(n-2)$-gaps of binary objects in arbitrary dimension. Discrete Appl. Math. **157**(3), 452–463 (2009)
15. Chamizo, F., Cristobal, E.: The sphere problem and the L-functions. Acta Math. Hungar. **135**(1–2), 97–115 (2012)
16. Cohen-Or, D., Kaufman, A.: Fundamentals of surface voxelization. Graphics Model. Image Process. **57**(6), 453–461 (1995)
17. Cohen-Or, D., Kaufman, A.: 3D line voxelization and connectivity control. IEEE Comput. Graph. Appl. **17**(6), 80–87 (1997)
18. Cox, D., Little, J., OShea, D.: Using Algebraic Geometry. Springer, New York (2005)
19. Debled-Rennesson, I., Domenjoud, E., Jamet, D.: Arithmetic discrete parabolas. In: Bebis, G., Boyle, R., Parvin, B., Koracin, D., Remagnino, P., Nefian, A., Meenakshisundaram, G., Pascucci, V., et al. (eds.) ISVC 2006. LNCS, vol. 4292, pp. 480–489. Springer, Heidelberg (2006)
20. Fiorio, C., Jamet, D., Toutant, J.L.: Discrete circles: an arithmetical approach with non-constant thickness. In: Vision Geometry XIV, Electronic Imaging, SPIE, vol. 6066, pp. 60660C.1–60660C.12 (2006)
21. Gouraud, H.: Continuous shading of curved surfaces. IEEE Trans. Comput. **20**(6), 623–629 (1971)
22. Hoffmann, C., Vermeer, P.: Eliminating extraneous solutions for the sparse resultant and the mixed volume. J. Symbolic Geom. Appl. **1**(1), 47–66 (1991)
23. Kaufman, A.: Efficient algorithms for 3d scan-conversion of parametric curves, surfaces, and volumes. SIGGRAPH Comput. Graph. **21**(4), 171–179 (1987)
24. Klette, R., Rosenfeld, A.: Digital Geometry: Geometric Methods for Digital Picture Analysis. Morgan Kaufmann, San Francisco (2004)
25. Maehara, H.: On a sphere that passes through n lattice points. European J. Combin. **31**(2), 617–621 (2010)
26. Montani, C., Scopigno, R.: Spheres-to-voxels conversion. In: Graphics Gems. Academic Press, pp. 327–334 (1990)
27. Mukhopadhyay, J., Das, P.P., Chattopadhyay, S., Bhowmick, P., Chatterji, B.N.: Digital Geometry in Image Processing. Chapman & Hall/CRC, Boca Ration, UK (2013)
28. Toutant, J.L., Andres, E., Roussillon, T.: Digital circles, spheres and hyperspheres: From morphological models to analytical characterizations and topological properties. Discrete Appl. Math. **161**(16–17), 2662–2677 (2013)

Optimal Consensus Set for nD Fixed Width Annulus Fitting

Rita Zrour[(✉)], Gaelle Largeteau-Skapin, and Eric Andres

Laboratoire XLIM, SIC, UMR CNRS 7252, Université de Poitiers, BP 30179,
86962 Futuroscope Chasseneuil, France
{rita.zrour,gaelle.largeteau.skapin,eric.andres}@univ-poitiers.fr

Abstract. This paper presents a method for fitting a nD fixed width spherical shell to a given set of nD points in an image in the presence of noise by maximizing the number of inliers, namely the consensus set. We present an algorithm, that provides the optimal solution(s) within a time complexity $O(N^{n+1} \log N)$ for dimension n, N being the number of points. Our algorithm guarantees optimal solution(s) and has lower complexity than previous known methods.

1 Introduction

Shape fitting is a general problem that is of very practical nature, namely extracting features from data with a high level of noise. This problem has been widely addressed in very different fields spanning from computer graphics and image processing to data mining in large dimensional databases [2]. In this paper we are looking into the problem of optimal fitting an nD annulus of fixed width in an image in the presence of outliers. The set of points which fits a model is called a *consensus set*. Note that an annulus in higher dimensions is sometimes referred to as *n-sphere shell*. We preferred to use the common denomination *annulus* for all dimensions.

This paper aims at finding the optimal consensus set (maximal number of inliers) inside a fixed width nD annulus, where the center and the radius are unknowns. Most annuli fitting methods try to minimize the thickness of the fitted annuli [1,5,17]. In our case, we are interested in digital circles and spheres and more specifically Andres digital circles and hyperspheres [3] or k-Flake digital circles-spheres [18]. In those cases, the thickness is directly linked with topological properties. The most common fitting methods are based on variants of the RANdom Sample Consensus (RANSAC) algorithm [8], which is a robust parameter estimation algorithm widely used in the field of computer vision. However RANSAC is inherently probabilistic in its approach and does not guarantee any optimality. In this paper we are looking for (all the) optimal solutions in order to generate base solutions to which we can compare and validate other, non exact, methods. It is also a problem when looking for multiple fittings in the same image.

In [4,20], brute force algorithms were proposed to compute the optimal consensus set respectively for Andres digital circles [3] (defined as digital points

© Springer International Publishing Switzerland 2015
R.P. Barneva et al. (Eds.): IWCIA 2015, LNCS 9448, pp. 101–114, 2015.
DOI: 10.1007/978-3-319-26145-4_8

inside a classical annulus of fixed width) and 0-Flake digital circles [18] (8-connected circles) with a time complexity of $O(N^4)$ where N is the number of points. A new method was proposed in [11] for fitting 0-Flake digital circles that has a complexity $O(N^3 \log N)$. Our main contribution in this article is the extension of the fitting problem proposed in [20] to nD and reformulating it using a space transformation similar to the one proposed by [6] which was used in [11]. However the major difference with [6] is that the width is fixed and the parameter we want to maximize is the number of inliers.

The simpler problem of fitting circles and spheres is a largely studied problem. Most common approaches are based on least squares [15] or adapted Hough Transforms [10]. Those are however not adapted for annulus fitting. The circle fitting method proposed by O'Rourke et al. [13,16] that transforms a circle separation problem into a plane separability problem, is also not well suited because the fixed width of the digital circles translates into non fixed vertical widths for the planes. In this case, the problem is complicated (See [14] for some ideas on how to handle this difficulty). For annuli detection, various approaches have been proposed. Most of these methods are probabilistic approaches that minimize the width of the annuli. Among these algorithms, some consider that no noise is present in the image, and concentrate only on the problem of recognition instead of the fitting problem [1,5,17]. However noise in real world is omnipresent in the input and so many algorithms dealing with outliers have been proposed [7,9,12]; in these algorithms, the number of outliers is usually predefined [9,12] which is not always realistic.

The idea of the 2D algorithm is the following: given a set S, we consider all the annuli that have two specific points of S on the border of the annulus. All the annuli centers with those two points on the border are then located on a straight line. This straight line is taken as a parametric axis. We then determine when a point of S enters and leaves the annulus while the center moves along the axis. This allows us to compute the intervals where the number of inliers is maximized. By considering all the combinations of points, we are able to compute the exhaustive set of all optimal consensus sets in $O(N^3 \log N)$. See [19] for a similar approach for line and plane fitting. The nD algorithm works in a similar way than the 2D algorithm [20]. However, we show that the annulus can de defined by n specific points that are all on the external border. This characterization allows to greatly simplify both the proof of the characterization of annuli and the computation of the parametric axis defined by annuli centers. The final algorithm has an $O(N^{n+1} \log N)$ time complexity and leads to the exact optimal solutions for the problem of fixed width annulus fitting.

The paper is organized as follows: in Sect. 2 we expose the problem of annulus fitting and present some properties of the annuli with fixed width. Section 3 presents the dual space we use and provide the algorithm for finding the optimal nD annuli. Section 4 presents some 2D and 3D experiments. Finally Sect. 5 states some conclusion and perspectives.

2   Annulus Fitting

A nD annulus $A(C, R, w)$ of width w and radius R centered at $C(c_1, c_2, ..., c_n)$, is defined by the set of points in $\mathbb{R}^n$ satisfying two inequalities:

$$A(C, R, w) = \left\{ (p_1, p_2, ..., p_n) \in \mathbb{R}^n : R^2 \leq \sum_{i=1}^{n} (p_i - c_i)^2 \leq (R + w)^2 \right\} \quad (1)$$

where $C(c_1, c_2, ..., c_n) \in \mathbb{R}^n$ and $R, w \in \mathbb{R}^+$.

Using the above annulus model, our fitting problem is then described as follows: given a finite set $S = \{P_i, i \in [1, N], \in \mathbb{R}^n\}$ of N nD points we would like to find the parameters (center and radius) of an annulus A of given width w that contains the maximum number of points in S. Points $P_i \in S \cap A$ in nD are called inliers; otherwise they are called outliers. We also say that the annulus A covers the set $S \cap A$. We denote $B_i(C, R)$ (respectively $B_e(C, R+w)$) the internal (resp. external) border of the annulus $A(C, R, w)$, i.e. the set of points located at distance R (resp. $R + w$) from C.

2.1   Annulus Characterization

Our approach is focused on inlier sets, also called consensus sets. Since S is finite, the number of different consensus sets for the annulus is finite as well although too big to consider them all. However, as we are looking for the biggest consensus set(s), only the annuli that contain a minimum number of $n+1$ points, uniquely defining an annuli, are considered. For all the different consensus sets $\mathcal{C}$, with at least $n + 1$ points from a given set S, we are going compute the size of each one. The center and radius are actually a side result of this search. We present in this section some characterization of nD annuli that will allow us to explore all those concensus sets and further on to build an optimal fitting algorithm for annuli.

2.2 nD Annular Characterizations

In [20], we proposed a brute force algorithm for fitting fixed width annuli in 2D. We showed that if an optimal solution exists then there exists an equivalent optimal solution (with the same set of inliers) having three points on the border (internal and/or external). Testing all the configurations of three points and counting the inliers leads therefore to all the possible optimal solutions. This brute force method leads however to an $O(N^4)$ complexity for N the number of points to fit. Its extension to nD is straight forward and needs $n + 1$ points to be on the border (either internal or external), but it also leads to an $O(N^{n+2})$ complexity.

In this article we need only n points instead of $n+1$ points in nD; however the restriction is that the n points must be on the external border of the annulus. The last point is found in $O(N \log N)$ using a dual space, which will be presented in

Sect. 3.1. Using such a dual space, the algorithm has an overall time complexity of $O(N^{n+1} \log N)$. Note that we are looking for the exact optimal solutions, this complexity is, so far, the best to our knowledge to solve this problem.

The following theorem states that given an annulus A of width w covering a set of points S, with $|S| \geq n$, there exists at least one other annulus A' of same width, that also covers S with at least n points of S on its external border B_e.

Theorem 1. *Let S be a finite set of N $(N \geq n)$ points in $\mathbb{R}^n$. Let $A = (C, R, w)$ be an annulus covering S. Then it exists $A' = (C', R', w)$ covering S such that:*

$$\exists! \, Q_1, Q_2, ..., Q_n \in S \cap B_e(C', R' + w).$$

Proof. Let S be a finite set of N $(N \geq n)$ points in $\mathbb{R}^n$. Let $A = (C, R, w)$ be an annulus that covers S. The theorem proof is given as follows:

A. If the internal radius of the annulus is 0, the problem is reduced to the problem of a hyper-sphere of fixed radius. The hypersphere can either be translated towards the closest point of S if there are no point on the external

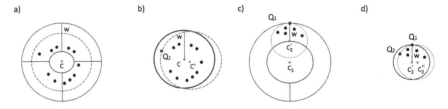

Fig. 1. Case A of the proof in 2D: (a) While decreasing the radius of the annulus (colored black) to reach a first point Q_1 we may arrive to an internal radius of 0 (annulus in red) without any point on B_e, in this case a translation is needed as shown in (b) to reach a point Q_1 on B_e. (c) While Decreasing the radius of the annulus (colored black) by moving the center along axis $C_2 Q_1$ to reach a second point Q_2, we may arrive to an internal radius of 0 and no point Q_2 on B_e (annulus in red), in this case a rotation is done as shown in (d) until reaching a second point Q_2 on B_e (Color figure online).

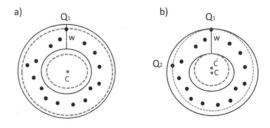

Fig. 2. Case B of the proof in 2D: (a) Decreasing the radius until reaching a point Q_1 on B_e. (b) Decreasing the radius by moving the center along axis CQ_1 until reaching a second point Q_2.

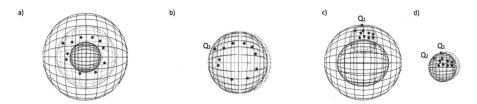

Fig. 3. Case A of the proof in 3D: (a) While decreasing the radius it may lead to an internal radius of 0 without any point on B_e; in this case a translation is needed as shown in (b) to reach a point Q_1 on B_e. (c) Decreasing the radius while maintaining Q_1 on the border until reaching an internal radius of 0; in this case a rotation is needed as shown in (d) in order to reach a point Q_2 on B_e.

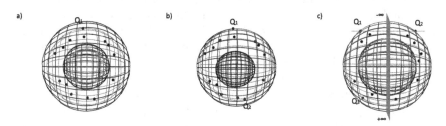

Fig. 4. Case B of the proof in 3D: (a) Decreasing the radius until reaching a first point Q_1 on B_e. (b) Decreasing the radius while maintaining Q_1 on the border until reaching a second point Q_2. (c) Case B of the proof in 3D: moving the center along the bisector of $Q_1 Q_2$ until reaching a third point Q_3.

border already (Figs. 1a,b and 3a,b) either it can be rotated around the axis formed by the already known points until reaching another point (Figs. 1c,d and 3c,d). This last step is done until n points lie on B_e.

B. If the internal radius is greater than 0, the building process consists in decreasing the radius while keeping the width fixed until reaching the points on the external border B_e:
  – If there is no point on the border, the radius is decreased while the center is fixed until reaching a point on B_e (Figs. 2a and 4a).
  – If there is one point Q_1 on the border, the radius is decreased while the center moves along the straight line CQ_1 towards Q_1 (which keep Q_1 on B_e) until reaching a second point on B_e (Figs. 2b, and 4b).
  – If there are already $k > 1$ points on B_e then, we consider the barycenter B of those k points. We can now consider the straight line Δ passing through B and C. By moving the center of the annulus along Δ towards B we can reduce the radius of B_e while keeping these k points on B_e (Fig. 4c). It should be noted that these steps are repeated until reaching n points on B_e or having an annulus with width 0 which leads to case A.

3   Fitting Algorithm

Let us define an equivalence class of all the annuli that cover the same consensus set. We suppose that there are more than n points in the image, otherwise all the points can be covered and the problem is somewhat trivial. The annulus characterization in Theorem 1 ensures that the optimal consensus set has always at least n points. It also shows that we can always find a representative of the equivalent class with n points of S on the outer border. Let us find now, among the representatives of the equivalent classes of consensus sets with at least n points, the annulus that cover(s) the maximum number of points of S.

The idea behind our fitting method is inspired by [6] where the authors maximize the width of an empty annulus in 2D. In [6], the authors look for the biggest empty annulus in a dual space based on the distance to the center. For each couple of points (Q_1, Q_2), the possible positions for the center of a 2D annulus passing through Q_1 and Q_2 is the bisector of both points which then forms the abscissa axis of the dual space. Each other point is associated to a curve that represents its distance to the possible centers. Using this specific space, they find when each point enters or leaves the annulus. A sorting of these intervals leads to the maximal empty annulus passing through Q_1 and Q_2. A comparison among all the possible couples of points leads to the general result. In [6], the authors do not represent an annulus in their parameter space but only circles and the width ω of the annulus is maximized.

Our purpose in this work is different since, in our problem, we tried to maximize the number of inliers inside an annulus with fixed width, so we have two concentric circles with a fixed distance ω between them. Moreover, we have adapted this method to nD. As we will see, their idea of taking the axes where the possible centers of the annuli are located can be adapted to our case. We first describe the dual space (Sect. 3.1) and then explain how we obtain the optimal consensus set(s) (Sect. 3.2).

3.1   Dual Space and Annulus Fitting in nD

According to Theorem 1, in nD, an annulus has at least n unique critical points $Q_1, .., Q_n$ located on its external border. Such an annulus has necessarily its center on a straight line, denoted Δ, that passes through the barycenter B of the points $Q_1, .., Q_n$ and that is orthogonal to the hyperplane H (of dimension $n - 1$) induced by those same points (see Fig. 5a for an example in 2D).

Let us define a 2D dual space as follows:

- The origin $O_{dual}(0, 0)$ corresponds to the barycenter B of the points $Q_1, .., Q_n$.
- The abscissa X axis represents the possible locations of the center (it is a representative of Δ): a center C is associated to $C_{dual}(dist(C, B), 0)$ (Fig. 5b).
- The ordinate Y axis represents the euclidean distance: each point $T(t_1, ... t_n)$ in the original nD space is associated in the dual space with a curve that represents the distance between T and every point of Δ.

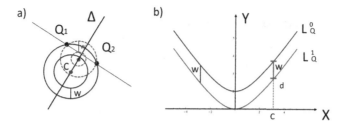

Fig. 5. (a) The 2D annulus having Q_1 and Q_2 as border point has its centers on the bisector Δ of the line segment Q_1Q_2. (b) All annuli for which Q_1 and Q_2 are on B_e correspond to the set of all the vertical line segments of length ω having one of its endpoints on L_Q^0 and the other on L_Q^1.

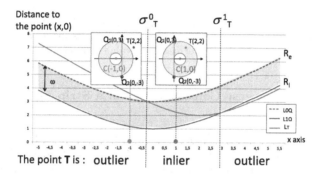

Fig. 6. 2D example: the point T represented in the dual space by the curve L_T intersects the curve L_Q^0 at σ_T^0 and the curve L_Q^1 at σ_T^1. The point T is thus inlier to the annulus having Q_1 and Q_2 on its external border between σ_T^0 and σ_T^1 and becomes outlier otherwise. The point T is outlier to the annulus centered at $C(-1,0)$ and inlier to the one centered at $C(1,0)$.

In this dual space, the points $Q_i, i \in [1, n]$ are all represented by the same curve L_Q^0 since they are all equidistant from each point of Δ (Fig. 5b). We consider that all the points $Q_i, i \in [1, n]$ are on the external border of the annulus, therefore L_Q^0 represents the external radius (R_e) variation with respect to the center position. An annulus of width ω passing through the points $Q_i, i \in [1, n]$ is represented by a vertical segment of length ω having one of its endpoints on L_Q^0. The X coordinate of the segment corresponds to the annulus center coordinate on Δ.

The translation of L_Q^0 by $(0, -\omega)$ is denoted by L_Q^1 which represents the internal radius variation with respect to the center position.

In the dual space, an annulus A of center $C(c', 0)$ with points $Q_i, i \in [1, n]$ on its external border corresponds to the vertical line segment $[L_Q^0(c'), L_Q^1(c')]$ of length ω.

For every point T in the image, it is possible to see if it is inlier or outlier to an annulus of width ω centered on C and having the n points Q_i on its external border by examining its associated curve L_T. The point T is inlier if, in the dual space, L_T intersects the vertical segment $[L_Q^0(c'), L_Q^1(c')]$ with $c' = dist(C, B)$ since in this case it is between $L_Q^0(c') = R_e$ and $L_Q^1(c') = R_i$ (Fig. 6).

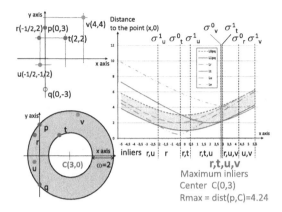

Fig. 7. 2D example: six points $p = Q_1, q = Q_2, r, u, v$, and t in the primal space and their corresponding curves. The maximum number of inliers for an annulus having p and q on B_e is reached when the center has an x-value around 3.

3.2 Finding the Largest Consensus Set in a Strip for a Given $(Q_1, Q_2, ..., Q_n)$

In order to know the number of inliers within any annulus defined by n points $Q_1... Q_n$, we check for every point T in the image, the intersections σ_T^0 and σ_T^1 of L_T with the strip boundaries L_Q^0 and L_Q^1. This check is important since any annulus corresponding to a vertical segment between the two intersections σ_T^0 and σ_T^1 in the strip always contains T as an inlier; outside this interval, T is always an outlier (Fig. 6).

When checking the intersections of every L_T with L_Q^0 and L_Q^1, we use two values f_T^i for $i = 0, 1$, which is set to 1 if L_T enters the strip from L_Q^i, and -1 if L_T leaves the strip from L_Q^i. Once the intersections σ_T^i, and the associated values f_T^i for $i = 0, 1$ are calculated, we sort all the intersections in increasing order. As for determining the location(s) of the maximum number of inliers, a function $F(x)$ is used; initially we set $F(x) = n$ for every x, since we already know that $Q_1,...Q_n$ are inliers. Then the value f_r^i is added to $F(x)$ following the sorted order. By looking for the maximum value of $F(x)$, we obtain the center location(s) in the dual space corresponding to the maximum optimal consensus set(s). Figure 7 shows an example in 2D of this algorithm; the annulus in the primal space having $p = Q_1$ and $q = Q_2$ on its external border is optimal in terms of inliers at a center around 3 between σ_v^0 and σ_t^1 when all the dual curves are inside L_Q^0 and L_Q^1 (i.e. when all the points are inliers).

This procedure is repeated for all the combination of n points in the image until finding the right center(s) of the annulus (annuli) having n points on B_e that maximizes the number of inliers. Since a sorting of complexity $O(N \log N)$ of the intersection is needed and since the algorithm is repeated for every couple of n points, the final complexity is $O(N^{n+1} \log N)$.

3.3 Algorithm

Algorithm 1 gives an example of the nD annulus fitting algorithm. The inputs are a set S of nD points and a width ω. Output is a set V of centers and radius of the best annuli.

Algorithm 1. nD Annulus Fitting

---

    input : A set **S** of N grid points and a width ω

    output: A list **V** of centers and radius of the best fitted annuli

1 **begin**

2   initialize $Max = 0$;

3   **foreach** n-uplet $\in$ **S** **do**

4   compute the barycenter of the n-uplet;

5   initialize the array $AR[k]$ for $k = 1, \ldots, 2N + 2$;

6   set $j = 0$;

7   **foreach** $T \in$ **S** **do**

8   calculate σ_T^i for $i = 0, 1$;

9   **if** $\sigma_T^0 < \sigma_T^1$ **then**

10   set $f_T^0 = 1$, $f_T^1 = -1$;

11   **else**

12   set $f_T^0 = -1$, $f_T^1 = 1$;

13   set the pair (σ_T^i, f_T^i), for $i = 0, 1$, in $AR[2j + i]$;

14             $j = j + 1$;

15   sort the pair elements (σ_k, f_k) for $k = 1, \ldots, 2j$ in AR with the values σ_k as keys;

16   initialize $F = 0$;

17   **for** $k = 1, \ldots, 2j$ **do**

18             $F = F + f_k$;

19   **if** $F > Max$ **then**

20   set $Max = F$, Erase **V** and set it to the interval $[\sigma_k, \sigma_{k+1}]$;

21             $[\sigma_k, \sigma_{k+1}]$ **if** $F = Max$ **then**

22   add the interval $[\sigma_k, \sigma_{k+1}]$ to **V**;

23   **return** **V**;

3.4 Degenerate Cases

There are degenerate cases that must be treated carefully when examining the number of inliers and outliers. They can be summarized as follows:

– Invalid radius: L_Q^1 represents the internal border of the annulus; when it is negative the annulus is not valid. An example of such invalid radius is seen in Fig. 8a, in this figure T becomes inlier at σ_T^1 and at σ_T^0, instead of σ_T^0 since at σ_T^0 the radius is negative.

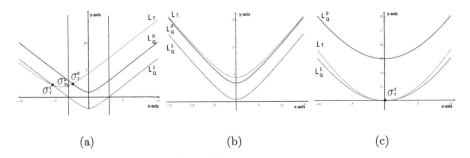

(a)   (b)   (c)

Fig. 8. Degenerate cases: (a) The value of L_Q^1 is negative between the two vertical dashed lines; the point T is inlier between σ_T^1 and $\sigma_{T'}^0$. (b) T is always outlier. (c) T is always inlier.

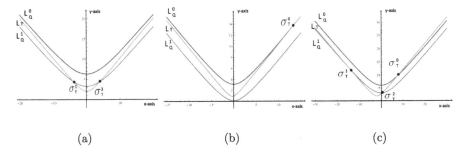

(a)   (b)   (c)

Fig. 9. Degenerate cases: (a) T is inlier before σ_T^1 and after σ_T^2. (b) T is inlier before σ_T^0. (c) T is inlier before σ_T^1 and between σ_T^2 and σ_T^0.

– Intersection of L_T with L_Q^0 and L_Q^1: When both L_Q^0 and L_Q^1 are each inter-sected once we have the regular case explained in the Algorithm 1. The degen-erate cases occur when one of the two curves is not intersected or when it is intersected twice. These cases can be explained as follows:

  • L_T and L_Q^0 have no common point: in this case the intersection of L_T with L_Q^1 must be verified. If L_Q^1 is not intersected, we must check if L_T is between the two curves, in this case T is always inlier, otherwise T is always outlier (Fig. 8b). If L_Q^1 is intersected once, L_T is inside the two curves and T is always inlier (Fig. 8c). If L_Q^1 is intersected twice, this means that L_T is between the two curves and thus the point T is inlier before the first intersection σ_T^1 of L_T and after the second intersection σ_T^2 as seen in Fig. 9a.
  • L_T has one intersection σ_T^0 with L_Q^0 : If L_T has no intersection with L_Q^1, and L_T is inside the two curves before the intersection σ_T^0, then T is inlier before the intersection σ_T^0 and becomes outlier after this intersection (Fig. 9b); otherwise T is outlier before the intersection σ_T^0 and becomes inlier after this intersection.
    If L_T has two intersections with L_Q^1, then we have two cases: T is inlier before the intersection σ_T^1 and between σ_T^2 and σ_T^0 and becomes outlier

otherwise (Fig. 9c) or T is inlier between the two intersections σ_T^0 and σ_T^1 and after the intersection σ_T^2 and is outlier otherwise.

4 Experiments

This section presents the 2D and 3D experiments.

4.1 Example for a 2D Real Image

We tested our method on a real test image, as shown in Fig. 10a, whose size is 140×69. Before applying our method, edge detection and mathematical morphological filtering have been performed on the image; the number of points in the image after this preprocessing is 646 points. Our method is then applied to fit an annulus to the set of points. Figure 10b shows the optimal consensus set, which includes 291 inliers. The width of the annulus is fixed to 3.

4.2 Example on 2D Noisy Images

We then applied our method for 2D noisy digital Andres circles (points for Andres circles and noise are generated randomly for 2D as well as for 3D experiments) as shown in Figs. 11a, b and 12a. For each of these set of points, an

Table 1. The number of points and the optimal consensus set size.

Figures	Number of points	Center	R	width	Opt. cons. set size
Fig. 10b	646	(104.992,31.017)	28.163	3	291
Fig. 11a	225	(40.871,41)	19.522	1	113
Fig. 11b	65	(31.109,31.109)	14.425	1	65
Fig. 12a	1127	(51,51.008)	26.992	6	1120
Fig. 12b	56	(0.878,1.244,1.976)	1.916	3	51
Fig. 13a	109	(-4,-4,-4)	2.742	1	90
Fig. 13b	151	(0,0,0)	6	3	116

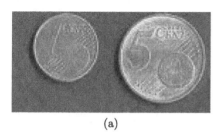 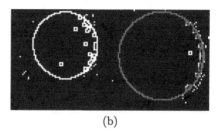

(a)   (b)

Fig. 10. An original image in (a) and its optimal consensus set colored red in (b) (Color figure online).

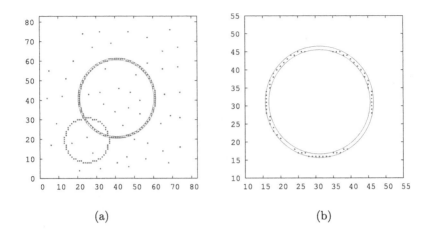

(a)   (b)

Fig. 11. (a) Annulus fitting for a noisy digital Andres circle of width 1. (b) Annulus fitting for a digital Andres circle of width 1.

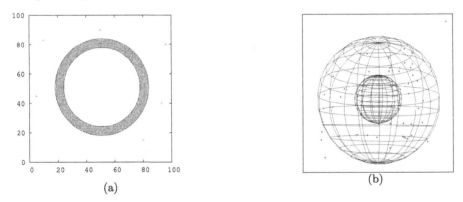

(a)   (b)

Fig. 12. (a) Annulus fitting for a noisy digital Andres circle of width 6. (b) Annulus fitting for noisy 3D data; a width $w = 3$ is used.

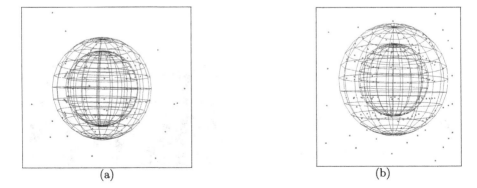

(a)   (b)

Fig. 13. (a) Annulus fitting for noisy 3D data; a width $w = 1$ is used. (b) Annulus fitting for noisy 3D data; a width $w = 3$ is used.

annulus of width $\omega = 1$, $\omega = 1$ and $\omega = 6$ is used respectively. Table 1 show the number of points, the optimal consensus set size as well as the center position and the radius R of the inner circle obtained after the fitting.

4.3   3D Noisy Images

We applied our method for 3D noisy digital Andres spheres as shown in Figs. 12a, 13a and b. For each of these set of points, an annulus of width $\omega = 3$, $\omega = 1$ and $\omega = 3$ is used respectively. The last three lines of Table 1 shows the number of points, the optimal consensus set size as well as the center position and the radius R of the inner sphere obtained after the fitting.

5   Conclusion and Perspectives

In this paper we have proposed a new approach for fitting nD annulus to a set of points while fixing the width of the annulus. The main advantage of our approach is that it guarantees optimal and exhaustive results from the point of view of the optimal (maximal) consensus set: we are guaranteed to fit an annulus with the least amount of outliers. We are also guaranteed to find all the optimal consensus sets. One of the future works concerns conic fitting such as ellipse, parabola and hyperbola. We also plan to implement a fast 2D/3D algorithm for fitting annulus; such algorithm does not guarantee optimality but guarantees local maximality of inliers in the sense of the set inclusion and thus has a lower time complexity. We also plan to adapt the nD characterization of annulus for $k - Flake$ digital hyperspheres [18] as an extension of [4,11].

Acknowledgments. The authors express their thanks to Mr. Pierre Boulenguez, who contributed in the implementation of a part of the 3D Fitting. The work for this paper was partly financed by Egide, franco-Japanese PHC Sakura project n° 27608XJ and by the Poitou Charentes region project n° 11/RPC-R-051.

References

1. Agarwal, P., Har-Peled, S., Varadarajan, K.: Approximating extent measures of points. J. ACM **51**(4), 606–635 (2004)
2. Agarwal, R., Gehrke, J., Gunopulos, D., Raghavan, P.: Automatic subspace clustering of high dimensional data for data mining applications. In: Proceeding SIGMOD 998 International Conference on Management of data. pp. 94–105 (1998)
3. Andres, E., Jacob, M.A.: The discrete analytical hyperspheres. IEEE Trans. Visual Comput. Graphics **3**, 75–86 (1997)
4. Andres, E., Largeteau-Skapin, G., Zrour, R., Sugimoto, A., Kenmochi, Y.: Optimal consensus set and preimage of 4-connected circles in a noisy environment. In: 21st International Conference on Pattern Recognition (ICPR 2012), pp. 3774–3777. IEEE Xplore (2012)

5. De Berg, M., Bose, P., Bremner, D., Ramaswami, S., Ramaswami, G., Wilfong, G.: Computing constrained minimum-width annuli of point sets. In: Rau-Chaplin, A., Dehne, F., Sack, J.-R., Tamassia, R. (eds.) WADS 1997. LNCS, vol. 1272, pp. 392–401. Springer, Heidelberg (1997)

6. Díaz-Báñez, J.M., Hurtado, F., Meijer, H., Rappaport, D., Sellares, T.: The largest empty annulus problem. In: Sloot, P.M.A., Tan, C.J.K., Dongarra, J., Hoekstra, A.G. (eds.) ICCS-ComputSci 2002, Part III. LNCS, vol. 2331, pp. 46–54. Springer, Heidelberg (2002)

7. Dunagan, J., Vempala, S.: Optimal outlier removal in high-dimensional spaces. J. Comput. Sys. Sci. **68**(2), 335–373 (2004)

8. Fischler, M., Bolles, R.: Random sample consensus: a paradigm for model fitting with applications to image analysis and automated cartography. Commun. ACM **24**, 381–395 (1981)

9. Har-Peled, S., Wang, Y.: Shape fitting with outliers. SIAM J. Comput. **33**(2), 269–285 (2004)

10. Kimme, C., Ballard, D., Sklansky, J.: Finding circles by an array of accumulators. Commun. ACM **18**(2), 120–122 (1975)

11. Largeteau-Skapin, G., Zrour, R., Andres, E.: $O(n^3\log n)$ time complexity for the optimal consensus set computation for 4-connected digital circles. In: Gonzalez-Diaz, R., Jimenez, M.-J., Medrano, B. (eds.) DGCI 2013. LNCS, vol. 7749, pp. 241–252. Springer, Heidelberg (2013)

12. Matousek, J.: On enclosing k points by a circle. Inf. Process. Lett. **53**(4), 217–221 (1995)

13. O'Rourke, J., Kosaraju, S., Megiddo, N.: Computing circular separability. Discrete Comput. Geom. **1**, 105–113 (1986)

14. Phan, M.S., Kenmochi, Y., Sugimoto, A., Talbot, H., Andres, E., Zrour, R.: Efficient robust digital annulus fitting with bounded error. In: Gonzalez-Diaz, R., Jimenez, M.-J., Medrano, B. (eds.) DGCI 2013. LNCS, vol. 7749, pp. 253–264. Springer, Heidelberg (2013)

15. Robinson, S.M.: Fitting spheres by the method of least squares. Commun. ACM **4**(11), 491 (1961). http://doi.acm.org/10.1145/366813.366824

16. Roussillon, T., Tougne, L., Sivignon, I.: On three constrained versions of the digital circular arc recognition problem. In: Brlek, S., Reutenauer, C., Provençal, X. (eds.) DGCI 2009. LNCS, vol. 5810, pp. 34–45. Springer, Heidelberg (2009)

17. Smid, M., Janardan, R.: On the width and roundness of a set of points in the plane. Int. J. Comput. Geom. **9**(1), 97–108 (1999)

18. Toutant, J., Andres, E., Roussillon, T.: Digital circles, spheres and hyperspheres: from morphological models to analytical characterizations and topological properties. Discrete Appl. Math. **161**(16–17), 2662–2677 (2013)

19. Zrour, R., Kenmochi, Y., Talbot, H., Buzer, L., Hamam, Y., Shimizu, I., Sugimoto, A.: Optimal consensus set for digital line and plane fitting. Int. J. Imaging Syst. Technol. **21**(1), 45–57 (2011)

20. Zrour, R., Largeteau-Skapin, G., Andres, E.: Optimal consensus set for annulus fitting. In: Debled-Rennesson, I., Domenjoud, E., Kerautret, B., Even, P. (eds.) DGCI 2011. LNCS, vol. 6607, pp. 358–368. Springer, Heidelberg (2011)

Number of Shortest Paths in Triangular Grid for 1- and 2-Neighborhoods

Mousumi Dutt[1]([⊠]), Arindam Biswas[2], and Benedek Nagy[3,4]

[1] Department of Computer Science and Engineering,
International Institute of Information Technology, Naya Raipur, India
duttmousumi@gmail.com
[2] Department of Information Technology,
Indian Institute of Engineering Science and Technology, Shibpur, India
barindam@gmail.com
[3] Department of Computer Science, Faculty of Informatics,
University of Debrecen, Debrecen, Hungary
nbenedek.inf@gmail.com
[4] Department of Mathematics, Faculty of Arts and Sciences,
Eastern Mediterranean University, Mersin-10, Famagusta, North Cyprus, Turkey

Abstract. This paper presents a novel formulation to determine the number of shortest paths between two points in triangular grid in 2D digital space. Three types of neighborhood relations are used on the triangular grid. Here, we present the solution of the above mentioned problem for two neighborhoods—1-neighborhood and 2-neighborhood. To solve the stated problem we need the coordinate triplets of the two points. This problem has theoretical aspects and practical importance.

Keywords: Triangular grid · Digital distances · Shortest paths · Combinatorics

1 Introduction

Digital geometry works on discrete space, i.e., points can have only integer coordinates. In digital geometry, two basic neighborhood relations are defined in the square grid [19]—cityblock and chessboard. The cityblock motion allows horizontal and vertical movements only, while at the chessboard motion the diagonal directions are also permitted. So, based on these motions, in this grid we have two kinds of distances. In [12,18], there is a short summary of examination of the square grid. Each coordinate of a point in square grid is independent of the other. Generalizing the concepts to n dimensions, n independent coordinates are used. In the n dimensional cubic grid, the structure of the nodes is isomorphic to the structure of the n dimensional cubes. The field called 'Geometry of Numbers' works on these grids [1,8,9,11]. The concept of 'lattice' and 'array' were used which have about the same meaning with 'grid', which we are using.

In digital geometry, hexagonal and triangular grids are also analyzed. There is a connection among the cubic, hexagonal and triangular grids [10,13,17,20],

© Springer International Publishing Switzerland 2015
R.P. Barneva et al. (Eds.): IWCIA 2015, LNCS 9448, pp. 115–124, 2015.
DOI: 10.1007/978-3-319-26145-4_9

and thus, coordinate systems with 3 coordinates are opt for these grids (see Fig. 1(left)). In this paper, the problem is formulated based on triangular grid, which has three neighborhoods. The three kinds of neighborhood criteria of the triangular grid can be found in [4], where thinning algorithms are shown in the three basic grids. The three coordinates used in triangular grid are not independent [14]. The digital distances of two points based on a neighborhood relation gives the length of a shortest path connecting the two points where in each step the path moves to next neighborhood point (given a neighborhood type). Weighted distance in triangular grid is discussed in [16], where three weights are used to define a distance function.

Shortest path is not unique in discrete space. Between two points there may exist more than one shortest path. The recursive formulation for the number of cityblock, chessboard, and octagonal shortest paths between two points in 2D digital plane was proposed in [3]. In [2], the number of minimal paths in a digital image between every pair of points with respect to a particular neighborhood relation is presented, where the image is considered as matrix and hence the algorithm contains matrix operations. Determination of shortest isothetic path (cityblock) between two points inside a digital object for a given grid size, is presented in [6,7]. Since a shortest isothetic path is not unique, finding the number of shortest isothetic paths between two points is essential. The corresponding problem is presented in [5]. Here, in this paper, we will discuss the number of shortest paths in triangular grid for 1- and 2-neighborhoods.

2   Preliminaries

In this section, we will discuss some definitions, that are required to understand the paper. Three types of neighborhood are used in triangular grid [4] as shown in Fig. 1(left). Each pixel of the triangular grid is a *triangle* and it is termed as a *point* of the grid. There are two orientations of the used triangles: there are triangles of shape $\triangle$ and there are triangles of shape $\triangledown$. There are three 1-neighbors, nine 2-neighbors (the 1-neighbors, and six more 2-neighbors), and twelve 3-neighbors (nine 2-neighbors, and three more 3-neighbors) w.r.t. each triangle. These relations are reflexive (i.e., the pixel marked with dark triangle is a 1-, 2-, and 3-neighbor of itself). In addition, all 1-neighbors of a pixel are its 2-neighbors and all 2-neighbors are 3-neighbors, as well (i.e., increasing and inclusion properties).

Three coordinate values are required to represent the triangles (and the term point is also used for the pixels). One point in triangular grid is selected as *origin* whose coordinate values are $(0,0,0)$. The three lines passing through the center of the origin triangle, which are orthogonal to its sides, are considered as three coordinate axes x, y, and z (see Fig. 1(right)). The coordinate values of a triangle can be determined from another triangle by considering the movement which must be parallel to one of the axes. If the step is in the direction of an axis, then the respective coordinate value is increased by 1, while in case when the step is in the opposite direction to an axis, the respective coordinate value is

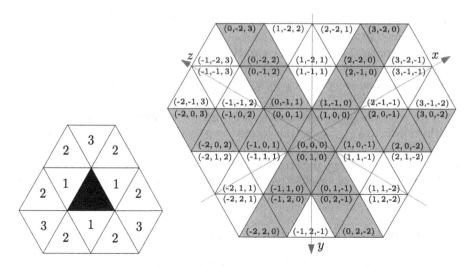

Fig. 1. Types of neighbors in triangular grid (**left**). The coordinate values and lanes in triangular grid (**right**).

decreased by 1. In this way, every triangle gets a unique coordinate triplet [15]. Since triangular grid is in 2-dimensional plane, the three coordinate values are dependent on each other. A point is termed as even (of shape $\triangle$) when the sum of the coordinate values is zero. When the corresponding sum is one, the point is termed as odd (of shape $\triangledown$).

The two grid points p and q of the triangular grid are m-neighbours ($m = 1, 2, 3$), if the following two conditions hold:

1. $|p(i) - q(i)| \leqslant 1$ for $i = 1, 2, 3$, and
2. $|p(1) - q(1)| + |p(2) - q(2)| + |p(3) - q(3)| \leqslant m$.

When for some value of m, the second condition holds equality, it is termed as strict m-neighbors.

The points having the same value as x, y, or z-coordinate, form a lane. Each lane is orthogonal to one of the coordinate axes. For the points of a lane a coordinate value is fixed. The other two values change by $+1/-1$ moving to neighbor points on the lane. If two points share a coordinate value, then they are in a common lane (according to this shared coordinate value). If there is no coordinate value that is shared by the two points, then the points can be connected by two lanes having angle $\frac{2}{3}\pi$ between them.

For technical purpose we name the sextants of the grid by terms direct and indirect sextants: a point (i, j, k) is in a direct sextant of the grid, if it has exactly 1 positive coordinate value and 2 negative coordinate values. A point is in an indirect sextant if it has two positive coordinate values (the third one is a negative value, necessarily, in this case). In this case, the grid consists of the three lanes through the origin and six pairwise disjoint sextants. Each sextant is bordered by two lanes (actually, only half parts of the lanes contain neighbor points of some of the points of the sextant).

It may happen that q is not origin. If q is an even point, let us say, (x, y, z) with $x + y + z = 0$, then a translation by vector $(-x, -y, -z)$ translates the grid in such a way, that every point (i, j, k) is mapped to $(i - x, j - y, k - z)$ [15]. It is also possible that q is an odd triangle. Any odd point q' with coordinates (x, y, z) $(x + y + z = 1)$ can be transformed to the origin by mirroring [15], and the point p' with coordinates (i, j, k) is transformed to p by the same transformation, e.g., assigning for any point (i, j, k) the point $(x - i, y - j, z - k)$. This transformation is also isometric, it keeps the structure of the grid. Thus further, it is enough to consider the number of shortest paths from the origin, i.e., we assume that q has coordinate triplet $(0, 0, 0)$.

3   Formulation for Number of Shortest Paths

The number of shortest paths between two points (say, p and q) in triangular grid depends on the neighborhood used on the path. Let q be the origin having coordinate triplet $(0, 0, 0)$ and p be any other point in the grid having coordinate triplet (i, j, k). Three lanes passes through q as shown in Fig. 1(right). Let $f_r(i, j, k)$ be the number of shortest paths from q to p in r-neighborhood It may happen that we need the number of shortest paths where one of the points (between p and q) is not the origin. In this scenario, we can translate the origin to q and calculate the number of shortest paths.

From the symmetry of triangular grid, it can be said that $f_r(i, j, k) = f_r(i, k, j) = f_r(j, i, k) = f_r(j, k, i) = f_r(k, j, i) = f_r(k, i, j)$. It is to be noted that in each sextant two values of the coordinate triplets have similar sign—either positive or negative, whereas the other one has opposite sign, which is called the *prime coordinate*. The other two, having the same sign, are called *secondary coordinates*. Consider the top-most sextant in Fig. 1(right), where $i \geqslant 0$ and $k \geqslant 0$ but $j \leqslant 0$. Thus, j is the prime coordinate and i and k are secondary coordinates. A sextant can also be named by its prime coordinate, therefore the mentioned sextant is called j-indirect sextant. The value of the prime coordinate can never be less than (by the absolute values of the coordinates) any of the secondary coordinates. To determine the number of shortest paths in 1-neighborhood, we need only secondary coordinates, as we will see.

The formulations for the number of shortest paths in 1-neighborhood and 2-neighborhood are discussed in the following sections (Sects. 3.1, 3.2).

3.1   Number of Shortest Paths Based on 1-Neighborhood

Let $f_1(i, j, k)$ be the number of shortest paths from (i, j, k) to $(0, 0, 0)$ considering 1-neighborhood in triangular grid. The number of shortest paths from p to q can be defined as follows.

$$f_1(i, j, k) = \binom{|a| + |b|}{|a|}, \quad \begin{array}{l} \text{where } a, b \geqslant 0 \text{ or } a, b \leqslant 0, \\ a \in \{i, j, k\} \text{ and } b \in \{i, j, k\} \setminus \{a\} \end{array} \tag{1}$$

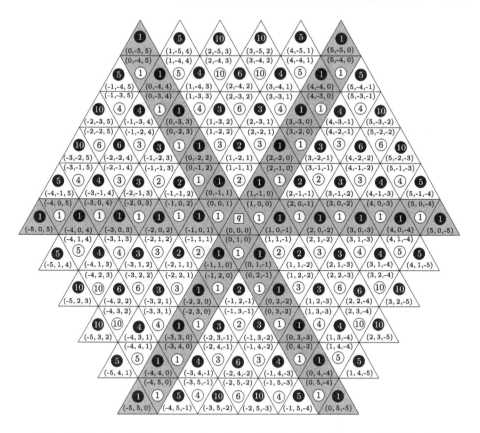

Fig. 2. Number of shortest paths (shown inside circles) in 1-neighborhood.

The number of shortest paths in 1-neighborhood for different coordinate triplets are shown in Fig. 2. It is to be noted that if p lies in any of the three lanes which intersect at q, then there will be only one shortest path between p and q in 1-neighborhood (see Fig. 2).

3.2   Number of Shortest Paths Based on 2-Neighborhood

Let $f_2(i, j, k)$ be the number of shortest paths between two points p and q in 2-neighborhood. The calculation of number of shortest paths in 2-neighborhood is dependent on the same in 1-neighborhood. Then, $f_2(i, j, k)$ can be defined as follows.

$$f_2(i,j,k) = f_1(i,j,k) \times \alpha, \quad \begin{array}{l} \alpha = 1, \text{ when } i + j + k = 0; \\ \alpha = \max(|i|, |j|, |k|), \text{ when } i \times j \times k \geqslant 0; \\ \alpha = \max(|i|, |j|, |k|) + 1, \text{ when } i \times j \times k < 0. \end{array} \quad (2)$$

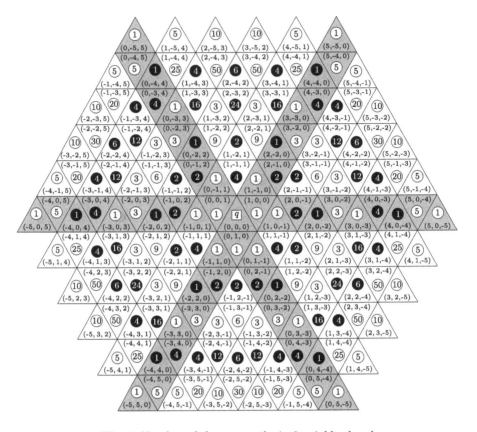

Fig. 3. Number of shortest paths in 2-neighborhood.

that is, the $f_2(i, j, k)$ remains same as $f_1(i, j, k)$ for even points and it is multiplied by the largest absolute value of the coordinates for odd points of the lanes containing the origin and for odd points of the direct sextants, and the sum of the absolute values of the secondary coordinates is used as coefficients for odd points of the indirect sextants (see also Fig. 3).

4   Proof of Correctness

Proof for 1-neighborhood: For 1-neighborhood, the proof goes by induction for the even and also for the odd points for both, points of the lanes and points of the (direct and indirect) sextants. Let us start by the lanes. It is clear that a shortest path from the origin q to any point having coordinates $(i, j, 0)$ or $(0, j, k)$ with $j > 0$ consists of $|i| + j$ or $j + |k|$ steps, respectively. Moreover, there is only one shortest path in these cases, stepping one by one, neighbor to neighbor on the given lane. Applying Eq. 1 for these cases we got $\binom{|i|}{0} = 1$ (or $\binom{|i|}{|i|}$, or $\binom{j}{0}$, or equivalently, $\binom{j}{j} = 1$) and $\binom{|k|}{0} = 1$ (or $\binom{|k|}{|k|}$), or $\binom{j}{0}$, or equivalently, $\binom{j}{j} = 1$), respectively. Thus, by symmetry, the formula is proven for the points of all the lanes containing the origin.

Now, let us consider the direct sextants, we show the proof for the j-direct sextant, but by symmetry it holds for the other direct sextants as well. We use an induction based on the sum of absolute values of the secondary coordinates w.r.t. a sextant. Now, let us consider an even point (i, j, k) such that it is inside the j-direct sextant, i.e., $j > 0$ and $i, k < 0$. Then in a shortest path from q one may arrive to the point (i, j, k) from $(i, j, k + 1)$ or from $(i + 1, j, k)$. There are no other possibilities. Therefore, $f_1(i, j, k) = f_1(i, j, k + 1) + f_1(i + 1, j, k) = \binom{|i|+|k|-1}{|i|} + \binom{|i|+|k|-1}{|i|-1} = \binom{|i|+|k|}{|i|}$. Notice that both points $(i, j, k+1)$, $(i+1, j, k)$ are odd.

Now, let us consider an odd point such that $j > 0$ and $i, k < 0$. Then, in a shortest path from q, the last step must be from the even point $(i, j - 1, k)$ to the point (i, j, k), and thus, $f_1(i, j, k) = f_1(i, j - 1, k)$.

Observe, that by using the already proven values for the lanes having a coordinate value 0, actually, the binomial coefficients and the Pascal's triangle are obtained, in a way that every value is repeated from an even point to an odd point right below.

Now, let us consider odd points in the j-indirect sextant where $j < 0$ and $i, k > 0$. The proof for the points of the lanes bordering the j-indirect sextant is already given. Now, let us consider the odd points of this sextant. In a shortest path from q one may arrive to the point (i, j, k) $(i+j+k = 1)$ from $(i, j, k-1)$ or from $(i-1, j, k)$. There are no other possibilities. Therefore $f_1(i, j, k) = f_1(i, j, k-1) + f_1(i - 1, j, k)$. Notice that both points $(i, j, k - 1)$, $(i - 1, j, k)$ are even. If we consider even points in the j-indirect sextant, then in a shortest path from q, the last step must go from the odd point $(i, j - 1, k)$ to the point (i, j, k), and thus, $f_1(i, j, k) = f_1(i, j-1, k)$. The proof is also inductive here and the binomial coefficients are obtained in a quite similar manner as at the j-direct sextant.

Proof for 2-neighborhood: Similarly, we will prove for 2-neighborhood first for even and odd points on the lanes containing the origin, and then for points in the j-direct and j-indirect sextants. By symmetry, proof for other sextants is immediately follows.

By using 2-neighborhood, in a shortest path from q to an even point only steps changing two of the coordinate values are used (they are the so-called strict 2-steps). From q to an odd point exactly one 1-step is used, all the other steps in a shortest path are strict 2-steps (see, e.g., [14]).

Let us start by the even points in lanes. It is clear that a shortest path from the origin q to any point having coordinates $(i, j, 0) = (-j, j, 0)$ or $(0, j, k) = (0, j, -j)$ with $j > 0$ consists of exactly j steps, respectively. Moreover, there is only one shortest path in these cases, stepping one by one, strict 2-neighbor to strict 2-neighbor on the given lane. Applying Eq. 2 for these cases we got $\binom{|i|}{0} = 1$ and $\binom{|k|}{0} = 1$, respectively, i.e. the same value as for 1-neighborhood.

Now consider an odd point in lanes containing the origin. Let it have the coordinates $(i, j, 0)$ or $(0, j, k)$ with $j > 0$, i.e., it is $(1 - j, j, 0)$ or $(0, j, 1 - j)$. In a shortest path from the origin to the point $(1 - j, j, 0)$ the last step could be either from $(1 - j, j - 1, 0)$ or from $(2 - j, j - 1, 0)$. Similarly, to $(0, j, 1 - j)$ one can arrive from the points $(0, j - 1, 1 - j)$ or $(0, j - 1, 2 - j)$ in a shortest path, there is no other choice. Thus, the number of shortest paths from q to these odd points can be written as, $f_2(0, j, k) = f_2(0, j - 1, k) + f_2(0, j - 1, k + 1)$ or $f_2(i, j, 0) = f_2(i, j - 1, 0) + f_2(i + 1, j - 1, 0)$, respectively. Here, the first terms are belonging to shortest paths for an even point and the second terms are belonging to the shortest paths for an odd point which will expand in similarly. By induction using Eq. 2, $f_2(0, j, k) = f_1(0, j - 1, k) + f_1(0, j - 1, k + 1) \times (j - 1)$ and $f_2(i, j, 0) = f_1(i, j - 1, 0) + f_1(i - 1, j - 1, 0) \times (j - 1)$, respectively. Thus, $f_2(0, j, k) = \binom{0 + |k|}{|k|} + \binom{0 + |k| - 1}{|k| - 1} \times (j - 1)$ or $f_2(i, j, 0) = \binom{|i| - 1 + 0}{|i| - 1} + \binom{|i| - 1 + 0}{|i| - 1} \times (j - 1)$, i.e., $f_2(0, j, k) = 1 + 1 \times (j - 1)$ or $f_2(i, j, 0) = 1 + 1 \times (j - 1)$. Hence, $f_2(0, j, k) = j$ and $f_2(i, j, 0) = j$. Thus, by symmetry, the formula is proven for the points of the lanes containing the origin.

Let us consider the shortest paths from q to an even point (i, j, k) in j-direct sextant ($j > 0$). Since only strict 2-steps, i.e., even points are used in these shortest paths, there are exactly to points that can precede (i, j, k) in a shortest path: they are $(i, j - 1, k + 1)$ and $(i + 1, j - 1, k)$. Thus, the number of shortest paths can be written as $f_2(i, j, k) = f_2(i, j - 1, k + 1) + f_2(i + 1, j - 1, k)$. The two terms represent the formula for two even points. Thus, $f_2(i, j, k) = f_1(i, j - 1, k + 1) + f_1(i + 1, j - 1, k)$, i.e., $f_2(i, j, k) = \binom{|i| + |k| - 1}{|i|} + \binom{|i| - 1 + |k|}{|i| - 1} = \binom{|i| + |k|}{|i|} = f_1(i, j, k)$.

In a shortest path from q one may arrive to the odd point (i, j, k) from $(i, j - 1, k + 1)$ or $(i, j - 1, k)$ or $(i + 1, j - 1, k)$, when (i, j, k) is an odd point in j-direct sextant. Subsequently, $f_2(i, j, k) = f_2(i, j - 1, k + 1) + f_2(i + 1, j - 1, k) + f_2(i, j - 1, k)$. The first two terms referring values for odd points and the third term does for an even point. Using Eq. 2, $f_2(i, j, k) = f_1(i, j - 1, k + 1) \times (j - 1) + f_1(i + 1, j - 1, k) \times (j - 1) + f_1(i, j - 1, k)$. Therefore, $f_2(i, j, k) = (j - 1) \times \left(\binom{|i| + |k| - 1}{|i|} + \binom{|i| - 1 + |k|}{|i| - 1} \right) + f_1(i, j - 1, k) = (j - 1) \times \binom{|i| + |k|}{|i|} + \binom{|i| + |k|}{|i|} = j \times \binom{|i| + |k|}{|i|} = j \times f_1(i, j, k)$. This is the formula that we wanted to proof in this case.

Let us consider a shortest path from q to an even point (i, j, k) in j-indirect sextant ($j < 0$). Their number can be written as $f_2(i, j, k) = f_2(i - 1, j + 1, k) + f_2(i, j + 1, k - 1)$ since one reach point (i, j, k) only from points $(i - 1, j + 1, k)$ and $(i, j + 1, k - 1)$ in a shortest path from the origin. The two terms represent values corresponding to two even points. Thus, $f_2(i, j, k) = f_1(i - 1, j + 1, k) + f_1(i, j + 1, k - 1)$, i.e., $f_2(i, j, k) = \binom{i - 1 + k}{i - 1} + \binom{i + k - 1}{k} = \binom{i + k}{i} = f_1(i, j, k)$.

Let us consider the last case. When (i, j, k) is an odd point in j-indirect sextant, then, in a shortest path, one may reach (i, j, k) from exactly one of the following points: $(i - 1, j, k)$, $(i, j, k - 1)$, $(i - 1, j + 1, k)$ and $(i, j + 1, k - 1)$. Consequently, $f_2(i, j, k) = f_2(i - 1, j, k) + f_2(i, j, k - 1) + f_2(i - 1, j + 1, k) + f_2(i, j + 1, k - 1)$. The first two terms are values for even points and the last two terms are vales for odd points. Using Eq. 2, $f_2(i, j, k) = f_1(i - 1, j, k) + f_1(i, j, k - 1) + f_1(i - 1, j + 1, k) \times (i - 1 + k) + f_1(i, j + 1, k - 1) \times (i + k - 1)$. Therefore,

$f_2(i,j,k) = \binom{i-1+k}{i-1} + \binom{i+k-1}{i} + \binom{i-1+k}{i-1} \times (i-1+k) + \binom{i+k-1}{i} \times (i+k-1)$. Thus, $f_2(i,j,k) = \binom{i+k}{i} + (i+k-1) \times \left(\binom{i-1+k}{i-1} + \binom{i+k-1}{i} \right) = \binom{i+k}{i} + (i+k-1) \times \binom{i+k}{i} = \binom{i+k}{i} \times (i+k-1+1) = f_1(i,j,k) \times (i+k)$. For odd points in j-indirect sextant, $i+j+k = 1$ and $|j|+1 = i+k$, since $j < 0$. Hence, $f_2(i,j,k) = f_1(i,j,k) \times (|j|+1)$.

The proof for other sextants, i.e., i-direct sextant, i-indirect sextant, k-direct sextant, and k-indirect sextant, can be done similarly.

5   Conclusions

Digital distances have various application in several research areas, specially in image processing. Many studies have already been proposed on it. The usage of non-traditional grids can have several advantages based on their better symmetric properties (e.g., they have more symmetry axes) than the traditional (square, cubic) grids. The three types of basic neighborhood relations give more flexibility in triangular grid compared to square and hexagonal grid. The number of shortest paths based on 1- and 2-neighborhoods in triangular grid is presented in this paper. Theoretical background, such as the formulation of the problem and proof of correctness are elaborated. In future, we can extend the problem by finding number of shortest paths based on 3-neighborhood.

References

1. Crawley, P., Dilworth, R.P.: Algebraic Theory of Lattices. Prentice-Hall Inc., Englewood Cliffs (1973)
2. Das, P.P.: An algorithm for computing the number of the minimal paths in digital images. Pattern Recogn. Lett. **9**(2), 107–116 (1989)
3. Das, P.P.: Counting minimal paths in digital geometry. Pattern Recogn. Lett. **12**(10), 595–603 (1991)
4. Deutsch, E.S.: Thinning algorithms on rectangular, hexagonal and triangular arrays. Commun. ACM **15**(3), 827–837 (1972)
5. Dutt, M., Biswas, A., Bhattacharya, B.B.: Enumeration of shortest isothetic paths inside a digital object. In: Kryszkiewicz, M., Bandyopadhyay, S., Rybinski, H., Pal, S.K. (eds.) PReMI 2015. LNCS, vol. 9124, pp. 105–115. Springer, Heidelberg (2015)
6. Dutt, M., Biswas, A., Bhowmick, P., Bhattacharya, B.B.: On finding shortest isothetic path inside a digital object. In: Barneva, R.P., Brimkov, V.E., Aggarwal, J.K. (eds.) IWCIA 2012. LNCS, vol. 7655, pp. 1–15. Springer, Heidelberg (2012)
7. Dutt, M., Biswas, A., Bhowmick, P., Bhattacharya, B.B.: On finding a shortest isothetic path and its monotonicity inside a digital object. Annals of Mathematics and Artificial Intelligence (2014) (in press)
8. Gruber, P.M.: Geometry of numbers. In: Gruber, P.M., Wills, J.M. (eds.) Handbook of Convex Geometry, vol. B, pp. 739–763. Elsevier, Amsterdam (1993)
9. Gruber, P.M., Lekkerkerker, C.G.: Geometry of Numbers. North-Holland Mathematical Library, vol. 37, 2nd edn. North-Holland Publishing Co., Amsterdam (1987)

10. Her, I.: Geometric transformations on the hexagonal grid. IEEE Trans. Image Process. **4**(9), 1213–1221 (1995)
11. Lekkerkerker, C.G.: Geometry of numbers. Bibliotheca Mathematica, vol. VIII. Wolters-Noordhoff Publishing, Groningen, North-Holland Publishing Co., Amsterdam, London (1969)
12. Melter, R.: A survey of digital metrics. Contemp. Math. **119**, 95–106 (1991)
13. Nagy, B.: A family of triangular grids in digital geometry. In: Proceedings of the 3rd International Symposium on Image and Signal Processing and Analysis, ISPA 2003, vol. 1, pp. 101–106. Italy, Rome (2003)
14. Nagy, B.: Shortest path in triangular grids with neighbourhood sequences. J. Comput. Inf. Technol. **11**, 111–122 (2003)
15. Nagy, B.: Isometric transformations of the dual of the hexagonal lattice. In: Proceedings of the 6th International Symposium on Image and Signal Processing and Analysis, ISPA 2009. pp. 432–437. IEEE, Salzburg, Austria (2009)
16. Nagy, B.: Weighted distances on a triangular grid. In: Barneva, R.P., Brimkov, V.E., Šlapal, J. (eds.) IWCIA 2014. LNCS, vol. 8466, pp. 37–50. Springer, Heidelberg (2014)
17. Nagy, B.: Generalized triangular grids in digital geometry. Acta Mathematica Academiae Paedagogicae Nyíregyháziensis. **20**, 63–78 (2004)
18. Rosenfeld, A., Melter, R.A.: Digital geometry. Math. Intelligencer **11**(3), 69–72 (1989)
19. Rosenfeld, A., Pfaltz, J.L.: Distance functions on digital pictures. Pattern Recogn. **1**, 33–61 (1968)
20. Wüthrich, C.A., Stucki, P.: An algorithmic comparison between square- and hexagonal-based grids. CVGIP: Graphical Models Image Process. **53**, 324–339 (1991)

Construction of 3D Orthogonal Convex Hull
of a Digital Object

Nilanjana Karmakar[(✉)] and Arindam Biswas

Department of Information Technology,
Indian Institute of Engineering Science and Technology, Shibpur, India
{nilanjana.nk,barindam}@gmail.com

Abstract. Orthogonal convex hull of a digital object in 3D domain is
defined as the minimum volume orthogonal polyhedron enclosing the
object such that its intersection with an axis-parallel face plane is either
empty or a collection of projection-disjoint convex polygons. A novel
and efficient algorithm for construction of 3D orthogonal convex hull of a
digital object is proposed. The algorithm is based on orthogonally slicing
the object into slab polygons followed by connecting all possible slab
polygons on a slicing plane and finding their 2D orthogonal convex hulls.
The regions belonging to the 2D orthogonal convex hulls are replaced
by the corresponding UGCs (unit grid cubes) and the exterior UGC-
faces are merged to give the 3D orthogonal convex hull. The algorithm
operates in integer domain and executes in time linear in the number
of voxels on the object surface. The algorithm operates in exactly two
passes irrespective of the object size or grid resolution. Experimentation
with a wide range of objects has provided accurate results, some of which
are presented here to demonstrate the effectiveness of the algorithm.

Keywords: 3D orthogonal convex hull · Orthogonal slicing · 3D orthogonal outer cover · 3D concavity · 2D orthogonal convex hull

1 Introduction

Determination of the convex hull of a digital object has been a well-studied
problem in the domain of computational geometry. Convex hull finds its use in
a wide range of applications including shape analysis, visual pattern matching,
intersection and collision detection, robot motion planning, obstacle detection,
etc. The standard algorithms in 2-dimensions include Gift wrapping or Jarvis
march [8] which executes in $O(nh)$, and Graham scan [7], Quickhull [5], and
other divide-and-conquer approaches [14] each of which executes in $O(n \log n)$
time, where n is the number of points or vertices on the object surface, and h
is the number of vertices of the convex hull. Compared to the 2D convex hull
in the general domain, the 2D orthogonal convex hull construction requires less
time at the same grid resolution [2].

Construction of 3D convex hull in the general domain has been attempted
since the Gift wrapping algorithm has been extended to 3-dimensions, thereby

© Springer International Publishing Switzerland 2015
R.P. Barneva et al. (Eds.): IWCIA 2015, LNCS 9448, pp. 125–142, 2015.
DOI: 10.1007/978-3-319-26145-4_10

replacing the edge-by-edge approach by a facet-by-facet approach [8,14]. A suggested improvement on the Gift wrapping algorithm [16] takes care of numerical computations and emphasizes on the topological condition that the boundary of the 3D convex hull should be isomorphic to a sphere. Variations of Quickhull [1,3] have been extended to the 3-dimensions by incrementally inserting points to the convex hull, as in the 2D case. A divide-and-conquer approach adopted in [13] is characterized by merging of two non-intersecting convex hulls as the procedure progresses. Both Quickhull and the divide-and-conquer approach execute in $O(n \log n)$ time. Other techniques based on Quickhull, Voronoi diagram, etc. and executing on GPU and PRAM architectures have been reported [4,6,15]. But there exists a sharp distinction among the two sets of techniques as the latter shifts from the standard CPU architecture to the fast and parallel processing GPU.

A novel and efficient algorithm for the construction of 3D orthogonal convex hull is reported. To the best of our knowledge, no work regarding construction of 3D orthogonal convex hull has been reported in literature. In [1], as also in many other works, the digital object is provided as a set of points in 3D space. As our algorithm accepts the input digital object as a triangulated data set, the space requirement is less than most other algorithms. Unlike [1], our algorithm does not follow an incremental approach. Rather, the orthogonal convex hull is determined in exactly two passes of the algorithm irrespective of the object size or the grid resolution. The algorithm executes in general CPU-based architecture.

2 Definitions and Preliminaries

The theoretical foundation in the context of this work is given below in brief. A *digital object* A is defined as a finite subset of $\mathbb{Z}^3$, with all its constituent points (i.e., voxels) having integer coordinates and connected in 26-neighborhood. Each voxel is equivalent to a *3-cell* [12] centered at the concerned integer point (Fig. 1 (Left)). In the current work, the digital object A is considered as a triangulated data set where each vertex of a triangle is an object voxel.

2.1 Digital Grid

The orthogonal cover of A is obtained w.r.t. a digital grid in 3D digital space. A *digital grid* $\mathbb{G}$ consists of three orthogonal sets of equi-spaced grid lines, $\mathbb{G}_{yz}$, $\mathbb{G}_{zx}$, and $\mathbb{G}_{xy}$, where $\mathbb{G}_{yz} = \{l_x(j \pm ag, k \pm bg) \mid a \in \mathbb{Z}, b \in \mathbb{Z}\}$. Similarly, $\mathbb{G}_{zx}$ and $\mathbb{G}_{xy}$ can be represented in terms of l_y and l_z for a grid size $g \in \mathbb{Z}^+$. Here, $l_x(j, k) = \{(x, j, k) : x \in \mathbb{R}\}$, $l_y(i, k) = \{(i, y, k) : y \in \mathbb{R}\}$, and $l_z(i, j) = \{(i, j, z) : z \in \mathbb{R}\}$ denote the *grid lines* (Fig. 1(Left)) along x-, y-, and z-axes respectively, where i, j, and k are integer multiples of g. The three orthogonal lines $l_x(j, k)$, $l_y(i, k)$, and $l_z(i, j)$ intersect at the point $(i, j, k) \in \mathbb{Z}^3$, which is called a *grid point*; a shift of $(\pm 0.5g, \pm 0.5g, \pm 0.5g)$ with respect to a grid point designates a *grid vertex*, and a pair of adjacent grid vertices defines a *grid edge* [12] (Fig. 1(Left)).

A grid, as defined above, is characterized by several elements (Fig. 1(Left)). A *unit grid cube* (UGC) is a (closed) cube of length g whose vertices are *grid*

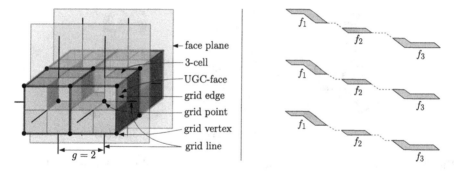

Fig. 1. Left: α-adjacent 3-cells for $g = 2$. **Right:** Possible ways of forming a convex polygon out of three projection-disjoint convex polygons (Color figure online).

vertices, edges constituted by *grid edges*, and faces constituted by *grid faces*. Each face of a UGC lies on a *face plane* (henceforth referred as a UGC-face), which is parallel to one of three coordinate planes. Clearly, each *face plane*, containing coplanar UGC-faces, is at a distance of integer multiple of g from its parallel coordinate plane. A UGC-face, f_k, has two adjacent UGCs, U_1 and U_2, such that $f_k = U_1 \cap U_2$. The *interior* of a UGC is the open cubical region lying strictly inside the UGC. A smaller (larger) value of g implies a denser (sparser) grid. For $g = 1$, the grid $\mathbb{G}$ essentially corresponds to $\mathbb{Z}^3$. As each grid point p is equivalent to a 3-cell c_p centered at p for $g = 1$, each face of c_p is a *grid face* lying on a *face plane*, which is parallel to a coordinate plane. A UGC consists of $g \times g \times g$ voxels and each UGC-face consists of $g \times g$ voxels.

2.2   3D Orthogonal Convex Hull

An *orthogonal polyhedron* is a 3D polytope with all its vertices as grid vertices, all its edges made of grid edges, and all its faces lying on face planes [11]. Each face of an orthogonal polyhedron is an *isothetic polygon* whose alternate edges are orthogonal and constituted by grid edges of $\mathbb{G}$. An *orthogonal convex polyhedron* is an orthogonal polyhedron whose intersection with a face plane parallel to any coordinate plane is either empty or a collection of projection-disjoint orthogonal convex polygons. The *3D orthogonal convex hull* of a digital object A, denoted by $3OH(A)$, is the minimum volume orthogonal polyhedron such that

i. each point $p \in A$ lies inside $3OH(A)$ and
ii. intersection of $3OH(A)$ with a face plane parallel to any coordinate plane is either empty or a collection of projection-disjoint orthogonal convex polygons.

Here, collection of projection-disjoint convex polygons defines a collection of orthogonal convex polygons (also known as "hv-convex" polygons in literature) whose intersection with every orthogonal line is either empty or a line segment [17]. Henceforth, the term "convex polygon" would refer to "orthogonal convex polygon".

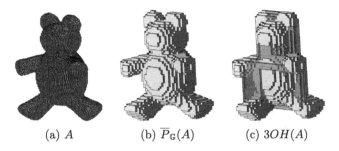

(a) A (b) $\overline{P}_{\mathrm{G}}(A)$ (c) $3OH(A)$

Fig. 2. A digital object, its 3D orthogonal outer cover, and its 3D orthogonal convex hull (Color figure online).

A minimal collection convex hull of a collection of polygons is defined as "the minimal area convex collection of disjoint polygons containing the given collection" [17]. The term "projection-disjoint" is used in conformance with the concept of minimal collection convex hull of a collection of polygons. If the intersection of a given collection of orthogonal convex polygons with an orthogonal line is either empty or a line segment, then an attempt to construct a "minimal area convex polygon containing the given collection" [17] results in a convex polygon which is not unique. In Fig. 1(Right), three possible ways of forming a minimal area convex polygon out of three such polygons f_1, f_2, and f_3 are displayed. In such cases, the concept of projection-disjoint convex polygons produces a unique collection of orthogonal convex polygons, i.e., the collection of f_1, f_2, and f_3 without the red dotted lines joining them.

3 Proposed Work

Given a 3D digital object A in the form of a triangulated data set, the construction of the 3D orthogonal convex hull is presented in this section. In Fig. 2(c), 3D orthogonal convex hull $3OH(A)$ of a digital object A (Fig. 2(a)) is shown as regions appended to the 3D orthogonal outer cover $\overline{P}_{\mathrm{G}}(A)$ in Fig. 2(b). The process involves orthogonal slicing of A with slicing planes parallel to yz-, zx-, and xy-planes, followed by the construction of the 2D orthogonal convex hull [2] of all the slab polygons on each slicing plane Π_i. Since on each edge of the triangulated surface exactly two triangles are incident, the set of triangles representing A truly captures the peripheral topology of the object. The triangulation contains neither degenerate triangles nor holes. The interiors of two triangles do not intersect. In this case, it is not necessary to represent the object as a set of object points. The object is imposed on a 3D background grid which is represented as a set of unit grid cubes (UGCs) (Sect. 2.1), each of grid size g. If a UGC U_k completely contains or is intersected by a triangle $T_{abc}(v_a, v_b, v_c)$, then U_k is considered as *object-occupied*.

Let $\Pi = \{\Pi_1, \Pi_2, ..., \Pi_n\}$ be a set of *slicing planes* (coinciding with face planes) separated by g and parallel to the zx-plane (or yz- or xy-plane) such

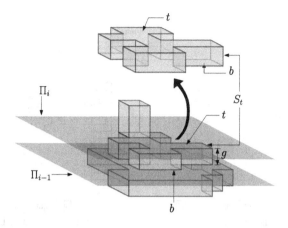

Fig. 3. Slab S_t of height g, bounded on top by slab polygon t lying on Π_i and in the bottom by b, the projection of t on Π_{i-1} (Color figure online).

that it extends upto the length of the 3D orthogonal outer cover $\overline{P}_{\mathbb{G}}(A)$ in a direction perpendicular to the zx-plane. A UGC-face f_k on Π_i is considered as object-occupied if the adjacent UGC below f_k is object-occupied. The boundary of the object intersected by Π_i is traversed orthogonally such that the object always lies left. Thus, a slab polygon on Π_i is obtained [9]. Let t be a slab polygon on Π_i and b be the projection of t on Π_{i-1}. A slab S_t is defined as the section of $\overline{P}_{\mathbb{G}}(A)$ of height g intercepted between Π_i and Π_{i-1} and bounded horizontally on top and bottom by t and b respectively (Fig. 3). Since b is the projection of t, their shapes are identical, that is, t does not vary from Π_i to Π_{i-1}. Hence the slab S_t can be conveniently represented by the slab polygon t.

If more than one slab polygon exists on a single slicing plane, then we minimally connect each pair of such polygons which are not projection-disjoint by a half-edge and its twin. The x and y ranges of each such slab polygon are stored in two indexed lists L_x and L_y by a single traversal of the slab polygons. A stack is used to find the overlapping slab polygons along each of x- and y-axes. Repetition of the process on L_x and L_y alternately for a finite number of times gives a connected slab polygon or more than one projection-disjoint slab polygons on a single slicing plane. Let F be a set of one or more 2D isothetic polygons representing a connected slab polygon. In Fig. 4(a), F is initially represented by $\{f_1, f_2, f_3, f_4\}$ and then connected minimally to form $F = f_1 \cup f_2 \cup f_3 \cup f_4$. The 2D orthogonal convex hull $OH(F)$ of the slab polygons on Π_i is determined as stated in [2]. $OH(F)$ is defined as the smallest area orthogonal polygon such that (i) each point $p \in F$ lies inside $OH(F)$ and (ii) intersection of $OH(F)$ with any horizontal or vertical line is either empty or exactly one line segment. To construct $OH(F)$ the concavities in F are resolved by applying a set of reduction rules to derive the edges of the orthogonal hull that maintain the property of orthogonal convexity (Fig. 4(b)).

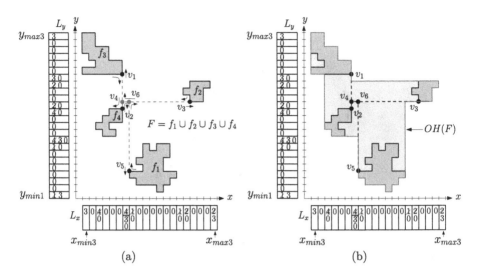

Fig. 4. Joining of coplanar slab polygons, which are not projection-disjoint, is shown here. In L_x and L_y, the indices corresponding to the maximum and minimum values of the x- and y-coordinates of each polygon are marked by polygon *ids*. (a) Each pair of slab polygons in $\{f_1, f_2, f_3, f_4\}$ are minimally connected by a half-edge and its twin to form the connected polygon F. (b) Resolving the concavities in F yields the 2D orthogonal convex hull $OH(F)$ (Color figure online).

Likewise, the construction of the 3D orthogonal convex hull $3OH(A)$ also requires a method of addressing the concavities in A. "Concavity" in 3-dimensions and its related terminology are explained next. A grid vertex in the 3D domain may be shared by at most eight UGCs. Depending on the object occupancy of these UGCs, the grid vertices may be classified into different types where each type is represented by a 3-tuple defined as (# incident UGCs, # incident edges, # incident faces) w.r.t. the grid vertex. In Fig. 5(Left), some instances of the possible concave vertices of types $(3, 3, 3)$, $(4, 4, 4)$, $(4, 6, 6)$, $(5, 3, 3)$, $(6, 2, 2)$, and $(7, 3, 3)$ are shown. It is to be noted that only those concave vertices that do not form an intersecting polyhedron (pseudo-polyhedron) are considered here.

3.1   Concavity in Three Dimensions

During the traversal of t (Fig. 5(Right)), if two consecutive concave vertices are encountered, then it implies a concavity in t. Since t represents the slab S_t, two consecutive concave vertices in t also represents a *concavity* or *concave region* in S_t. Hence, in this case, the concavity consists of two consecutive concave vertices in each of t and b. In Fig. 5(Right)(a), the concave vertices in t and b are marked in red. If t lies on slicing plane Π_i and b on Π_{i-1}, then this concavity is referred to as a concavity on Π_i. The minimum volume polyhedron that is plugged into a concave region to resolve the concavity is defined as a *polyhedral*

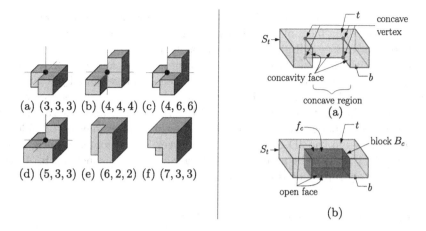

(a) $(3,3,3)$ (b) $(4,4,4)$ (c) $(4,6,6)$

(d) $(5,3,3)$ (e) $(6,2,2)$ (f) $(7,3,3)$

Fig. 5. Left: Types of concave vertex (that do not belong to intersecting polyhedron). **Right:** (a) A concavity on slab S_t containing two consecutive concave vertices (red) in each of the slab polygons t and b. The concave region contains a pair of parallel concavity faces. (b) The concavity in S_t is resolved by plugging in a polyhedral block which contains three open faces (Color figure online).

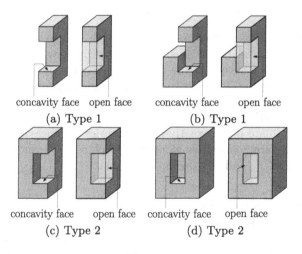

concavity face   open face
(a) Type 1

concavity face   open face
(b) Type 1

concavity face   open face
(c) Type 2

concavity face   open face
(d) Type 2

Fig. 6. Types of concavity: Type 1 where (a) three faces of the concavity are open and (b) two faces are open and one face is partially open, and Type 2 where (c) two faces, and (d) one face of the concavity is open (Color figure online).

block, henceforth referred to as a block (Fig. 5(Right)(b)). For simplicity, the polyhedral block is shown as a tetrahedron in Fig. 5(Right)(b). Here, t and b are horizontal faces. The faces of the block, B_c, that coincide with vertical faces of S_t are called *concavity faces*. If two concavity faces are parallel to each other, then they are referred to as parallel concavity faces. The concavity in Fig. 5(Right) contains three concavity faces out of which two are parallel to each other. Each face of the block that does not coincide with a face of S_t is defined as *open face*. Henceforth, the open faces of a block will also be referred to as open faces of the concave region. In Fig. 5(Right), the block has three open faces.

The concavities may be classified into two types, Type 1 and Type 2, as shown in Fig. 6. Type 1 (Fig. 6(a, b)) has a pair of parallel open faces and Type 2 (Fig. 6 (c, d)) has no parallel open faces. For simplicity, the concave vertices involved in the concavities in Fig. 6 are only of types $(3, 3, 3)$, $(4, 4, 4)$, and $(7, 3, 3)$. Based on the object occupancy of the UGCs the concavities in A may be considered as the concavities in the 3D orthogonal outer cover $\overline{P}_\mathbb{G}(A)$ for sufficiently small grid sizes [10,11]. It is to be noted that for larger grid sizes a concavity in A may not be accurately represented by a concavity in $\overline{P}_\mathbb{G}(A)$.

3.2   Finding 3D Orthogonal Convex Hull

Let a concavity in the slab polygon t be resolved by appending the polygon f_c to t (Fig. 5(Right)(b)). Hence, the concavity in the slab S_t should be resolved by plugging in the block, B_c, which consists of the UGC(s) corresponding to f_c. The polygon f_c is an open face of B_c. B_c may contain three, two, or one open face(s), as mentioned in Fig. 6.

The following features are observed in a concavity in a slab S_t.

    i. A concavity in S_t is viewed as a concave slab polygon t and/or as a collection of coplanar slab polygons (not necessarily convex) along a direction perpendicular to t, such that the polygons are not projection-disjoint. W.l.o.g. a concavity $C_{x,y}$ refers to a concavity along yz-plane. $C_{x,y}$ is viewed as a collection of coplanar slab polygons that are not projection-disjoint along zx-plane (Fig. 7(b)).

   ii. Along a coordinate plane exactly two categories of concavities exist in S_t. $C_{x,z}$ and $C_{x,y}$ are concavities along the yz-plane, $C_{y,z}$ and $C_{y,x}$ along the zx-plane, and $C_{z,y}$ and $C_{z,x}$ along the xy-plane (Fig. 7).

3.2.1   Induced Concavities

The introduction of B_c in S_t may induce a new concavity. A concavity is characterized by a pair of parallel concavity faces separated by a distance of integer multiple of g which defines the concave region. A concavity is induced when a concavity face is introduced such that it is parallel to an existing concavity face.

A Type 1 concavity C_1 contains a pair of parallel open faces, say, f_1 and f_2, as shown in Fig. 8(a). Let f_1 lies on the slicing plane Π_i parallel to the zx-plane. If a block B_c is used to address this concavity, then f_1 and f_2 are two parallel open faces of B_c. f_1 and f_2 are introduced simultaneously and none of them lies between two existing concavity faces. Let there exists a polytope face f_4 on some slicing plane Π_l parallel to the zx-plane such that its projection on Π_{l-1} is the concavity face f_3. Let f_1 and the projection of f_3 on Π_i have a non-empty intersection. Then f_1 induces a new concavity C_1' with f_3 where the concave region lies between f_1 and f_3. On the other hand, a Type 2 concavity C_2 does not contain parallel open faces. In Fig. 8(b), an open face f_1 of C_2 is parallel

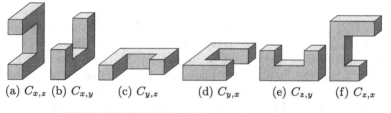

(a) $C_{x,z}$ (b) $C_{x,y}$ (c) $C_{y,z}$ (d) $C_{y,x}$ (e) $C_{z,y}$ (f) $C_{z,x}$

Plane	Concavity	Induced concavity		
		yz-plane	zx-plane	xy-plane
yz	$C_{x,z}$ $C_{x,y}$	$C_{y,z}, C_{z,y}$	-	-
zx	$C_{y,z}$ $C_{y,x}$	-	$C_{x,z}, C_{z,x}$	-
xy	$C_{z,y}$ $C_{z,x}$	-	-	$C_{x,y}, C_{y,x}$

Fig. 7. Two categories of possible concavities w.r.t. a slab along each coordinate plane. (a) $C_{x,z}$ concavity, induces $C_{y,z}$ and $C_{z,y}$ along yz-plane, (b) $C_{x,y}$ concavity, induces $C_{y,z}$ and $C_{z,y}$ along yz-plane, (c) $C_{y,z}$ concavity, induces $C_{x,z}$ and $C_{z,x}$ along zx-plane, (d) $C_{y,x}$ concavity, induces $C_{x,z}$ and $C_{z,x}$ along zx-plane, (e) $C_{z,y}$ concavity, induces $C_{x,y}$ and $C_{y,x}$ along xy-plane, and (f) $C_{z,x}$ concavity, induces $C_{x,y}$ and $C_{y,x}$ along xy-plane.

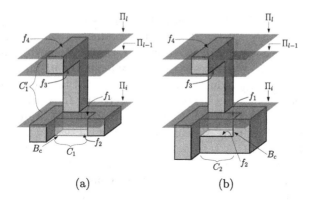

(a)   (b)

Fig. 8. Induced concavities: (a) Addressing a Type 1 concavity C_1 induces a concavity C_1' and (b) addressing a Type 2 concavity C_2 does not induce any new concavity (Color figure online).

to one of its concavity faces f_2. It is evident that f_1 is introduced between two concavity faces f_2 and f_3. Earlier f_2 and f_3 had defined a concavity. The same concavity is now defined by f_1 and f_3. Hence, no new concavity is induced. This implies that resolving a Type 1 concavity induces a new concavity but resolving a Type 2 concavity does not induce new concavity.

W.r.t. a slab S_t, a concavity $C_{x,y}$ ($C_{x,z}$) is viewed as a concave slab polygon along the yz-plane and as a set of slab polygons which are not projection-disjoint along the zx-plane (xy-plane) (Fig. 7(a) and (b)). In this case, two of the open faces must be parallel to the yz-plane. Such a face may act as a concavity face only for a $C_{y,z}$ or a $C_{z,y}$ concavity (Fig. 7(c) and (e)). Hence, addressing the concavities $C_{x,y}$ and $C_{x,z}$ while traversing slab polygons along yz-plane may induce concavities $C_{y,z}$ and $C_{z,y}$. Concavity $C_{x,y}$ ($C_{x,z}$), as shown in Fig. 7, is not detected along the xy-plane (zx-plane). Along the zx-plane (xy-plane) $C_{y,z}$ and $C_{z,y}$ are induced and resolved simultaneously, as evident later from Fig. 9(c) and (d). Hence, addressing $C_{x,y}$ and $C_{x,z}$ along zx- and xy-planes induces no concavity. Such properties are true for concavities $C_{y,x}$ and $C_{y,z}$ (Fig. 7(c) and (d)), and $C_{z,x}$ and $C_{z,y}$ (Fig. 7(e) and (f)) along the corresponding coordinate planes, as summarized in tabular form in Fig. 7.

3.2.2   Two Passes of the Algorithm

Since addressing a concavity may induce a new concavity as stated in Sect. 3.2.1, a second pass of the algorithm may be required in order to deal with the induced concavities, if any. W.r.t. a slicing plane Π_i parallel to a given coordinate plane, the algorithm is applicable if it performs at least one of the operations – connecting at least one pair of slab polygons which are not projection-disjoint, or resolving at least one concavity.

Let us consider that $\overline{P}_{\mathbb{G}}(A)$, intersected by the given slicing plane Π_i, is subjected to the algorithm along the coordinate planes in the given order of yz-, zx-, and xy-planes.

Pass 1, Step yz-plane: Let the set of slab polygons on the slicing plane Π_i be denoted as $F = \{f_1, f_2, ..., f_r\}$. The method addresses concavities $C_{x,z}$, $C_{x,y}$, $C_{y,x}$, and $C_{z,x}$ which already existed on $\overline{P}_{\mathbb{G}}(A)$ (Fig. 7(a), (b), (d), and (f)). Addressing $C_{y,x}$ and $C_{z,x}$ induces no concavity along the yz-plane (Sect. 3.2.1). Addressing $C_{x,z}$ and $C_{x,y}$ induces $C_{y,z}$ and $C_{z,y}$ such that the block describing the concave region of $C_{y,z}$ or $C_{z,y}$ is not intersected by Π_i, as shown in Fig. 9(c) and (d) (Sect. 3.2.1). In this case, concavities $C_{y,z}$ and $C_{z,y}$ are not identified along the yz-plane (Fig. 9(a)). But in some cases, $C_{y,z}$ or $C_{z,y}$ may get resolved while addressing $C_{y,x}$ or $C_{z,x}$ w.r.t. one or more slicing planes along the yz-plane. In Fig. 9(b), $C_{z,y}$ gets resolved while addressing $C_{z,x}$ w.r.t. slicing planes Π_i and Π_{i+1}. No concavity is induced in the process. Thus, only concavities $C_{y,z}$ and $C_{z,y}$ (both originally present in $\overline{P}_{\mathbb{G}}(A)$ and induced here) are carried over to the next step.

Pass 1, Step zx-plane: Once the yz-plane is exhausted, $\overline{P}_{\mathbb{G}}(A)$ is subjected to the algorithm along the zx-plane. The concavities $C_{y,z}$ and $C_{z,y}$ which are originally present in $\overline{P}_{\mathbb{G}}(A)$ and those which are induced in the last step are carried over to this step. The recently induced $C_{y,z}$ and $C_{z,y}$ (Fig. 9(c) and (d)), in turn, induces concavities which do not belong to Π_i. Therefore, in this step $C_{y,z}$ and $C_{z,y}$ will refer to the original ones (Fig. 9(a)). When $C_{y,z}$ is addressed along the zx-plane, $C_{x,z}$ and $C_{z,x}$ may be induced (Fig. 9(e)). Removal of $C_{z,y}$ does not induce any concavity along the zx-plane, as mentioned in Sect. 3.2.1. Hence, induced concavities $C_{x,z}$ and $C_{z,x}$ are carried over to the next step.

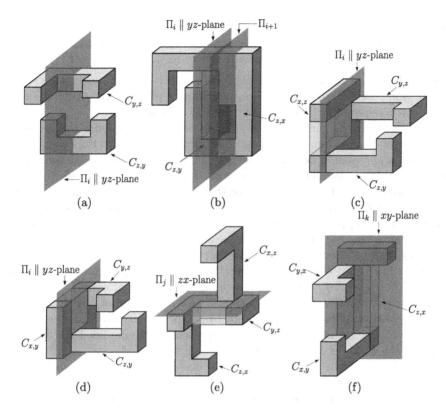

Fig. 9. (a) Concavities $C_{y,z}$ and $C_{z,y}$ are not identified along the yz-plane. (b) $C_{z,y}$ gets resolved while addressing $C_{z,x}$ w.r.t. Π_i and Π_{i+1} along yz-plane. (c, d) Addressing $C_{x,z}$ and $C_{x,y}$ in Step yz-plane induces $C_{y,z}$ and $C_{z,y}$, (e) addressing $C_{y,z}$ in Step zx-plane induces $C_{x,z}$ and $C_{z,x}$, and (f) addressing $C_{z,x}$ in Step xy-plane induces $C_{y,x}$ and $C_{x,y}$ (Color figure online).

Pass 1, Step xy-plane: As $C_{x,z}$ is addressed in this step along the xy-plane, no concavity is induced. As the remaining concavity $C_{z,x}$ is addressed along the xy-plane, concavities $C_{x,y}$ and $C_{y,x}$ are induced (Fig. 9(f)).

Thus, at the end of the first pass of the algorithm, induced concavities $C_{x,y}$ and $C_{y,x}$ are left to be resolved w.r.t. the slicing plane Π_i. Let the set of UGCs P represents $\overline{P}_\mathbb{G}(A)$. When $\overline{P}_\mathbb{G}(A)$ is subjected to the algorithm along yz-, zx-, and xy-planes, the UGCs comprising the blocks that resolve the concavities are appended to P. Thus, $\overline{P}_\mathbb{G}(A)$ is modified. Orthogonal slicing of $\overline{P}_\mathbb{G}(A)$ may give rise to a new set of slab polygons $f_{r+1}, f_{r+2}, ..., f_n$ on the slicing plane Π_i. In the next pass, all the slab polygons $f_1, f_2, ..., f_r, f_{r+1}, ..., f_n$ are connected and the resultant concavities are addressed. Let the second pass of the algorithm be started with the yz-plane just as the first pass.

Pass 2, Step yz-plane: $C_{x,y}$ and $C_{y,x}$ are the only concavities present in this step w.r.t. Π_i. Addressing $C_{y,x}$ does not induce any concavity along the yz-plane

(Sect. 3.2.1). Addressing $C_{x,y}$ induces concavities $C_{y,z}$ and $C_{z,y}$ (Fig. 9(d)) but the blocks describing these concavities in any further step are not intersected by Π_i. Hence, no induced concavity is carried over to the next step w.r.t. to Π_i.

Pass 2, Step zx-plane and Step xy-plane: Since no induced concavity is carried over to Steps zx and xy from the previous Step yz-plane of the second pass, no concavity is required to be resolved in these two steps.

Therefore, all the concavities in Π_i are resolved by applying the proposed algorithm in at most four steps along the coordinate planes in the order of yz-, zx-, xy-, and yz-planes, i.e., order starting and ending with the same coordinate plane. In other words, all the concavities on Π_i are resolved within the first pass and the first step of the second pass of the algorithm; hence the last two steps of the second pass are redundant w.r.t. Π_i. These two steps may be useful for some other slicing plane. Intuitively, the concavities on any other slicing plane Π_j parallel to zx- or xy-plane can also be addressed in the above process where the steps may be taken in a different order (say, zx, xy, yz, zx or xy, zx, yz, xy, etc.). The set of convex and projection-disjoint slab polygons formed from $f_1, f_2, ..., f_r, f_{r+1}, ..., f_n$ represents the final 2D orthogonal convex hull on Π_i that contributes in the construction of $3OH(A)$. At the end of each step the 3D orthogonal outer cover $\overline{P}_{\mathbb{G}}(A)$ is modified by appending the concavity resolving UGCs to P. The orthogonal polyhedron representing the modified $\overline{P}_{\mathbb{G}}(A)$ is the 3D orthogonal convex hull $3OH(A)$ of the digital object A.

A single pass of the proposed algorithm consists of three steps along yz-, zx-, and xy-planes and the concavities w.r.t. a slicing plane are resolved in at most four steps. Hence, it is evident that two passes of the algorithm are required to construct $3OH(A)$. As mentioned in Sect. 3.2.1, resolving a concavity C_1 may induce another concavity C_2 if C_1 is a Type 1 concavity. Hence, if a Type 2 concavity is encountered in any step of the process, then the construction of 2D orthogonal convex hull on a slicing plane may even be completed in less than four steps, i.e., in a single pass of the algorithm. Thus, all the concavities of $\overline{P}_{\mathbb{G}}(A)$ will be addressed if the proposed algorithm is executed along the full cycle of yz-, zx-, and xy-planes at most twice.

It is to be noted that the resultant 3D orthogonal convex hull is independent of the order of application of the algorithm. But the intermediate modifications of $\overline{P}_{\mathbb{G}}(A)$ may vary with the order. According to Fig. 7, concavities belonging to exactly two categories are resolved along each coordinate plane and concavities of two specific categories are induced in the process. Hence, the concavities addressed and induced in each step are pre-defined. But the concavities carried over to each step may vary with the order. Since the initial and final steps are parallel to the same coordinate plane and the slicing plane Π_i in question is also parallel to the same coordinate plane, the process leads to the same result for any order of the steps.

3.3   Algorithm

Given an object A in the form of triangulated data set, its 3D orthogonal convex hull $3OH(A)$ is constructed by the two-pass algorithm presented in Fig. 10. A is

```
01.  set U ← 0                              ▷ empty set of UGCs
02.  set count ← 0                          ▷ # passes
03.  do
04.     for each coordinate plane j    ▷j ∈ {yz, zx, xy}
05.        slice P̄_G(A) with Π_j        ▷Π_j = {Π_{j1}, Π_{j2}, ..., Π_{jn}}
06.        for each Π_{jk}               ▷1 ≤ k ≤ n
07.           for each slab polygon f
08.              find coordinate ranges along x- and y-axes
09.              connect slab polygons which are not projection-disjoint
10.           set K ← set of connected slab polygons
11.           set s ← # connected slab polygons
12.           for i = 1 to s
13.              set H[i] ← 2D orthogonal convex hull OH(K[i])
14.              R[i] ← H[i] − K[i]
15.              set C[i] ← set of UGCs corresponding to R[i]
16.              U ← U ∪ C[i]
17.        increment count by 1
18.  while(count < 2)
19.  set U_M ← polyhedron formed by merging the exterior
                   UGC-faces of U
20.  report P̄_G(A) ∪ U_M
```

Fig. 10. Brief outline of the proposed algorithm.

sliced orthogonally by a set of slicing planes $\Pi_j = \{\Pi_{j1}, \Pi_{j2}, ..., \Pi_{jn}\}$ parallel to each of the coordinate planes yz, zx, and xy (Steps 4 and 5) in a method similar to that stated in [9]. One or more slab polygons may exist on a slicing plane Π_{jk}, where $1 \leq k \leq n$. The slab polygons on Π_{jk} which are not projection-disjoint are minimally connected by a half-edge and its twin using their maximum and minimum coordinate values along the x- and y-axes (Steps 7–9). Each connected slab polygon $K[i]$ is subjected to a procedure for finding its 2D orthogonal convex hull $H[i]$ which is based on the algorithm stated in [2] (Step 13). The region $R[i]$ intercepted between $K[i]$ and $H[i]$ and its corresponding set of UGCs $C[i]$ are determined in Steps 14 and 15. The process is repeated exactly once for each slicing plane parallel to each of the three coordinate planes (Steps 3–18). W.r.t. each slicing plane Π_{jk}, the set of UGCs $C[i]$ corresponding to $R[i]$ are accumulated as U (Step 16) and their exterior UGC-faces are merged to construct the orthogonal polyhedron U_M (Step 19). Finally, the 3D orthogonal convex hull $3OH(A)$ is reported as the orthogonal polyhedron formed by merging $\overline{P}_G(A)$ and U_M (Step 20).

3.4   Time Complexity

Let n be the number of voxels on the object surface connected in 26-neighborhood. Since the object is triangulated, n refers to the total number of voxels that approximate all the triangles representing the object surface. A UGC is a cube of length g which contributes a maximum of five faces to the cover. Therefore, the number of UGCs on the object surface containing object voxels is $O(n/g)$ in the worst case, which implies that the number of UGC-faces on the object surface is given by $O(n/g)$. As the area of a UGC-face is g^2, the object occupancy of a UGC is determined in $O(g^2)$ time.

W.r.t. each slicing plane, orthogonal slicing involves traversal of the grid vertices on the slicing plane exactly once. Therefore, considering all the slicing planes, the UGCs on the object surface are traversed exactly once. This traversal requires $O(n/g)$ time. Since object-occupancy of a UGC-face is checked in $O(g^2)$ time, the direction of traversal at each grid vertex is determined in $O(g^2)$ time. Hence, the orthogonal slicing along all the coordinate planes is completed in $O(n/g) \times O(g^2) = O(ng)$ time.

For all the slicing planes, finding the slab polygons which are not projection-disjoint requires a single traversal of all the slab polygons in $O(n/g)$ time. The stack operations and indexed list operations are done in constant time. The slab polygons are traversed once more along with the introduced half-edges to retrieve the connected slab polygon, which requires $O(n/g)$ time. Hence, the slab polygons which are not projection-disjoint are minimally connected in $O(n/g) + O(1) + O(1) + O(n/g) = O(n/g)$ time.

Let there be k number of slicing planes along a given coordinate plane. The time complexity for traversing a connected slab polygon is $O((n/k)/g) \times O(g^2) = O(ng/k)$, where the number of grid vertices traversed is bounded by $O((n/k)/g)$ and object-occupancy at each vertex is checked in $O(g^2)$ time. Time required for detection and removal of concavity is given by $(O(n/kg) - 4).O(1) = O(n/kg)$. Therefore, the total time complexity for finding the 2D orthogonal convex hull of a connected slab polygon is given by $O(ng/k) + O(n/kg) = O(ng/k)$ time. The process is repeated for k slicing planes which gives a collective complexity of $O(ng)$. The process is repeated to determine the 2D orthogonal convex hull on each slicing plane parallel to each of the three coordinate planes.

Therefore, the total time complexity for a single pass of the algorithm is given by $O(ng) + O(n/g) + O(ng) = O(ng)$. Exterior UGC-faces of the set of UGCs which are used to resolve concavities are merged by a traversal of the UGC-faces exactly once in $O(n/g)$ time. Thus, the total time complexity for finding the 3D orthogonal convex hull $3OH(A)$ of the given object A is given by $(O(ng) \times 2) + O(n/g) = O(ng)$.

4   Experimental Results and Conclusion

The proposed algorithm has been implemented in C in Linux Fedora Release 7, Dual Intel Xeon Processor 2.8 GHz, 800 MHz FSB. The experimental results in

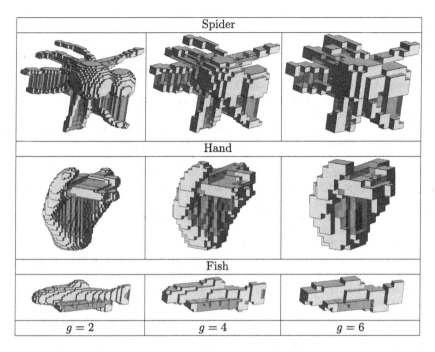

Fig. 11. 3D orthogonal convex hull for Spider, Hand, and Fish for different grid sizes. The regions added to the 3D orthogonal outer cover to construct the 3D orthogonal hull are shown in green color (Color figure online).

Table 1. CPU time of construction of 3D orthogonal convex hull of different digital objects.

Object	CPU time (in secs.)		
	$g = 2$	$g = 4$	$g = 6$
Spider	0.297	0.083	0.053
Hand	0.253	0.101	0.082
Fish	0.119	0.045	0.036

Fig. 11 for different objects like `Spider`, `Hand`, and `Fish` display the 3D orthogonal convex hull for various grid sizes. The regions added to the outer cover to construct the 3D orthogonal hull are shown in green color. The CPU time of hull construction increases with decrease in the grid size for different digital objects as shown in Table 1.

Acknowledgement. A part of this research is funded by CSIR, Govt. of India under SRF (File No. 08/03(0091)/2012-EMR-1) and Sponsored project (Scheme No. 22(0568)/12/EMR-II).

References

1. Barber, C.B., Dobkin, D.P., Huhdanpaa, H.: The quickhull algorithm for convex hulls. ACM Trans. Math. Softw. **22**(4), 469–483 (1996)
2. Biswas, A., Bhowmick, P., Sarkar, M., Bhattacharya, B.B.: A linear-time combinatorial algorithm to find the orthogonal hull of an object on the digital plane. Inf. Sci. **216**, 176–195 (2012)
3. Clarkson, K.L., Shor, P.W.: Applications of random sampling in computational geometry, II. Discrete Comput. Geom. **4**(1), 387–421 (1989)
4. Dehne, F., Deng, X., Dymond, P., Fabri, A., Khokhar, A.A.: A randomized parallel 3D convex hull algorithm for coarse grained multicomputers. In: Proceedings of the Seventh Annual ACM Symposium on Parallel Algorithms and Architectures, SPAA 1995, pp. 27–33. ACM, New York (1995)
5. Eddy, W.F.: A new convex hull algorithm for planar sets. ACM Trans. Math. Softw. **3**(4), 398–403 (1977)
6. Gao, M., Cao, T.T., Tan, T.S., Huang, Z.: gHull: A three-dimensional convex hull algorithm for graphics hardware. In: Symposium on Interactive 3D Graphics and Games, I3D 2011, pp. 204–204. ACM, New York (2011)
7. Graham, R.L.: An efficient algorithm for determining the convex hull of a finite planar set. Inf. Process. Lett. **1**(4), 132–133 (1972)
8. Jarvis, R.: On the identification of the convex hull of a finite set of points in the plane. Inf. Process. Lett. **2**(1), 18–21 (1973)
9. Karmakar, N., Biswas, A., Bhowmick, P.: Fast slicing of orthogonal covers using DCEL. In: Barneva, R.P., Brimkov, V.E., Aggarwal, J.K. (eds.) IWCIA 2012. LNCS, vol. 7655, pp. 16–30. Springer, Heidelberg (2012)
10. Karmakar, N., Biswas, A., Bhowmick, P., Bhattacharya, B.B.: Construction of 3D orthogonal cover of a digital object. In: Aggarwal, J.K., Barneva, R.P., Brimkov, V.E., Koroutchev, K.N., Korutcheva, E.R. (eds.) IWCIA 2011. LNCS, vol. 6636, pp. 70–83. Springer, Heidelberg (2011)
11. Karmakar, N., Biswas, A., Bhowmick, P., Bhattacharya, B.B.: A combinatorial algorithm to construct 3D isothetic covers. Int. J. Comput. Math. **90**(8), 1571–1606 (2013)
12. Klette, R., Rosenfeld, A.: Digital Geometry: Geometric Methods for Digital Picture Analysis. Morgan Kaufmann, San Francisco (2004)
13. Preparata, F.P., Hong, S.J.: Convex hulls of finite sets of points in two and three dimensions. Commun. ACM **20**(2), 87–93 (1977)
14. Preparata, F.P., Shamos, M.I.: Computational Geometry–An Introduction, 3rd edn. Springer, New York (1985)
15. Stein, A., Geva, E., El-Sana, J.: Applications of geometry processing: CudaHull: fast parallel 3D convex hull on the GPU. Comput. Graph. **36**(4), 265–271 (2012)
16. Sugihara, K.: Robust gift wrapping for the three-dimensional convex hull. J. Comput. Syst. Sci. **49**(2), 391–407 (1994)
17. Wood, D.: An Isothetic View of Computational Geometry. University of Waterloo, Computer Science Department (1984)

Appendix

If more than one slab polygon exists on a single slicing plane, then we minimally connect each pair of such polygons which are not projection-disjoint by a half-edge and its twin. The procedure is demonstrated in Fig. 12. Let f_1, f_2, f_3, and

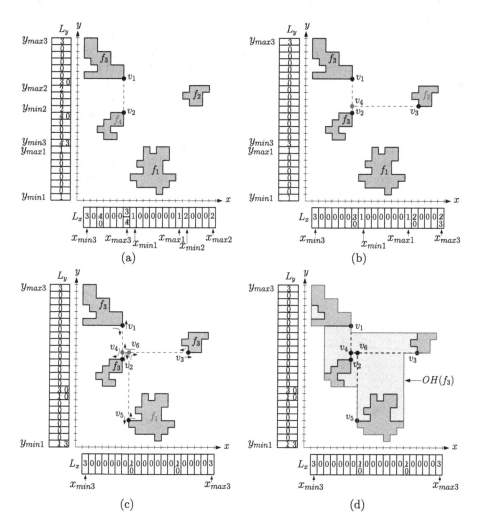

Fig. 12. Connecting coplanar slab polygons which are not projection-disjoint. In L_x and L_y, the start and end indices of the x and y ranges of each polygon are marked by polygon id. (a) As slab polygons f_3 and f_4 are not projection-disjoint, f_4 is minimally connected with f_3 by a half-edge and its twin. (b) New f_3 is minimally connected with f_2. (c) Finally f_1 is appended to the new f_3 to form the current polygon f_3. In each step, L_x and L_y are updated to reflect the changes in x- and y-ranges of the polygons. (d) Resolving the concavities in f_3 yields the 2D orthogonal convex hull $OH(f_3)$ (Color figure online).

f_4 be four coplanar slab polygons. Two lists L_x and L_y are maintained where the start and end of x and y ranges of each slab polygon are marked by the slab polygon id. For example, the start and end indices of f_3 are populated with '3' in L_x and L_y. Initially only f_3 and f_4 are not projection-disjoint. L_x is traversed and f_3 and f_4 are identified using a stack. f_4 is connected to f_3 (Fig. 12(a)) by a half-edge and its twin between vertices v_1 and v_2. L_x and L_y are updated to reflect the new polygon f_3 thus formed. A traversal of L_y reveals that the new polygon f_3 is no more projection-disjoint with f_2. f_3 and f_2 are connected to form the new polygon f_3 as shown in Fig. 12(b). The changes are reflected in L_y and L_x. L_x is traversed again and f_1 is appended to f_3 (Fig. 12(c)). The procedure is continued by traversing L_x and L_y alternately until we have a single slab polygon or more than one projection-disjoint slab polygons. The connected slab polygon f_3 is retrieved by traversing the original polygons in the given order, i.e., f_1, f_2, f_3, and f_4, connected by the introduced edges such that at each of the vertices v_1, v_2, v_3, v_4, v_5, and v_6 the direction of traversal is selected keeping the object to the left, as shown by the arrows in Fig. 12(c). Thereafter, the 2D orthogonal convex hull of f_3 is constructed as stated in [2] (Fig. 12(d)).

Efficient Dominant Point Detection Based on Discrete Curve Structure

Phuc Ngo[1,2], Hayat Nasser[1,2], and Isabelle Debled-Rennesson[1,2(✉)]

[1] Université de Lorraine, LORIA, UMR 7503, F-54506 Vandoeuvre-lès-Nancy, France
[2] CNRS, LORIA, UMR 7503, F-54506 Vandoeuvre-lès-Nancy, France
{hoai-diem-phuc.ngo,hayat.nasser,isabelle.debled-rennesson}@loria.fr

Abstract. In this paper, we investigate the problem of dominant point detection on digital curves which consists in finding points with local maximum curvature. Thanks to previous studies of the decomposition of curves into sequence of discrete structures [5–7], namely maximal blurred segments of width ν [13], an initial algorithm has been proposed in [14] to detect dominant points. However, an heuristic strategy is used to identify the dominant points. We now propose a modified algorithm without heuristics but a simple measure of angle. In addition, an application of polygonal simplification is as well proposed to reduce the number of detected dominant points by associating a weight to each of them. The experimental results demonstrate the efficiency and robustness of the proposed method.

Keywords: Dominant point · Polygonal simplification · Discrete structure

1 Introduction

Dominant points of discrete curves are identified by Attneave [2] as the local maximum curvature points on a curve. Such points content a rich information which is sufficient to characterize this curve. They play a critical role in curve approximation, image matching and in other domains of computer vision. Many works have been conducted regarding the dominant point detection [1,3,4,8–12,14,17–20] and surveys are presented in [1,12,14]. Several problems have been identified in the different approaches: time computation, number of parameters, selection of starting point, bad results with noisy curves, ...

Nguyen *et al.* proposed in [14] a new sequential method issued from theoretical results of discrete geometry, it only requires to set one parameter, it is invariant to the choice of the starting point and it naturally works with general curves: possibly being noisy or disconnected. It relies on the geometrical structure of the studied curve, in particular the decomposition of the curve into maximal blurred segments for a given width [6,13]. However at the end of the proposed method, the choice of dominant points is done with an heuristic strategy.

In this paper, an improvement of this dominant point detection algorithm is proposed with an efficient choice of dominant points by considering a simple

© Springer International Publishing Switzerland 2015
R.P. Barneva et al. (Eds.): IWCIA 2015, LNCS 9448, pp. 143–156, 2015.
DOI: 10.1007/978-3-319-26145-4_11

measure of angle. Furthermore, an algorithm of polygonal simplification is also proposed to reduce the number of detected dominant points while preserving the principal angular deviations in the discrete curve of the shape border.

The paper is organized as follows: in Sect. 2, we recall results of discrete geometry used in this paper to analyze a curve. Then, in Sect. 3, we describe the previous dominant point detection algorithm [14] and propose improvements. In Sect. 4, we present experimental results, comparisons with previous methods and an application to polygonal simplification.

2 Decomposition of a Curve into Maximal Blurred Segments

We recall hereafter several notions concerning discrete lines [15], blurred segments [5] and maximal blurred segments [13] which are used throughout the paper.

Definition 1. *A **discrete line** $\mathcal{D}(a, b, \mu, \omega)$, with a main vector (b, a), a lower bound μ and an arithmetic thickness ω (with a, b, μ and ω being integer such that $gcd(a, b) = 1$) is the set of integer points (x, y) verifying $\mu \leq ax - by < \mu + \omega$. Such a line is denoted by $\mathcal{D}(a, b, \mu, \omega)$.*

Let us consider $\mathcal{S}_f$ as a sequence of integer points.

Definition 2. *A discrete line $\mathcal{D}(a, b, \mu, \omega)$ is said to be **bounding** for $\mathcal{S}_f$ if all points of $\mathcal{S}_f$ belong to $\mathcal{D}$.*

Definition 3. *A bounding discrete line $\mathcal{D}(a, b, \mu, \omega)$ of $\mathcal{S}_f$ is said to be **optimal** if the value $\frac{\omega-1}{max(|a|,|b|)}$ is minimal, i.e. if its vertical (or horizontal) distance is equal to the vertical (or horizontal) thickness of the convex hull of $\mathcal{S}_f$.*

This definition is illustrated in Fig. 1 and leads to the definition of the blurred segments.

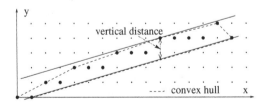

Fig. 1. $\mathcal{D}(2, 7, -8, 11)$, the optimal bounding line of the set of points (vertical distance $= \frac{10}{7} = 1.42$).

Definition 4. *A set $\mathcal{S}_f$ is a **blurred segment of width** ν if its optimal bounding line has a vertical or horizontal distance less than or equal to ν i.e. if $\frac{\omega-1}{max(|a|,|b|)} \leq \nu$.*

The notion of maximal blurred segment was introduced in [13]. Let C be a discrete curve and $C_{i,j}$ a sequence of points of C indexed from i to j. Let us suppose that the predicate "$C_{i,j}$ is a blurred segment of width ν" is denoted by $BS(i,j,\nu)$.

Definition 5. *$C_{i,j}$ is called a **maximal blurred segment of width** ν and noted $MBS(i,j,\nu)$ iff $BS(i,j,\nu)$, $\neg BS(i,j+1,\nu)$ and $\neg BS(i-1,j,\nu)$.*

An incremental algorithm was proposed in [13] to determine the sequence of all maximal blurred segments of width ν of a discrete curve C. The main idea is to maintain a blurred segment when a point is added (or removed) to (from) it. The two following important properties were proved.

Property 1. *Let $MBS_\nu(C)$ be the sequence of width ν maximal blurred segments of the curve C. Then,*
$MBS_\nu(C) = \{MBS(B_0, E_0, \nu), MBS(B_1, E_1, \nu), \ldots, MBS(B_{m-1}, E_{m-1}, \nu)\}$
and satisfies $B_0 < B_1 < \ldots < B_{m-1}$. So we have: $E_0 < E_1 < \ldots < E_{m-1}$ (see Fig. 2).

Property 2. *Let $L(k)$, $R(k)$ be the functions which respectively return the indexes of the left and right extremities of the maximal blurred segments on the left and right sides of the point C_k in a discrete curve C. So, for each index k such that $E_{i-1} < k \le E_i$, we have $L(k) = B_i$, and, for each k such that $B_i \le k < B_{i+1}$, we have $R(k) = E_i$.*

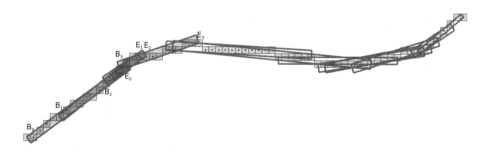

Fig. 2. A sequence of width 1.4 maximal blurred segments of the curve with grey pixels (each red box bounds the pixels of a maximal blurred segment of width 1.4). The indexes of the begin (B_i) and end (E_i) points of the four first maximal blurred segments are indicated ($B_0 = 0$ and $E_0 = 15$, $B_1 = 5$ and $E_1 = 17$, ...). The points with blue frame are points in the common zone – intersecting zone – defined by these four first maximal blurred segments (Color figure online).

For a given width ν, the sequence of the maximal blurred segments of a curve C entirely determines the structure of C. It can be considered as an extension to thick curves of the notion of **tangential cover** introduced by Feschet and

Tougne in [7] and used to obtain curvature profile or fast polygon approximation [8]. In [14], a method for dominant point detection, constructed from the
sequence of maximal blurred segments, was given. In the next section, we propose
an improvement of this method which implies new applications.

3 Dominant Point Detection

The sequence of maximal blurred segments of a curve provides important information about the geometrical structure of the studied curve. The width of the
maximal blurred segments permits to work in different scales and to consider
the noise eventually present in a curve C.

In [2], Attneave defined a dominant point (corner point) on a curve as a point
of local maximum curvature. In this way, the method proposed in [14] uses a
notion of region of support (ROS) of a point in a curve.

Definition 6. *Width ν maximal left and right blurred segments of a point constitute its region of support (ROS). The angle between them is called the ROS
angle of this point.*

The method first chooses a width ν and computes $MBS_\nu(C)$, the sequence of
maximal blurred segments of C. The main idea of the method is that dominant
points of C are localised in the common zones of successive maximal blurred
segments of C (see Fig. 2). Hereafter, we recall the properties and the heuristic
strategy used in [14]. We show that problems can occur and propose a solution.

3.1 Heuristic Strategy of Method [14]

Let us now consider the common zone of more than 2 successive maximal blurred
segments.

Proposition 1. *The smallest common zone of successive width ν maximal blurred
segments, of which slopes are either increasing or decreasing, contains a candidate
as dominant point.*

Property 3. *A maximal blurred segment contains at most 2 dominant points.*

Based on the previous propositions and properties, the algorithm proposed
by Nguyen *et al.* in [14] consists in finding the smallest common zone induced
by successive maximal burred segments of width ν which have either increasing
or decreasing slopes. It is stated that such a zone contains the candidates of
dominant point, since the points in this zone have the smallest ROS angle. By
an heuristic strategy, the dominant point is identified as the middle point of this
zone.

This heuristic is very effective, but sometimes leads to a non optimal solution
for the polygonal simplification problem, as shown in the next section.

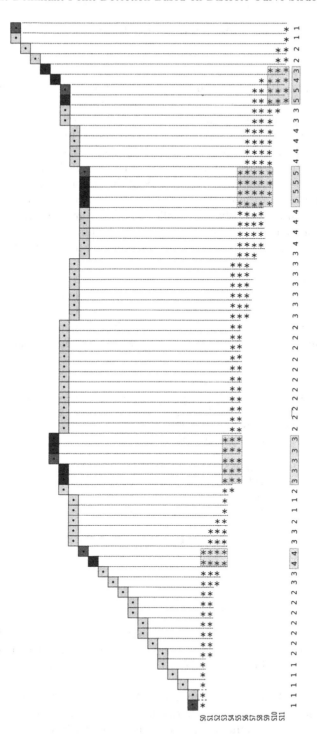

Fig. 3. Interest regions (common zones) obtained with the sequence of width 1.4 maximal blurred segments of the curve with grey pixels (Color figure online).

3.2   New Dominant Point Detection Algorithm

In this section, we present a modified algorithm of [14] for detecting the dominant points of discrete curves with high accuracy.

Since the purpose of dominant point detection is to detect significant points with curvature extreme, we propose a new decision-making strategy of dominant point in common zones. For this, we use a simple measure, as a pseudo-curvature, estimated for each point located in the interest regions. In particular, we consider the k-cosine measure [16] adapted to the decomposition into width ν blurred segments of the input curve. More precisely, it is the angle between the considered point and the two left and right extremities of the left and right maximal blurred segments, involved in the localisation of the studied common zone and passing through the point. Obviously, the smaller angle, the greater curvature and thus the higher dominant character of the point. Accordingly, the dominant point is identified as the point, in the common zone, having a local minimum measure of angle.

The modified algorithm is given in Algorithm 1. It is similar to the one in [14] except for the selection of dominant points as described above. Moreover we simplify the process because we don't need to decompose the slopes of the maximal blurred segments located in the common zones into monotone sequences. Indeed, the angle computation permits to well localize the dominant point in each common zone.

Algorithm 1: Dominant point detection.

 Input : C discrete curve of n points, ν width of the segmentation
 Output: D set of dominant points

1 **begin**
2   Build $MBS_\nu = \{MBS(B_i, E_i, \nu)\}_{i=0}^{m-1}$;
3   $n = |C|$; $m = |MBS_\nu|$;
4   $q = 0$; $p = 1$; $D = \emptyset$;
5   **while** $p < m$ **do**
6   **while** $E_q > B_p$ **do** $p + +$;
7    $D = D \cup \min\{Angle(C_{B_q}, C_i, C_{E_{p-1}}) \mid i \in [\![B_{p-1}, E_q]\!]\}$;
8    $q = p - 1$;

9 **end**

Particularly, in this algorithm, each common zone is computed by Line 6, where q is the index of the first maximal blurred segment involved in the common zone and $p - 1$ is the index of the last. The points in the common zone are points C_i for $i \in [\![B_{p-1}, E_q]\!]$.

Moreover, at Line 7, the function *Angle* calculates, for each point in the common zone, the angle between the point and its two extremities, then in a common zone, the dominant point is the point with local minimum angle. Afterwards, the next common zone is computed by considering the last segment (index $p-1$) of the previous zone as the first one (index q) of the next common zone. This process permits to detect the points of the curve through which a maximum number of maximal blurred segments is passing (see Fig. 3).

If the curve C is not closed, the first and the last points of C are dominant points and they must be added to D.

The algorithm is illustrated in Fig. 3; a curve C is analysed by the algorithm, the sequence $MBS_{1.4}(C)$ is computed and 12 maximal blurred segments (named $s0$ to $s11$ in the figure) are obtained. Their localisation in the curve is indicated by one star ($*$) for each point belonging to the considered maximal blurred segment. Four common zones (dark pixels on the curve) are detected by the algorithm: C_{14} to C_{15}, C_{22} to C_{26}, C_{49} to C_{52} and C_{59} to C_{62}. The pink boxes in the figure permit to count the number of maximal blurred segments passing through each point of common zones. The angles are calculated by considering the starting point of the first maximal blurred segment involved in the common zone and the end point of the last involved maximal blurred segment. The Table 1 shows the angles at each point of common zones and permits to deduce the dominant points (minimal values are in bold in the table and the corresponding points are in red in Fig. 3). In the common zone 4, the selected dominant point is not located at the middle of the common zone.

Table 1. Angles at each point of common zones (see Fig. 3).

Common Zone	Points	Angle
Common Zone 1 B_{S0} and E_{S3}	C_{14}	162.9
	C_{15}	**159**
Common Zone 2 B_{S3} and E_{S5}	C_{22}	155.6
	C_{23}	157.6
	C_{24}	**152**
	C_{25}	153.6
	C_{26}	154.9
Common Zone 3 B_{S5} and E_{S9}	C_{49}	158.6
	C_{50}	157.4
	C_{51}	156
	C_{52}	**154.3**
Common Zone 4 B_{S9} and E_{S11}	C_{59}	155.7
	C_{60}	**150.5**
	C_{61}	155.3
	C_{62}	160.2

The complexity of the proposed algorithm is the same as the one in [14], which depends on the complexity to decompose a curve of n points into maximal blurred segments. We can use the technique proposed in [6] to obtain the tangential cover (corresponding to the sequence of maximal blurred segments for a given width) and the complexity of this method is in $O(n \log n)$.

4   Evaluation Results and Applications

We now present some experimental results of dominant point detection using the proposed method, and evaluate by widely used error criteria [17,18] described in Sect. 4.1. Using these criteria, in Sect. 4.2 we show the improvement results of the modified algorithm regarding the original of Nguyen [14]. Afterwards, in Sect. 4.3, we present an application in polygonal simplification with detected dominant points. The results are then compared with other popular methods in Sect. 4.4.

4.1   Evaluation Criteria

In order to assess the effectiveness of the proposed method, we consider the following five evaluation criteria:

1. Number of dominant points (nDP).
2. Compression ratio (CR) is defined as the ratio between number of curve points and number of detected dominant points. The larger CR, the more effective data simplification.

$$CR = \frac{n}{nDP}$$

3. Integral sum of square errors (ISSE) is the sum of squared distances of the curve points from approximating polygon. The smaller ISSE, the better description of the shape by the approximating polygon.

$$ISSE = \sum_{i=1}^{n} d_i^2$$

where d_i is distance from i^{th} curve point to approximating polygon.

4. Maximum error (L_∞) is the maximum distance of the curve points from approximating polygon. The smaller L_∞, the better fitness of polygonal approximation.

$$L_\infty = max\{d_i\}_{i=1}^{n}$$

5. Figure of merit (FOM) is estimated as ratio between CR and ISSE. FOM a compromise between the low approximation error and the benefit of high data reduction.

$$FOM = \frac{CR}{ISSE}$$

4.2   Effectiveness Compared to Nguyen's Algorithm

In this section, we present the experimental results of the modified algorithm (see Sect. 3.2) and compare them with Nguyen's algorithm [14]. The experiments are carried out on the data with and without noise. For each input curve, a fixed width $\nu = 1.5$ of maximal blurred segments is used for both algorithms, the results are shown in Fig. 4 and Table 2. It can be seen that selecting the middle point of common zone is not always a relevant strategy, in particular in the high-pass zones.

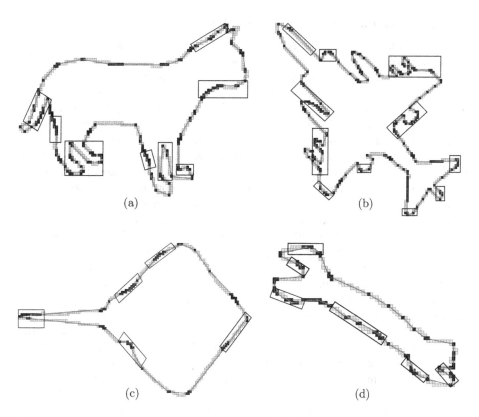

(a)

(b)

(c)

(d)

Fig. 4. Improved results with respect to Nguyen [14] (see also Table 2). The effectiveness of the modified algorithm is more significant in the *highly derivative zones* of the shape as highlighted by the black boxes. Green (resp. red) points are dominant points detected by Nguyen's (resp. modified) algorithm. Lines in green (resp. red) denote the polygonal approximation from detected dominant points. Blue points are candidates of dominant points in the interest regions (Color figure online).

Table 2. Comparison with Nguyen's method [14] on Fig. 4 using significant measures described in Sect. 4.1.

Curve	Method	nDP	CR	ISSE	L_∞	FOM
(a) $n = 536$	Nguyen	68	7.882	149.15	1	0.0523
	Ours	68	7.882	**125.763**	1	**0.0627**
(b) $n = 722$	Nguyen	105	6.876	202.809	1	0.0339
	Ours	105	6.876	**166.74**	1	**0.0412**
(c) $n = 404$	Nguyen	20	20.2	236.806	3.536	0.0853
	Ours	20	20.2	**150.314**	**1.539**	**0.1344**
(d) $n = 252$	Nguyen	43	5.86	68.896	1	0.0851
	Ours	43	5.86	**57.582**	1	**0.1018**

4.3   Application in Polygonal Simplification

The goal of finding the dominant points is to have an approximate description of the input curve, called *polygonal simplification* or *approximating polygon*. On the one hand, due to the nature of the maximal blurred segment sequence defined on a discrete curve, we observe that the common zones – contain candidates of dominant points – may be close to each others. On the other hand, using the algorithm proposed in Sect. 3.2, the dominant point is selected according to its angle with the extremities of maximal segments. As a consequence, detected dominant points are sometimes redundant or stay very near, which is presumably undesirable in particular for polygonal simplification. More precisely, this leads to an overmuch dominant points and thus the polygon induced by such points is not an optimal solution of curve simplification. Therefore, we can eliminate certain dominant points to achieve a high compression of approximating polygon of the input curve. To this end, we associate to each detected dominant point a weight indicating/describing its importance with respect to the approximating polygon of the curve. Such a weight must: (1) be related to some error criteria of approximating polygon, and (2) not induce costly computation.

From Sect. 4.1, the criterion ISSE describes the distortions caused by the approximated polygon of a curve. More precisely, ISSE allows to evaluate how much the approximated polygon is similar to the curve, thus smaller ISSE means better descriptive of the curve by the polygon. This error evaluation is suitable to our criterion of weight associated to dominant points since: (1) we would like to eliminate points that less affect the possible error (i.e. less ISSE), and (2) in particular the ISSE can be calculated *locally* by using their two neighbouring dominant points. Indeed, the value of ISSE between the approximated polygon and the curve before and after deleting a dominant point differs only at the part related to the point, in particular its two neighbours as illustrated in Fig. 5. The ISSE induced by the two neighbours of a dominant point thus characterizes the important of the point; the greater ISSE, the more important the point.

Still in Fig. 5, we remark that dominant point may have a small ISSE, however regarding its neighbours, in particular the angular relationship, it is more important than the others. In other words, this angle with the neighbours plays an important role in the decision of suppressing a dominant point. This leads to a consideration of weight associated to each dominant point detected that not only involves ISSE but also the angle to its two neighbours. In particular, this weight is determined by the ratio of ISSE and angle, *i.e.*, $ISSE/angle$. Then, fixing a desired number m of dominant points on the approximated polygon, roughly speaking, the process of polygonal simplification is performed by removing one by one the dominant points of small weight until reaching m.

In the next section, we test the proposed method and compare it to the other polygonal simplification techniques based on dominant points extraction.

4.4   Comparison with Other Methods

The experiments are first carried out on three benchmarks: chromosome, leaf and semicircle shown in Figs. 6, 7 and 8 respectively. Table 3 compares the proposed

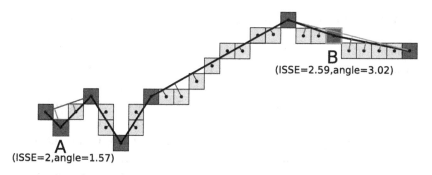

Fig. 5. Illustration of calculating the weights associated to detected dominant points. In grey (resp. red) are points (resp. dominant points) of the curves. The continuous black line connecting the dominant points denotes the approximated polygon, while dashed lines are perpendicular distances from points on the curve to the polygon, the sum of these distances defines the ISSE. Considering the point A in blue (resp. B in green) frame, deleting this point leads to a *local* modification in the approximated polygon, and thus ISSE induced by blue (resp. green) dashed lines. One remarks that ISSE of the point in blue frame is smaller than the green, however according to the neighbours – in angle – it is more important than the green one. Therefore, the weight associated to each dominant point involves both ISSE and angle criteria, in particular $\frac{ISSE}{angle}$. Then, the weight of A and B are 1.27 and 0.86, respectively (Color figure online).

method with other popular algorithms such as: Masood [11], Marji [9,10] and Teh [19], on the benchmarks by the evaluation criteria described in Sect. 4.1. Overall the experiments, the proposed method presents an improvement more than of 38 % on FOM, and has a better approximation error in ISSE and L_∞.

Figure 9 shows polygonal approximation results of the other proposed digital curves with noise. The results demonstrate the effectiveness and robustness of the proposed method on noisy data. Note that the experiments are performed by using Algorithm 1 (see Sect. 3.2 with a process of polygonal simplification as described in Sect. 4.3).

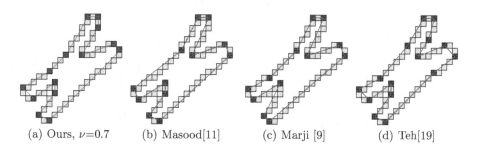

(a) Ours, $\nu=0.7$ (b) Masood[11]   (c) Marji [9]   (d) Teh[19]

Fig. 6. Dominant points of the chromosome curve.

Table 3. Results of the proposed method and of the other methods.

Curve	Method	nDP	CR	ISSE	L_∞	FOM
Chromosome $n = 60$	Ours, $\nu = 0.7$	14	4.286	**5.116**	0.8	**0.838**
	Masood [11]	12	5	7.76	0.88	0.65
	Marji [9]	12	5	8.03	0.895	0.623
	Teh [19]	15	4	7.2	**0.74**	0.556
Leaf $n = 120$	Ours, $\nu = 0.7$	23	5.218	**9.065**	**0.671**	**0.576**
	Masood [11]	23	5.217	10.61	0.74	0.49
	Marji [10]	22	5.45	13.21	0.78	0.413
	Teh [19]	29	4.14	14.96	0.99	0.277
Semicircle $n = 102$	Ours, $\nu = 0.7$	23	4.435	**7.639**	0.724	**0.581**
	Masood [11]	22	4.64	8.61	**0.72**	0.54
	Marji [10]	26	3.92	9.01	0.74	0.435
	Teh [19]	22	4.64	20.61	1	0.225

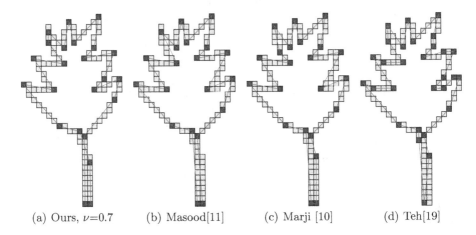

(a) Ours, ν=0.7   (b) Masood[11]   (c) Marji [10]   (d) Teh[19]

Fig. 7. Dominant points of the leaf curve.

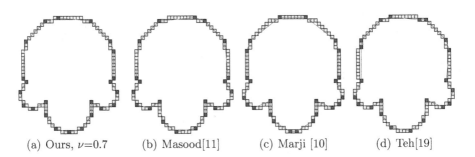

(a) Ours, ν=0.7   (b) Masood[11]   (c) Marji [10]   (d) Teh[19]

Fig. 8. Dominant points of the semicircle curve.

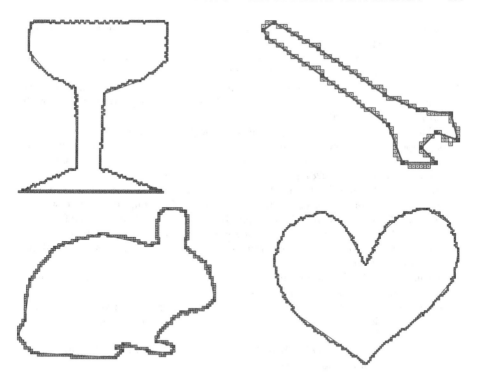

Fig. 9. Polygonal simplification results (in red) on noisy data using width parameter $\nu = 1.5$, and a reduction of 10 % of dominant points detected by Algorithm 1 (Color figure online).

5   Conclusion and Future Work

We present in this paper a dominant point detection algorithm, improvement of a previous algorithm which uses an heuristic strategy. This algorithm relies on the structure of the discrete curve and the width, parameter of the algorithm, permits to work on data with or without noise. For a given width, our method associates to each detected dominant point a weight which permits to evaluate the importance of the dominant point in the structure of the studied curve. We then deduce a method to reduce the number of dominant points and to obtain the smallest possible error according to a given number of dominant points.

The proposed method opens numerous perspective of future work, for example the study of the method behaviour in a multiscale approach to automatically detect the more appropriate width for dominant point selection.

References

1. Aguilera-Aguilera, E.J., Poyato, Á.C., Madrid-Cuevas, F.J., Carnicer, R.M.: The computation of polygonal approximations for 2D contours based on a concavity tree. J. Vis. Comm. Image Represent. **25**(8), 1905–1917 (2014)

2. Attneave, E.: Some informational aspects of visual perception. Psychol. Rev. **61**(3), 183–193 (1954)
3. Backes, A.R., Bruno, O.M.: Polygonal approximation of digital planar curves through vertex betweenness. Inf. Sci. **222**, 795–804 (2013)
4. Bhowmick, P., Bhattacharya, B.B.: Fast polygonal approximation of digital curves using relaxed straightness properties. IEEE Trans. Pattern Anal. Mach. Intell. **29**(9), 1590–1602 (2007)
5. Debled-Rennesson, I., Feschet, F., Rouyer-Degli, J.: Optimal blurred segments decomposition of noisy shapes in linear time. Comput. Graph. **30**(1), 30–36 (2006)
6. Faure, A., Buzer, L., Feschet, F.: Tangential cover for thick digital curves. Pattern Recogn. **42**(10), 2279–2287 (2009)
7. Feschet, F., Tougne, L.: Optimal time computation of the tangent of a discrete curve: application to the curvature. In: Bertrand, G., Couprie, M., Perroton, L. (eds.) DGCI 1999. LNCS, vol. 1568, pp. 31–40. Springer, Heidelberg (1999)
8. Feschet, F.: Fast guaranteed polygonal approximations of closed digital curves. In: Kalviainen, H., Parkkinen, J., Kaarna, A. (eds.) SCIA 2005. LNCS, vol. 3540, pp. 910–919. Springer, Heidelberg (2005)
9. Marji, M., Siy, P.: A new algorithm for dominant points detection and polygonization of digital curves. Pattern Recogn. **36**(10), 2239–2251 (2003)
10. Marji, M., Siy, P.: Polygonal representation of digital planar curves through dominant point detection - nonparametric algorithm. Pattern Recogn. **37**(11), 2113–2130 (2004)
11. Masood, A.: Dominant point detection by reverse polygonization of digital curves. Image Vis. Comput. **26**(5), 702–715 (2008)
12. Masood, A.: Optimized polygonal approximation by dominant point deletion. Pattern Recogn. **41**(1), 227–239 (2008)
13. Nguyen, T.P., Debled-Rennesson, I.: Curvature Estimation in Noisy Curves. In: Kropatsch, W.G., Kampel, M., Hanbury, A. (eds.) CAIP 2007. LNCS, vol. 4673, pp. 474–481. Springer, Heidelberg (2007)
14. Nguyen, T.P., Debled-Rennesson, I.: A discrete geometry approach for dominant point detection. Pattern Recogn. **44**(1), 32–44 (2011)
15. Reveillès, J.P.: Géométrie discrète, calculs en nombre entiersgorithmique, thèse d'état. Université Louis Pasteur, Strasbourg (1991)
16. Rosenfeld, A., Johnston, E.: Angle detection on digital curves. IEEE Trans. Comput. **22**, 875–878 (1973)
17. Rosin, P.L.: Techniques for assessing polygonal approximations of curves. IEEE Trans. Pattern Anal. Mach. Intell. **19**(6), 659–666 (1997)
18. Sarkar, D.: A simple algorithm for detection of significant vertices for polygonal approximation of chain-coded curves. Pattern Recogn. Lett. **14**(12), 959–964 (1993)
19. Teh, C., Chin, R.: On the detection of dominant points on the digital curves. IEEE Trans. Pattern Anal. Mach. Intell. **2**, 859–872 (1989)
20. Wang, B., Brown, D., Zhang, X., Li, H., Gao, Y., Cao, J.: Polygonal approximation using integer particle swarm optimization. Inf. Sci. **278**, 311–326 (2014)

Thoughts on 3D Digital Subplane Recognition and Minimum-Maximum of a Bilinear Congruence Sequence

Eric Andres[✉], Dimitri Ouattara, Gaelle Largeteau-Skapin, and Rita Zrour

Laboratoire XLIM, SIC, UMR CNRS 7252, Université de Poitiers, BP 30179,
86962 Futuroscope Chasseneuil, France
{eric.andres,jean.ouattara,gaelle.largeteau.skapin,
rita.zrour}@univ-poitiers.fr

Abstract. In this paper we take first steps in addressing the 3D Digital Subplane Recognition Problem. Let us consider a digital plane $P : 0 \leq ax + by - cz + d < c$ (w.l.o.g. $0 \leq a \leq b \leq c$) and a finite subplane S of P defined as the points (x, y, z) of P such that $(x, y) \in [x_0, x_1] \times [y_0, y_1]$. The Digital Subplane Recognition Problem consists in determining the characteristics of the subplane S in less than linear (in the number of voxels) complexity. We discuss approaches based on remainder values $\left\{ \frac{ax+by+d}{c} \right\}, (x, y) \in [x_0, x_1] \times [y_0, y_1]$ of the subplane. This corresponds to a bilinear congruence sequence. We show that one can determine if the sequence contains a value ϵ in logarithmic time. An algorithm to determine the minimum and maximum of such a bilinear congruence sequence is also proposed. This is linked to leaning points of the subplane with remainder order conservation properties. The proposed algorithm has a complexity in, if $m = x_1 - x_0 < n = y_1 - y_0$, $O(m \log (\min(a, c - a)))$ or $O(n \log (\min(b, c - b)))$ otherwise.

Keywords: Digital planes · Digital subplane recognition problem · Congruence sequence

1 Introduction

Since J-P. Reveilles, among other previous authors [4,5], proposed an analytical description of a Digital Straight Line (DSL) $0 \leq ax - by + c < \omega$ [14], many papers have been devoted to its study. Indeed, the structure of DSL is rich, with immediate links to word theory, the Stern-Brocot tree, the Farey sequence, etc. See [10] for an historical perspective. The natural extension to higher dimensions has opened new venues for arithmeticians [2].

Lately, the problem of characterizing a Digital Straight Segment (DSS), segment of a DSL with known characteristics, has gained some traction [11,13,15,16]. This problem is linked to multiscale shape analysis [11,15,19]. When considering geometrical features at multiple scales, it is important to be able to recompute the new, scaled, characteristics as rapidly as possible. In this paper we are interested

© Springer International Publishing Switzerland 2015
R.P. Barneva et al. (Eds.): IWCIA 2015, LNCS 9448, pp. 157–171, 2015.
DOI: 10.1007/978-3-319-26145-4_12

in the extension of this problem to dimension three: the Digital Subplane (DSP) Recognition Problem. Check the following papers for a general approach on Digital Plane Recognition [3,6,7,9,12]. In Sect. 2, we propose a recall on 2D results and the unsolved problems in 3D. Right now, minimal characteristics of a subplane are chosen as representative of the equivalence class formed by all the characteristics that fit the Digital SubPlane (DSP) [3,6,9]. In particular, we conjecture the existence of a class of Digital SubPlanes characteristics for which the remainder order property is respected. This means that by searching for the minimum and maximum of the remainders on a DSP, we can identify leaning points easily which would lead to the characteristics (a leaning point is a point of extremum remainder in the DSP that allows to compute its characteristics). In Sect. 3, we propose an algorithm for computing the minimum and maximum of a simple bilinear congruence sequence corresponding to the remainders of a DSP. This represents a first step towards solving a particular subclass of the general problem of characterization of a Digital Plane Subsegment Recognition Problem. We conclude in Sect. 4.

2   Recalls on the 2D Problem and State of the 3D Problem

2.1   Recalls of the 2D Digital Straight Subsegment Recognition Problem

An 8-connected Digital Straight Line (DSL) in the first octant, is defined by analytical inequalities $0 \leq ax - by + c < c$, with $0 \leq a \leq b$ and $\omega = b$, $\gcd(a, b) = 1$, $a, b, c \in \mathbb{Z}^3$. There is a unique DSL with a given set of characteristics but there are an infinite number of finite Digital Straight Lines that contain a same, finite connected, Digital Straight Segment (DSS). All these DSL containing a DSS form an equivalence class. There is therefore a question of the unique characterization of a DSS. If one simply takes the known characteristics of a DSL containing a DSS, then we may end up with different characteristics for the DSS and have the problem of comparing them. There is a unique DSL among the class that has a minimal parameter b [6]. The parameters (a, b, c), with minimal parameter b among all the DSL containing a same DSS, are chosen as characteristics for the DSS. These characteristics are called *minimal characteristics* of the DSS. The DSS can then be fully characterized by those parameters and two points A and B corresponding to the end points of the segment. These parameters happen also to be those that are given by the analytical recognition algorithm proposed by I. Debled-Renesson [6].

The problem of the characterization of a Digital Straight Subsegment contained in a known DSL is different from the regular recognition problem of a DSS since we already know that all the points of the subsegment belong to a known DSL. Various approaches have been proposed [11,15,16] such as considering the Stern-Brocot tree or the Farey fans. These methods are logarithmic in the coefficients of the input slope or the length of the segment. The main problem with these methods is that they do not offer an obvious extension to higher dimensions. In [13], the authors have proposed an alternative algorithm based on the remainder values. For a DSL $\mathcal{D}(a, b, c)$ of characteristics (a, b, c), if a point (x, y) belongs to the DSL, then the remainder $ax - by + c$ is equal to $\left\{\frac{ax+c}{b}\right\}$ (where $\left\{\frac{n}{m}\right\}$ stands for $n \mod m$). The integer $\mathcal{R}_{(a,b,c)} = \left\{\frac{ax+c}{b}\right\}$ is called a remainder in x as remainder of a Euclidean division where $\left\lfloor\frac{ax+c}{b}\right\rfloor$ is the ordinate of the point of abcissa x belonging to the DSL. The remainders define a simple congruence sequence noted $\mathcal{R}_{(a,b,c)}$. One of the main results that we showed is that for a DSL of characteristics (a, b, c) and a DSS of minimal characteristics (α, β, γ) defined on $x \in [u, v]$, the remainder order is conserved on the DSS [13]:

$$\forall x, x' \in [u, v], |x - x'| \leq b : \mathcal{R}_{a,b,c}(x) < \mathcal{R}_{a,b,c}(x') \Rightarrow \mathcal{R}_{\alpha,\beta,\gamma}(x) \leq \mathcal{R}_{\alpha,\beta,\gamma}(x')$$

This has some direct consequences such as the fact that the minimal and maximal values of the DSL remainders on the subsegment are leaning points. By computing a third minimal or maximal remainder, it allows to determine the minimal characteristics of the subsegment. Furthermore, the computation of these minimal and maximal values of the congruence sequence can be done in logarithmic time with a simple characteristic substitution and sequence reduction scheme (DSL collapse) akin to Euclid's Algorithm. The method is faster than previous ones [11,15,16] and offers a possible extension to higher dimensions. It is this extension we start exploring in the present paper.

2.2 State of the 3D Problem

This paper is interested in the exploration of the Digital SubPlane (DSP) Recognition (characterization) Problem in dimension three. The paper is meant as a first step as there are some significant differences with the problem in dimension two and, as we will see, many questions that remain open and that require future investigations. First of all, Digital Plane (DP) recognition problems in 3D can be way more difficult than in 2D. For instance, decomposing a 2D closed curve into a minimal number of 2D DSS can be performed in linear time [8] while the equivalent problem in 3D is NP-hard [17]. The preimage of a 2D DSS is a polygon of a maximum of four vertices while there is no limit to the number of vertices for the preimage polytope of a DSP in 3D [3]. There are also some similarities: as in 2D, a finite connected digital subplane belongs to an infinite number of digital planes containing the DSP (and thus defining an equivalence class). There exists a unique DP of characteristics (a, b, c, d) such that c is minimal among the characteristics in the equivalence class. This is called the *minimal characteristics* of a DSP. The DSP Recognition algorithms provide the minimal characterictics [3,6,7,9,12]. For the problem that interests us, the problem of characterizing a plane subsegment of a known Digital Plane in dimension three, the Stern-Brocot exploration approach followed by Said and Iachaud [11,15] and the Farey fan walk approach followed by Sivignon [16] are not easily extended to dimension three, while our remainder approach seems more appropriate [13].

The extension of the 2D remainder sequence to 3D is straightforward and constitutes the starting point of this investigation. Let us consider a digital plane (DP) $\mathcal{P}(a,b,c,d) = \{(x,y,z) \in \mathbb{Z}^3; 0 \le ax + by - cz + d < c\}$ of known characteristics (a,b,c,d), with, w.l.o.g. $\gcd(a,b,c) = 1$ and $0 \le a \le b \le c$. For all the points of the DP, one can define a simple bilinear congruence sequence $\mathcal{R}_{a,b,c,d}(x,y) = ax + by - cz + d = \left\{\frac{ax+by+d}{c}\right\}$ [1,14].

One could think that the 2D remainder properties extend naturally to 3D, but they do not, at least not always. The 2D remainder order conservation property (recalled in the previous subsection) is not verified anymore in 3D (see the conclusion of [13] for an example). It is verified quite often but not systematically. The consequences are immediate: there are leaning points of the DP that may not be leaning point anymore in the DSP and vice-versa. It means also that the minimum and maximum remainder value on a DSP remainder sequence are not necessarily leaning points for the minimal characteristics of the DSP. However, it seems that there always exists characteristics for the DSP such that the minimum and maximum remainder are leaning points. Let us express this in the form of a conjecture:

Conjecture 1. Let us consider a DP $P = \mathcal{D}(a,b,c,d)$, with $\gcd(a,b,c) = 1$ and $0 \le a \le b \le c$ and a finite DSP of P defined on $(x,y) \in [x_0, x_1] \times [y_0, y_1]$, then there exists characteristics $(\alpha, \beta, \gamma, \delta)$ such that the points among those with the minimum and maximum values of the bilinear remainder sequence $\mathcal{R}_{(a,b,c,d)}(x,y)$ on $[x_0, x_1] \times [y_0, y_1]$ are leaning points of the DSP for the characteristics $(\alpha, \beta, \gamma, \delta)$.

Actually, what that means is that the minimal characteristics for a DSP may not be the only choice as DP equivalence class characteristics' representative. While the remainder conservation property is not always verified for minimal characteristics, it seems that there are actually always characteristics (sometimes not minimal) that have this property. This needs to be looked upon more closely before we state it as a conjecture or simply prove it, but it opens the way to a subclass of DSP characteristics with some very interesting properties. On the opposite, the remainder conservation property seems to be the norm. For instance, when we consider a DSP with minimal characteristics, most DP that contain this DSP seems to verify the conservation property but not necessarily all. Are there always DP that do not verify this property and always some that do? How are those groups characterized? They represent after all the very same voxels so what makes them behave differently?

So, why considering characteristics that are not necessarily minimal? First of all, this may lead to characterization algorithms that are sublinear in the DSP number of voxels (for the Digital Subplane Recognition Problem). The insight in the characteristics classes may lead to new and better general understanding of Digital Planes. In this paper, we provide the first algorithm with sublinear complexity to determine the minimum and maximum of a simple bilinear congruence sequence (DSP remainder sequence) and provide some thoughts on particular classes of Digital Plane collapses.

After some notations, we will present in Sect. 3 some extensions of the results presented in [13] on linear congruence sequences. We show that one can determine in logarithmic time if a bilinear congruence sequence contains a given value ϵ. We provide an algorithm to compute the minimum and maximum of a bilinear congruence sequence. We conclude in Sect. 4 and give some clues on the computation of DSP characteristics and future work.

3 Finding the Minimum and Maximum of a Simple Bilinear Congruence Sequence

We are looking for the minimum and maximum of a bilinear congruence sequence $\left\{ \frac{ax+by+d}{c} \right\}$ for $(x, y) \in [x_0, x_1] \times [y_0, y_1]$. We suppose that $\gcd(a, b, c) = 1$ and that $0 \le a \le b \le c$. After the presentation of some notations, we will discuss properties of linear congruence sequences and especially linear congruence sequence collapses that preserve minimum or maximum values. This leads to a first algorithm for the search of a minimum and maximum in a bilinear congruence sequence. We will end this section with some thoughts on digital plane collapses in order to obtain even better complexities.

3.1   Notations

A *Digital Plane* (DP for short) $\mathcal{P}(a, b, c, d)$ of *integer characteristics* (a, b, c, d) is the set of digital points $(x, y, z) \in \mathbb{Z}^3$ such that $0 \le ax + by - cz + d < max(|a|, |b|, |c|)$ with $\gcd(a, b, c) = 1$. This digital plane is 18-connected and called a *naive* digital plane [3]. The value d is sometimes called the translation constant. In this paper, without loss of generality, we assume that $0 \le a \le b \le c$. In this case, we have one and only one point, denoted $P_D(x, y)$, in P with abscissa x and ordinate y. The z-coordinate is then $z = \left\lfloor \frac{ax+by+d}{c} \right\rfloor$.

A *Digital SubPlane* (DSP for short) $\mathcal{S}(P, x_0, x_1, y_0, y_1)$ associated to the DP $P = \mathcal{P}(a, b, c, d)$ is the subset of P with points of abscissa and ordinate in $[x_0, x_1] \times [y_0, y_1]$. A DSP is a finite 18-connected subset of a DP.

We will use the notation $\left\{ \frac{n}{m} \right\}$ for $n \bmod m$ [14]. In 3D, the *remainder* at abscissa and ordinate (x, y) is the value $\mathcal{R}_{a,b,c,d}(x, y) = ax + by - cz + d$. For a point of the DP, we have $\mathcal{R}_{a,b,c,d}(x, y) = \left\{ \frac{ax+by+d}{c} \right\}$. The bilinear remainder sequence $\mathcal{R}_{a,b,c,d}(x_0, x_1, y_0, y_1)$ is a set of remainders $\mathcal{R}_{a,b,c,d}(x, y)$ for $(x, y) \in [x_0, x_1] \times [y_0, y_1]$. In 2D, the *remainder* for a DSL of characteristics (a, b, c) is the value $\mathcal{R}_{a,b,c}(x) = \left\{ \frac{ax+c}{b} \right\}$ at abscissa x. The linear remainder sequence $\mathcal{R}_{a,b,c}(u, v)$ corresponds to the values $\mathcal{R}_{a,b,c}(x)$ for $u \le x \le v$.

Let us first note that $\mathcal{R}_{a,b,c,d}(x, y) = \mathcal{R}_{a,c,d+by}(x) = \mathcal{R}_{b,c,d+ax}(y)$. The simpliest way of looking at a bilinear congruence sequences is to look at them as sequences of linear congruence sequences. The results presented here are slight extensions of properties already presented in [13] for sequences of type $\mathcal{R}_{a,b,0}(x)$ with parameter $c = 0$. Here we are looking at the same properties for $\mathcal{R}_{a,b,c}(x)$. As we will see, the extensions are pretty straightforward.

3.2   Linear Sequence Collapse

Let us look at collapsed linear congruence sequences to $\mathcal{R}_{a,b,c}(u,v) = \left\{ \left\{ \frac{ax+c}{b} \right\} : u \le x \le v \right\}$ that preserve the minimum and maximum values of the sequence. We suppose that $0 \le b$ and $\gcd(a,b) = 1$. Let us remark that the sequence of remainders $\mathcal{R}_{a,b,c}(x)$ corresponds to a naive DSL of slope $\frac{a}{b}$ in the first octant. When one looks at such a sequence as a DSL, one can see that the minimal values and maximal values are located at specific places on the DSL. Let us call a *span*, a set of pixels with same ordinate. The remainders between 0 and $a-1$ are located at the beginning of a complete span while the remainders between $b-a$ and $b-1$ are located at the end of a complete span. Let us note as well that, in Fig. 1, the first span on the bottom left is not complete and the upper top span neither.

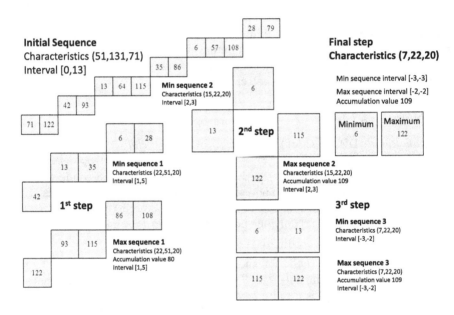

Fig. 1. DSS of characteristics $(51, 131, 71)$, $0 \le x \le 13$. Minimum remainders are at the beginning of a span while maximum remainders are at the end of a span. Three steps are needed in this case to determine the minimum and maximum.

The following proposition states that the span start and end remainders form linear congruence sequences as well:

Proposition 1. *Let us consider the remainder subsequence* $\zeta = \mathcal{R}_{a,b,c}(u,v)$, *with* $0 \le a \le b$ *and* $\gcd(a,b) = 1$.

- *if* $\left\lfloor \frac{au+c}{b} \right\rfloor = \left\lfloor \frac{av+c}{b} \right\rfloor$ *then* $\min(\zeta) = \left\{ \frac{au+c}{b} \right\}$ *and* $\max(\zeta) = \left\{ \frac{av+c}{b} \right\}$;
- *otherwise* $\min(\zeta) \in \zeta'$ *where* $\zeta' = \mathcal{R}_{\left\{ \frac{-b}{a} \right\}, a, c}\left(1 + \left\lfloor \frac{a(u-1)+c}{b} \right\rfloor, \left\lfloor \frac{av+c}{b} \right\rfloor \right)$;

– and $\max(\zeta) \in \zeta''$ where $\zeta'' = b - a + \mathcal{R}_{\{\frac{-b}{a}\},a,c}\left(1 + \lfloor\frac{au+c}{b}\rfloor, \lfloor\frac{a(v+1)+c}{b}\rfloor\right)$.

Proof. In [13] a similar result has been presented but with $c = 0$. We have therefore simply to prove that the result stands with $c \neq 0$. For the first line, $\lfloor\frac{au+c}{b}\rfloor = \lfloor\frac{av+c}{b}\rfloor$ corresponds to the ordinate of the points of abscissa u and v. If the ordinates are equal, both points are on the same span and the minimum is located at abscissa u and the maximum at abscissa v regardless if the span is complete or not.

Let $m = \min(\zeta)$. Now, let us consider the Bezout coefficient (α, β) of (a, b) such that $a\alpha - b\beta = 1$. It is easy to see that $\mathcal{R}_{a,b,c}(u, v) = \mathcal{R}_{a,b,0}\left(u - \{\frac{c\alpha}{b}\}, v - \{\frac{c\alpha}{b}\}\right)$ [13]. We know already that if $m \in \mathcal{R}_{a,b,0}$ $\left(u - \{\frac{c\alpha}{b}\}, v - \{\frac{c\alpha}{b}\}\right)$ then $m \in \mathcal{R}_{\{\frac{-b}{a}\},a,0}\left(1 + \lfloor\frac{a(u-\{\frac{c\alpha}{b}\}-1)}{b}\rfloor, \lfloor\frac{a(v-\{\frac{c\alpha}{b}\})}{b}\rfloor\right)$ [13]. It is now easy to see that this is the same as $m \in \mathcal{R}_{\{\frac{-b}{a}\},a,c}$ $\left(1 + \lfloor\frac{a(u-1)+c}{b}\rfloor, \lfloor\frac{av+c}{b}\rfloor\right)$. The same goes for the maximum. $\square$

Proposition 1 means that we can build two linear congruence sequences that maintain the minimum and maximum values respectively. Although the maximum value is only conserved indirectly via an accumulator value. The interesting aspect is that we replace a sequence of $(v - u)$ values by a sequence with $\left(\lfloor\frac{av+c}{b}\rfloor - \lfloor\frac{a(u-1)+c}{b}\rfloor - 1\right)$ values. However if the slope is close to 1, the number of points is equal to the number of spans and we do not gain much by replacing one sequence by the other. This is solved by performing the following substitution:

Lemma 1. *Let us consider the remainder subsequence* $\zeta = \mathcal{R}_{a,b,c}(u, v)$, *with* $0 \leq a \leq b$ *and* $\gcd(a, b) = 1$. *Let us suppose that* $2a > b$ *then:*

$$\min(\zeta) \in \zeta' \text{ and } \max(\zeta) \in \zeta' \text{ where } \zeta' = \mathcal{R}_{b-a,b,c}(-v, -u)$$

The proof is similar to the one that can be found in [13]. With Lemma 1, we transform a sequence with a spans into a sequence with $b - a$ spans and a DSS of slope $\frac{a}{b} > \frac{1}{2}$ into a DSS of slope $\frac{b-a}{b} < \frac{1}{2}$. The spans are bigger and the computation time is reduced.

Lemma 2. $\mathcal{R}_{a,b,c}(u, v) = \mathcal{R}_{a,b,\{\frac{c}{b}\}}(u, v)$

This result is obvious since the remainder sequence has a periodicity of b. This lemma can help if c is big compared to b.

Right now we have supposed that for a DSL characteristics (a, b, c), we have $\gcd(a, b) = 1$. This is reasonable since the DSL of characteristics (a, b, c), for $g = \gcd(a, b) > 1$, is the same than the DSL of characteristics $\left(\frac{a}{g}, \frac{b}{g}, \lfloor\frac{c}{g}\rfloor\right)$. However, if we are simply looking at the remainder sequence for $\gcd(a, b) > 1$, the values in the sequence are different although related to the remainder values obtained by dividing the characteristics by the gcd, as the following lemma shows (note that the Algorithm 1 works even if the GCD is not equal to one):

Lemma 3. *Let us consider a DSL of characteristics* (a, b, c) *such that* $\gcd(a, b) = g > 1$, *then:* $\mathcal{R}_{a,b,c}(u, v) = \left\{ \frac{c}{g} \right\} + g\mathcal{R}_{\frac{a}{g}, \frac{b}{g}, c}(u, v)$

Algorithm 1. ComputeMinMax2D (In: a, b, c, u, v. Out: $mini, maxi$) - - a, b, c: characteristics of the DSL; u, v: interval of definition of the DSS; mini, maxi: minimal, maximal remainder).

---

begin
    $minifound \leftarrow False$; $maxifound \leftarrow False$; $cumul \leftarrow 0$;
    (* (u', v') min sequence interval and (u'', v'') max sequence interval *)
    $u' \leftarrow u$; $v' \leftarrow v$; $u'' \leftarrow u$; $v'' \leftarrow v$;
    while *not (minifound and maxifound)* **do**
        if $2a > b$ **then**
            (* Dealing with longer spans reduce computation time *)
            $(a, b, c, u', v', u'', v'') \leftarrow (b - a, b, c, -v', -u', -v'', -u'')$;
        $(a', b') \leftarrow (\{ \frac{-b}{a} \}, a)$;
        $c' = \{ \frac{c}{b'} \}$;
        if *not(minifound)* **then**
            $y_u \leftarrow \lfloor \frac{au' + c}{b} \rfloor$; $y_v \leftarrow \lfloor \frac{av' + c}{b} \rfloor$;
            if $y_u = y_v$ *(* only one span *)* **then**
                $mini \leftarrow \{ \frac{au' + c}{b} \}$;
                $minifound \leftarrow True$; (* We have our minimal remainder *)
            else
                $(u', v') \leftarrow (1 + \lfloor (a'(u' - 1) + c')/b' \rfloor, \lfloor (a'v' + c')/b' \rfloor)$;
        if *not(maxifound)* **then**
            $y_u \leftarrow \lfloor \frac{au'' + c}{b} \rfloor$; $y_v \leftarrow \lfloor \frac{av'' + c}{b} \rfloor$;
            if $y_u = y_v$ *(* only one span *)* **then**
                $maxi \leftarrow cumul + \{ \frac{au'' + c}{b} \}$;
                $maxifound \leftarrow True$; (* We have our maximal remainder *)
            else
                $(u'', v'') \leftarrow (1 + \lfloor (a'u'' + c')/b' \rfloor, \lfloor (a'(v'' + 1) + c')/b' \rfloor)$;
        $(a, b, c) \leftarrow (a', b', c')$;

We now have all we need for a complete 2D algorithm: see Algorithm 1 for the search of the minimum in a 2D sequence. This algorithm is an extension of the one proposed in [13] as it computes the minimum and the maximum at the same time. Note however that we do not check if the value 0 or $b - 1$ belong to the sequence for algorithm ComputeMinMax2D parameters (a, b, c, u, v). In 3D, there is an overall check for the presence of those values in the complete bilinear sequence. If the reader wants to use the algorithm ComputeMinMax2D to solve 2D cases, he may add these checks although it is not necessary. It requires to compute the Bezout coefficients and thus it adds the complexity of this computation to the general case and substitutes it to the complexity of the algorithm (equivalent to the complexity of the Euclidean algorithm).

Example: Figure 1 shows an example of simple linear congruence collapse. The DSS is defined by $0 \leq 31x - 151y + 71 < 151$ with $0 \leq x \leq 13$. The first step transforms the DSS characteristics in $\left(\left\{\frac{-151}{31}\right\}, 31, 71\right) = (22, 51, 20)$. The translation constant is $\left\{\frac{71}{51}\right\} = 20$ (Lemma 2). We have now two sequences: the one that contains the minimum values and the one with the maximum values. The minimum sequence is defined on the interval $\left[1 + \left\lfloor \frac{31(0-1)+20}{151} \right\rfloor, \left\lfloor \frac{31 \cdot 13+20}{151} \right\rfloor\right] = [1, 5]$. The formula $1 + \left\lfloor \frac{a(u-1)+c}{b} \right\rfloor$ ensures that the span considered is the first complete span. The first value in the minimum sequence will be 42 and not 71. The maximum sequence is defined on the interval $\left[1 + \left\lfloor \frac{a \cdot 0+20}{151} \right\rfloor, \left\lfloor \frac{31(13+1)+20}{151} \right\rfloor\right] = [1, 5]$ with accumulation value $131 - 51 = 80$. Note that the interval for the minimum sequence and the maximum sequence are not necessarily identical as can be seen in the last step. The second step is similar to the first but applied on the minimum and maximum sequence: the DSS $0 \leq 22x - 51y + 20 < 51$ is collapsed into the DSS $0 \leq 15x - 22y + 20 < 22$ with both intervals $2 \leq x \leq 3$, and accumulation value $80 + 51 - 22 = 109$. The third step corresponds to an inversion on the sequence: since $15 \cdot 2 > 22$, the DSS is transformed into $7x - 22y + 20 < 7$, with intervals $-3 \leq x \leq -2$. The accumulation value does not change. As can be seen in the figure, the values are now, for both minimum and maximum sequence, on a same span, and the minimum is given by the first value in the span piece while the maximum value is given by the last value in the span piece.

3.3   Efficient Search for a Given Value in a Bilinear Congruence Sequence

We have now almost all we need for a first 3D algorithm. There is however a last problem that we are going to address. Let us consider a bilinear congruence sequence $\zeta = \mathcal{R}_{a,b,c,d}(x_0, x_1, y_0, y_1)$. We know that the minimum value in ζ cannot be smaller than 0 and the maximum not greater than $c - 1$. So, by providing an efficient method that determines if a given value ϵ belongs to the sequence (in our case $\epsilon = 0$ or $\epsilon = c - 1$), we will not have to search further for a minimum or a maximum. Of course, one can check row by row or column by column but one can actually do better than that using the following theorems (see Fig. 2 for an example):

Theorem 1. *Let us consider the bilinear congruence sequence $\zeta = \mathcal{R}_{a,b,c,d}(x_0, x_1, y_0, y_1)$ and a value ϵ, with $0 \leq \epsilon < c$. Let us suppose that $\gcd(a, c) = 1$ and (α, β) their Bezout coefficients verifying $a\alpha - c\beta = 1$.*

Let us define the sequence $x_\epsilon(y), y \in [y_0, y_1]$ of the smallest abscissa greater or equal to x_0 with remainder $\mathcal{R}_{a,b,c,d}(x,y) = \epsilon$. Then:

$$x_\epsilon \text{ is given by the sequence } x_0 + \mathcal{R}_{-b\alpha, c, \alpha(\epsilon - d - ax_0)}(y_0, y_1).$$

Proof. Let us consider a DSP defined by $0 \leq ax + by - cz + d < c$ with $(x,y) \in [x_0, x_1] \times [y_0, y_1]$. Let us suppose that $\gcd(a,c) = 1$ and (α, β) their Bezout coefficients verifying $a\alpha - c\beta = 1$. First, let us note that $0 \leq ax + by - cz + d < c$ with $(x,y) \in [x_0, x_1] \times [y_0, y_1]$ is equivalent to $0 \leq ax' + by - cz + d + ax_0 < c$ with $(x', y) \in [0, x_1 - x_0] \times [y_0, y_1]$ and $x' = x - x_0$. For a given ordinate, we are searching for the abscissa x' greater or equal to 0 with a remainder equal to ϵ. For a given ordinate y, the abscissa $x'(y)$ with remainder $\mathcal{R}_{a,b,c,d}(x', y) = ax' + by - cz + d + ax_0 = \epsilon$ verifies $ax' - cz = \epsilon - d - by - ax_0$. With the Bezout coefficients (α, β), we have $(\alpha(\epsilon - d - by - ax_0) + kc)a - (\beta(\epsilon - d - by - ax_0) + ka)c = \epsilon - d - by - ax_0, k \in \mathbb{Z}$. This means that $x' \in \{(\epsilon - d - by - ax_0)\alpha + kc : k \in \mathbb{Z}\}$. The smallest abscissa $x = x' + x_0$, greater or equal to x_0 with remainder equal to ϵ is then given by $x_0 + \left\{\frac{-b\alpha y + \alpha(\epsilon - d - ax_0)}{c}\right\}$. □

12	17	3	8	13	18	4	9	14	0	5	10	15	1	6	11	16	2	7	12	17
0	5	10	15	1	6	11	16	2	7	12	17	3	8	13	18	4	9	14	0	5
7	12	17	3	8	13	18	4	9	14	0	5	10	15	1	6	11	16	2	7	12
14	0	5	10	15	1	6	11	16	2	7	12	17	3	8	13	18	4	9	14	0
2	7	12	17	3	8	13	18	4	9	14	0	5	10	15	1	6	11	16	2	7
9	14	0	5	10	15	1	6	11	16	2	7	12	17	3	8	13	18	4	9	14

Fig. 2. Bilinear Congruence Sequence $\left\{\frac{4x+12y+4}{17}\right\}$ on $[3,15] \times [0,5]$. The blue rectangle shows the DSP subsequence. In Pink, the values $\epsilon = 0$ with abscissa greater than $x_0 = 3$. In Dark Blue, the values 0 with abscissa smaller than $x_0 = 3$ (Colour figure online).

In Theorem 1, we have supposed that $\gcd(a,c) = 1$ which is not necessarily the case. Let us now examine what happens when $gcd(a,c) = g > 1$.

Theorem 2. *Let us consider the bilinear congruence sequence $\zeta = \mathcal{R}_{a,b,c,d}(x_0, x_1, y_0, y_1)$ and a value ϵ, with $0 \leq \epsilon < c$. Let us suppose that $\gcd(a,c) = g > 1$. Let us suppose that (α, β) are the Bezout coefficients for (b, g) such that $b\alpha - g\beta = 1$. Then, the bilinear congruence sequence ζ contains the value ϵ iff the sequence $\mathcal{R}_{\frac{a}{g}, \left\{\frac{b}{g}\right\}, \frac{c}{g}, \frac{d + b \cdot y_i - e}{g}}\left(x_0, x_1, 0, \left\lfloor \frac{y_1 - y_i}{g} \right\rfloor\right)$ with $y_i = y_0 + \left\{\frac{\alpha(e - d - by_0)}{g}\right\}$, contains the value 0.*

Proof. Let us consider the bilinear congruence sequence $\zeta = \mathcal{R}_{a,b,c,d}(x_0, x_1, y_0, y_1)$ and a value ϵ, with $0 \leq \epsilon < c$. Let us suppose that $\gcd(a,c) = g > 1$. Let us suppose that (α, β) are the Bezout coefficients for (b, g) such that $b\alpha - g\beta = 1$. Of course, here, $\gcd(b, g) = 1$ or otherwise we would not have $\gcd(a, b, c) = 1$. It is easy to see that $\mathcal{R}_{a,b,c,d}(x_0, x_1, y_0, y_1)$ for $y \in [0, y1 - y0]$ is the same as $\mathcal{R}_{a,b,c,d+by_0}(x_0, x_1, 0, y_1 - y_0)$. We know that $ax + by' - cz + d + by_0 = \epsilon$, with $y' = y - y_0$, is only possible if $\left\{ \frac{by' + d + by_0 - \epsilon}{g} \right\} = 0$. The smallest value $y' \geq 0$ verifying this is given by $y' = \left\{ \frac{\alpha(\epsilon - d - by_0)}{g} \right\}$. Let us denote $y_i = y_0 + \left\{ \frac{\alpha(\epsilon - d - by_0)}{g} \right\}$. The ordinate y_i is the first ordinate between y_0 and y_1 for which the sequence ζ may contain ϵ. The other ordinates where we may find ϵ are then all the $y_i + kg$ for $k \in \left[0, \left\lfloor \frac{y1 - y_i}{g} \right\rfloor \right]$. Now we need to replace y by gy'' in order to have steps of 1 on the ordinates. We also need to start with the ordinate 0 as y_i is not necessarily divisible by g. It is easy to see that $\zeta = \mathcal{R}_{a,b,c,d+by_i}(x_0, x_1, 0, y_1 - y_i)$. Since the value ϵ can only be found on the lines with ordinate $y_i + kg$, it is easy to see that ϵ can be found in ζ iff it can be found in $\mathcal{R}_{a,bg,c,d+by_i}\left(x_0, x_1, 0, \left\lfloor \frac{y1 - y_i}{g} \right\rfloor \right)$. If we denote $(a', c') = (a/g, c/g)$ then we have $a'gx + bgy - c'gz + d + by_i = \epsilon$ if $a'gx + bgy - c'gz + d + by_i - \epsilon = 0$ or $a'x + by - c'z + \frac{d + by_i - \epsilon}{g} = 0$ (note that $d + by_i - \epsilon$ is divisible by g). Here, if b is bigger than c, then it is easy to see that it can be replaced by $\left\{ \frac{b}{c} \right\}$. □

Theorem 3. *Deciding if a value belongs to a bilinear congruence sequence can be decided in logarithmic time.*

The proof is obvious. With Theorems 1 and 2, we exhibit linear congruence sequences. We can search for its minimum with Algorithm 1. If the minimum is smaller or equal to $x_1 - x_0$ then the sequence ζ contains the value ϵ. This search for the values 0 or $c - 1$ can thus be done in logarithmic time.

3.4 First Algorithm for the Minimum and Maximum Search in a Bilinear Congruence Sequence

Let us consider a bilinear congruence sequence $\zeta = \mathcal{R}_{a,b,c,d}(x_0, x_1, y_0, y_1)$. The previous section let us check if the values 0 or $c - 1$ belong to ζ. If both values are in ζ then the search is over. Otherwise, let us consider the smallest of the intervals $n = x_1 - x_0$ and $m = y_1 - y_0$. W.l.o.g., let us consider that we have $m < n$. The first 3D algorithm consists simply in applying algorithm ComputeMinMax2D$(a, c, d + ay)$, for $y \in [y_0, y_1]$. We keep the minimum and maximum over all these 2D sequences.

Proposition 2. *Let us consider a bilinear congruence sequence $\zeta = \mathcal{R}_{a,b,c,d}$ (x_0, x_1, y_0, y_1) with $n = x_1 - x_0 < m = y_1 - y_0$. The complexity of the search for the minimum and maximum value in ζ is bounded by $O(n \cdot \log(\min(a, c - a)))$.*

The proposition is a direct consequence of the complexity of the 2D algorithm [13]. Figure 3 shows an example. As one can see, the collapse line by line does not produce a rectangle on (x, y). Also, one can see that the final minimum or maximum values do not necessarily form a connected final set. The problem comes from the sequences of the values (u, v) over the different lines.

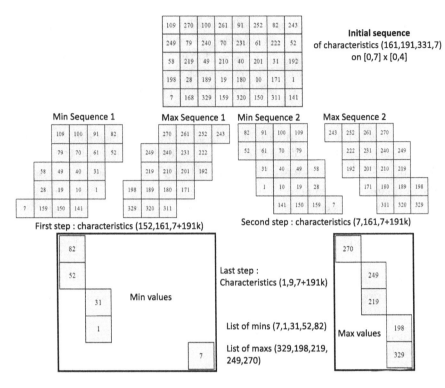

Fig. 3. Bilinear Congruence Sequence $\left\{ \frac{161x+191y+7}{331} \right\}$ on $[0,7] \times [0,4]$.

4   Discussion, Conclusion and Perspectives

In this paper we were interested in the Digital SubPlane (DSP) Recognition Problem. We tried to extend our remainder approach for the recognition of straight line segments to the recognition of subplanes. The extension is not immediate. In particular, the remainder order property that is verified in 2D is not always verified in 3D. As a consequence, a point may be a Leaning Point for the Digital Plane but not for the Digital SubPlane and vice-versa. There seems however to be classes of subplanes for which the remainder order property are conserved. The characterization of this subclass is an open question. It could represent an interesting candidate as representative of the equivalent class of Digital Planes containing a SubPlane. From this starting point, we proposed an extension of the search of a minimum and maximum of a linear congruence sequence to the third dimension. We showed in particular that one can determine if a given value belongs to a bilinear congruence sequence in logarithmic time. The minimum and maximum value in a bilinear congruence sequence $\left\{ \frac{ax+by+d}{c} \right\}$, $(x,y) \in [x_0, x_1] \times [y_0, y_1]$ can be found, if $m = x_1 - x_0 < n = y_1 - y_0$, in $O(m \log(\min(a, c-a)))$ or $O(n \log(\min(b, c-b)))$ otherwise.

This paper is only a very first step in the investigation of the SubPlane Recognition Problem. As already discussed, we would like to prove that there are always subplanes that verify the remainder order property, namely, for a Digital Plane of characteristics (a, b, c, d) and a SubPlane of characteristics $(\alpha, \beta, \gamma, \delta)$ defined on $[x_0, x_1] \times [y_0, y_1]$ $\forall (x, y)$ and $(x', y') \in [x_0, x_1] \times [y_0, y_1],: \mathcal{R}_{a,b,c,d}(x, y) < \mathcal{R}_{a,b,c}(x', y') \Rightarrow \mathcal{R}_{\alpha,\beta,\gamma,\delta}(x, y) \leq \mathcal{R}_{\alpha,\beta,\gamma,\delta}(x', y')$. The question comes actually down to a construction problem. Starting from the Digital Plane, it is possible to erode it in such a way that we obtain a sequence of SubPlanes that verify the remainder order property and vice-versa?

The algorithm for the search of a minimum and a maximum in a bilinear congruence sequence searches for the minimum and maximum line by line (or column by column). This is possible because a line of a bilinear congruence sequence is simply a linear congruence sequence. One could try to improve this by alternatively searching for a minimum in a line or a column. This corresponds to a form of 3D digital plane collapse (see [18] for other forms of plane collapses). See Fig. 4 when this is performed on an infinite Digital Plane and Fig. 5 where it is performed on a Digital SubPlane. Again, one can see that the collapse on a DSP does not preserve a simple shape. One would be able to achieve a logarithmic search for the minimum and maximum if one is able to characterize the shape of the collapsed DSP (See Fig. 5).

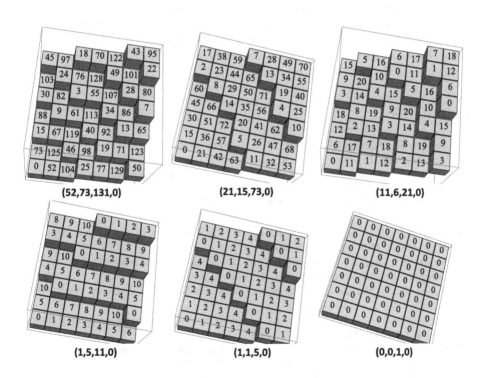

(52,73,131,0)   (21,15,73,0)   (11,6,21,0)

(1,5,11,0)   (1,1,5,0)   (0,0,1,0)

Fig. 4. Plane collapses.

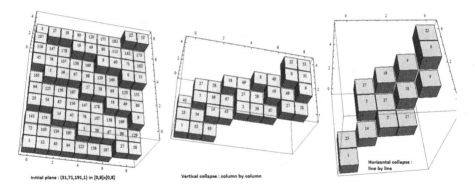

Fig. 5. DSP of characteristics $(31, 71, 191, 1)$ on $[0, 8] \times [0, 8]$. Two successive collapses are shown, a first vertical and then an horizontal one.

There are however possibilities of improvement in terms of complexity. Firstly, one can see that the shape after one collapse (see Figs. 3 and 5) is not defined on a rectangle anymore but the sides that are not parallel to an axis form actually a Digital Straight Line (when projected on 2D). Another possible improvement could be done by repetitively searching for specific values in the bilinear congruence sequence: looking for values 0, 1, .. when a value is found in the sequence it corresponds then to the minimum. The complexity is then c times a log. This works best when the size of the DSP is important compared to the characteristics values. A finer study needs to be conducted to check when doing one is more efficient than the other, or mixing both.

References

1. Andres, E., Acharya, R., Sibata, C.: Discrete analytical hyperplanes. Graphical Models Image Process. **59**(5), 302–309 (1997)
2. Berthé, V., Labbé, S.: An Arithmetic and combinatorial approach to three-dimensional discrete lines. In: Debled-Rennesson, I., Domenjoud, E., Kerautret, B., Even, P. (eds.) DGCI 2011. LNCS, vol. 6607, pp. 47–58. Springer, Heidelberg (2011)
3. Brimkov, V.E., Coeurjolly, D., Klette, R.: Digital planarity - a review. Discrete Appl. Math. **155**(4), 468–495 (2007)
4. Brons, R.: Linguistic methods for the description of a straight line on a grid. Comput. Graphics Image Process. **3**(1), 48–62 (1974)
5. Coven, E.M., Hedlund, G.: Sequences with minimal block growth. Math. Syst. Theory **7**(2), 138–153 (1973)
6. Debled-Rennesson, I., Reveilles, J.P.: A linear algorithm for segmentation of digital curves. IJPRAI **09**(04), 635–662 (1995)
7. Dexet, M., Andres, E.: A generalized preimage for the digital analytical hyperplane recognition. Discrete Appl. Math. **157**(3), 476–489 (2009)
8. Feschet, F., Tougne, L.: On the min DSS problem of closed discrete curves. Discrete Appl. Math. **151**(1–3), 138–153 (2005)

9. Gérard, Y., Debled-Rennesson, I., Zimmermann, P.: An elementary digital plane recognition algorithm. Discrete Appl. Math. **151**(1–3), 169–183 (2005)
10. Klette, R., Rosenfeld, A.: Digital straightness - a review. Discrete Appl. Math. **139**(1–3), 197–230 (2004)
11. Lachaud, J.O., Said, M.: Two efficient algorithms for computing the characteristics of a subsegment of a digital straight line. Discrete Appl. Math. **161**(15), 2293–2315 (2013)
12. Largeteau-Skapin, G., Debled-Rennesson, I.: Outils arithmétiques pour la géométrie discrète. In: Géométrie discrète et images numériques, pp. 59–74. Traité IC2 - Traitement du signal et de l'image, Hermès - Lavoisier (2007)
13. Ouattara, J.D., Andres, E., Largeteau-Skapin, G., Zrour, R., Tapsoba, T.M.: Remainder approach for the computation of digital straight line subsegment characteristics. Discrete Appl. Math. **183**, 90–101 (2015). http://dx.doi.org/10.1016/j.dam.2014.06.006
14. Reveillès, J.P.: Calcul en Nombres Entiers et Algorithmique. Ph.D. thesis, Université Louis Pasteur, Strasbourg, France (1991)
15. Said, M., Lachaud, J.-O., Feschet, F.: Multiscale discrete geometry. In: Brlek, S., Reutenauer, C., Provençal, X. (eds.) DGCI 2009. LNCS, vol. 5810, pp. 118–131. Springer, Heidelberg (2009)
16. Sivignon, I.: Walking in the Farey fan to compute the characteristics of a discrete straight line subsegment. In: Gonzalez-Diaz, R., Jimenez, M.-J., Medrano, B. (eds.) DGCI 2013. LNCS, vol. 7749, pp. 23–34. Springer, Heidelberg (2013)
17. Sivignon, I., Coeurjolly, D.: Minimal decomposition of a digital surface into digital plane segments Is NP-hard. In: Kuba, A., Nyúl, L.G., Palágyi, K. (eds.) DGCI 2006. LNCS, vol. 4245, pp. 674–685. Springer, Heidelberg (2006)
18. Fernique, T.: Generation and recognition of digital planes using multi-dimensional continued fractions. Pattern Recogn. **10**(42), 2229–2238 (2009)
19. Vacavant, A., Roussillon, T., Kerautret, B., Lachaud, J.O.: A combined multi-scale/irregular algorithm for the vectorization of noisy digital contours. Comput. Vis. Image Underst. **117**(4), 438–450 (2013)

Construction of Sandwich Cover
of Digital Objects

Apurba Sarkar[1]([⊠]) and Mousumi Dutt[2]

[1] Department of Computer Science and Technology,
Indian Institute of Engineering Science and Technology, Shibpur, India
as.besu@gmail.com
[2] Department of Computer Science and Engineering,
International Institute of Information Technology, Naya-Raipur, India
duttmousumi@gmail.com

Abstract. An algorithm to construct a minimum vertex cover of a digital object from its inner and outer isothetic covers such that it lies within the annular region bounded by its outer and inner isothetic covers is presented here which has $O(\frac{n}{g}\log(n/g))$ time complexity, where n being the number of pixels on the contour of the digital object and g is the grid size. After constructing inner and outer covers [2,3], a combinatorial technique is used to construct a sandwich cover. Sandwich cover reduces the storage complexity of the given digital object as it contains less number of vertices compared to inner or outer isothetic cover while preserving the shape of the object. Sandwich cover can be used as shape descriptor by generating several metrics on it.

Keywords: Isothetic covers · Sandwich cover · Shape analysis · Shape descriptor · Combinatorial technique

1 Introduction

Finding tightest possible covers both inner and outer of a digital object has diverse applications in many areas of computer science such as shape analysis, document image analysis, computer vision, VLSI layout design, robot motion planning, grasping object by robot [5,7], inner and outer approximation of polytopes [1], computing minimum area safety zone [8], and rough sets. The covers can also be used to extract shape based features for shape analysis. Shape based features can then be used to design OCR systems [9]. Shape analysis on the other hand can be used for road-sign detection [6]. These covers are of importance in the sense that they preserve the shape of the digital object and they are less complex than the actual object. Few works on finding optimal covers in different grid have been found in the literature. Sloboda et al. presented a work that deals with boundary approximation of objects in [11]. There is a work by Sklansky [10] that deals with minimum-perimeter polygons of digitised silhouettes. An algorithm to find the inner and outer isothetic cover has been proposed

© Springer International Publishing Switzerland 2015
R.P. Barneva et al. (Eds.): IWCIA 2015, LNCS 9448, pp. 172–184, 2015.
DOI: 10.1007/978-3-319-26145-4_13

in [2,3] which utilises combinatorial properties and relative arrangements of the digital object and the underlying grid lines. The algorithm proposed there could also control the relative error of the polygonal cover by varying the grid size. B. Das et al. [4] proposed a combinatorial algorithm to find the optimal covers of digital object in triangular grid. Although, these polygonal covers give an approximation about the shape of the digital object, for some applications a mere rough approximation about the shape of the object may be enough.

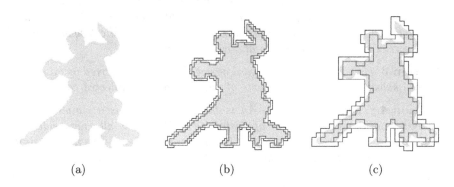

(a)   (b)   (c)

Fig. 1. (a) A digital object, its inner (blue), outer (black), and sandwich (red) covers are shown for $g = 4$ (b) and $g = 8$ (c) (Color figure online)

The algorithm presented in this paper computes a rough cover (henceforth called sandwich cover) that lies within the annular region bounded by the inner and outer isothetic cover of the given digital object[1] and the resulting cover has minimum number of vertices. The algorithm can also be applied to find the minimum vertex isothetic polygon, given one isothetic polygon inside another isothetic polygon. Figure 1 shows an object with its inner (blue), outer (black), and sandwich (red) covers for grid size $g = 4$ and $g = 8$.

The rest of the paper is organized as follows. All the required definitions and preliminaries are presented in Sect. 2. The construction of outer isothetic cover (OIC) and inner isothetic cover (IIC) are explained in brief in Sects. 2.1 and 2.2 respectively. The method for deriving a sandwich cover is presented in Sect. 3. Estimation of running time of the proposed algorithm is explained in Sect. 3.3. Section 4 presents the experimental results with analysis and the conclusion is presented in Sect. 5.

2   Definitions and Preliminaries

This section contains some definitions required to understand the paper. In Sects. 2.1 and 2.2, the construction of outer and inner isothetic covers are discussed in brief which are given in details in [2,3].

---

[1] It is to be noted that the construction of these two covers are given in [2,3].

Definition 1 *(k-connectedness). Two points p and q are said to be k-connected ($k = 4$ or 8) in a set S if and only if there exists a sequence $p = \langle p_0, p_1, \ldots, p_n = q \rangle \subseteq S$ such that $p_i \in N_k(p_{i-1})$ for $1 \leqslant i \leqslant n$. The 4-neighborhood of a point (x, y) is given by $N_4(x, y) = \{(x', y') : |x - x'| + |y - y'| = 1\}$ and its 8-neighborhood by $N_8(x, y) = \{(x', y') : \max(|x - x'|, |y - y'|) = 1\}$.*

Definition 2 *(Digital grid). A digital grid is given by $\mathcal{G} := (\mathcal{H}, \mathcal{V})$, where $\mathcal{H}$ and $\mathcal{V}$ represent the respective sets of (equi-spaced) horizontal grid lines and vertical grid lines. A grid size, g is defined as the distance between two consecutive horizontal/vertical grid line. A grid point is the point of intersection of a horizontal and a vertical grid line. A unit grid block (UGB) is the smallest square having its four vertices as four grid points and edges as grid edges.*

Definition 3 *(Digital object). A digital object (henceforth referred as an object) is a finite subset of $\mathbb{Z}^2$, which consists of one or more k-connected components.*

Definition 4 *(Isothetic polygon). An isothetic polygon P is a simple polygon (i.e., with nonintersecting sides) of finite size in $\mathbb{Z}^2$ whose alternate sides are subsets of the members of $\mathcal{H}$ and $\mathcal{V}$. The polygon P, hence given by a finite set of UGBs, is represented by the (ordered) sequence of its vertices, which are grid points. The border B_P of P is the set of points belonging to its sides. The interior of P is the set of points in the union of its constituting UGBs excluding the border of P.*

An isothetic cover has two type of vertices 90° (type **1**) and 270° (type **3**).

Definition 5 *(Outer isothetic cover). The outer (isothetic) cover (OIC), denoted by $\overline{P}(S)$, is a set of outer polygons and (outer) hole polygons, such that the region, given by the union of the outer polygons minus the union of the interiors of the hole polygons, contains a UGB if and only if it has object occupancy (i.e., has a non-empty intersection with S).*

Definition 6 *(Inner isothetic cover). The inner (isothetic) cover (IIC), denoted by $\underline{P}(S)$, is a set of inner polygons and (inner) hole polygons, such that the region, given by the union of the inner polygons minus the union of the interiors of the hole polygons, contains a UGB if and only if it is a subset of S.*

Definition 7 *(Sandwich cover). The sandwich (isothetic) cover (SC) is defined to be the minimum vertex simple isothetic polygon that lies completely within the annular region bounded by inner isothetic cover (IIC) and outer isothetic cover (OIC).*

Definition 8 *(Convexity and Concavity). An edge of P defined by two consecutive vertices of type **1** is termed as a convex edge, as it gives rise to a convexity. Similarly, an edge defined by two consecutive type **3** vertices gives rise to a concavity, and hence termed as a concave edge.*

The line segment, along a concavity, contained within the annular region defined by the OIC and the IIC, is called the *concavity line segment* (shown in red in Fig. 4(b)). Similarly, the portion of the line, along a convexity, within the annular region, is called the *convexity line segment* (shown in violet in Fig. 4(a)).

2.1   Deriving the Outer Isothetic Cover (OIC)

The outer isothetic cover is the minimum area isothetic polygon A_{in} that covers a digital object A registered with the grid, $\mathcal{G}$. The algorithm presented in [2,3] constructs an isothetic cover by computing ordered list of vertices of A_{in} using a combinatorial classification (type) of the grid points lying on/inside/outside the object boundary. The class or type of a grid point p in $\mathcal{G}$ is determined by checking object occupancy of four neighboring cells of size $g \times g$ incident at p. If i ($i \in [0\ldots4]$) is the number of UGBs fully/partially occupied by the object incident at p then type of p is determined to be C_i as shown in Fig. 2.

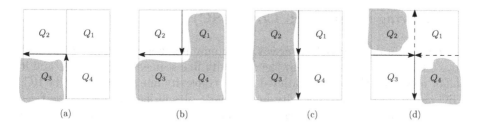

Fig. 2. Different vertex types for outer isothetic cover

The significance of a class is as follows. (i) C_0: p is not a vertex, since none of Q_i's has object containment; (ii) C_1: Q_i is a 90° vertex of A_{in} (Fig. 2(a)); (iii) C_2: (a) if two adjacent UGB has object containment, then p is an edge point (Fig. 2(c)); (b) if diagonally opposite cells contain object, then p is a 270° vertex of A_{in} (Fig. 2(d)); (iv) C_3: p is classified as a 270° vertex (Fig. 2(b)); (v) C_4: p is not a vertex of A_{in} and lies inside A_{in}.

To obtain the cover an anticlockwise traversal is made around the boundary of object and the start point of this traversal is determined to be the first 90° vertex found in the row wise scan of the grid points. Let v_i be the current vertex and v_{i-1} be the previous vertex in A_{in}, also let d_{i-1} be the direction of v_{i-1} and t_i be the type of v_i, then the next direction of traversal from v_i is obtained by the formula $d_i = (d_{i-1} + t_i) \bmod 4$. Where $d_i \in [0,3]$ indicating the direction along right, top, left and downward respectively. The next grid point of traversal will be the point along d_i. This procedure continues until it reaches the start point thereby finding the outer isothetic cover.

2.2   Deriving the Inner Isothetic Cover (IIC)

The inner isothetic cover, A'_{in}, is the maximum-area orthogonal polygon that inscribes the digital object A, registered with the background grid $\mathcal{G}$. Derivation of inner isothetic cover follows the same procedure as the outer isothetic cover, except the consideration of the grid point, q. The classification of the grid point q in this case is based on the full occupancy of UGBs incident at q as shown in Fig. 3. The algorithm in [2,3] constructs the inner isothetic cover using these classification of vertices.

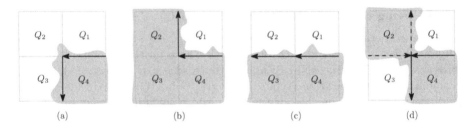

Q_2	Q_1	Q_2	Q_1
Q_3	Q_4	Q_3	Q_4

(a)   (b)   (c)   (d)

Fig. 3. Different vertex types of inner isothetic cover

3   Deriving a Sandwich Cover (SC)

In order to obtain sandwich cover with minimum number of vertices, SC cannot move into the region bounded by the concavity line segment and the OIC and it cannot move into the region bordered by the convexity line segment and the IIC. If SC moves beyond the concavity line segment, it will introduce extra vertices because SC has to traverse back to the line of concavity as it must be inside the annular region bounded by the OIC and the IIC. Same is true if SC moves inside the region bounded by a convexity line segment and IIC, new vertices will be introduced. The objective of the algorithm is to minimize the number of vertices, which can be achieved when there are less number of changes in direction during the traversal. Thus the traversal is made along concavity or convexity line segment.

The algorithm maintains four lists L_{out}, L_{in}, L_x, and L_y which are explained below. L_{out} contains list of all vertices (Types 1 and 3) of the outer cover in anti-clockwise order. This list is constructed during the construction of outer cover. L_{in} contains list of all the vertices (Types 1 and 3) of the inner cover in anti-clockwise order. L_{in} is constructed during the construction of inner cover. The lists L_x and L_y are global and contain vertices (Type 1 and 3) and edge points (Type 2) from both inner and outer cover. List L_x is lexicographically sorted in a non-decreasing order w.r.t. x (primary key) and y (secondary key), and L_y is lexicographically sorted in a non-decreasing order w.r.t. y (primary key) and x (secondary key). Lists L_x and L_y are used to determine concavity and convexity lines. For all the vertices present in these lists, following information are stored; coordinates of the point, type (1 or 2 or 3), outgoing direction, information about whether it belongs to inner cover or outer cover, information about whether it lies on a concavity (convexity) line segment (boolean), direction of the concavity (convexity) (if present).

3.1   Determining Concavity and Convexity Line Segments

It should be mentioned here that the algorithm requires the concavity line segment due to the outer cover and the convexity line segment due to the inner cover. To determine concavity line segment, the list L_{out} is traversed. During

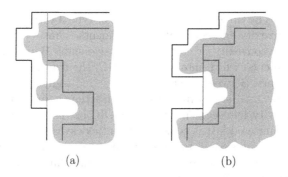

Fig. 4. (a) Convexity line segment (in red) and (b) Concavity line segment (in violet) (Color figure online)

the traversal if two or more Type 3 vertices are encountered it implies the presence of a concavity. To determine the concavity line two consecutive Type 3 vertices are considered at a time. The concave edge incident at two type 3 vertices are extended and the immediate next vertices along the direction of edge are found out and they are the end points of the concavity line formed by the concave edge. The end points are determined by consulting L_x or L_y and their appropriate fields (presence on concavity line segment and its direction) are set. For example, if the concave edge is vertical then L_y is consulted and if the concave edge is horizontal then L_x is consulted. The same procedure is followed to determine the convexity line segment except that traversal is made along inner cover and two or more consecutive Type 1 vertices are considered. It is to be noted that direction of concavity (convexity) line is set to be the direction associated with the first vertex of a concave (convex) edge. The concavity and convexity lines are shown in Fig. 4.

3.2   Obtaining Sandwich Cover

From top-left vertex of inner isothetic cover, an anticlockwise traversal is made to obtain sandwich cover. It is to be noted that the top-left point of IIC always lies on the convexity line segment. Let v_0 be a point on the sandwich cover. When the traversal reaches v_0, it checks three immediate next vertices along its front (v_f), left (v_l), and right (v_r) direction, which may lie either on IIC or OIC. The three vertices mentioned above are found out by searching in L_x or L_y depending on the direction of v_f, v_l, v_r w.r.t. v_0. The next direction of traversal is determined based on the following three simple rules.

– **R1:** If v_0 is on the concavity or convexity line segment, the traversal proceeds along that line in anticlockwise manner. The corresponding segment ends either on IIC or OIC. The traversal continues from that end point in anticlockwise manner. For example, as shown in Fig. 5(a), from v_0 it proceeds along the convexity line segment since v_0 is on a convexity line defined by the convex edge v_0v_1.

- **R2:** Let v_0, v_1, v_2 be three consecutive type **1** or type **3** vertices, producing two consecutive convexity and concavity line segments. Thus, v_1 belongs to either two consecutive convexity or concavity line segments. Between the two convexity line segments, which intersects the OIC, the traversal starts from that intersection points. Between the two concavity line segments, which intersects the IIC, the traversal starts from that intersection points. If both the convexity or concavity line segments intersect OIC or IIC respectively, then the point that encountered last is considered (anticlockwise direction). As shown in Fig. 5(b), as the traversal proceeds from v_0 along the convexity line and reaches v_1, it finds that there is another convexity line that passes through v_1. The new convexity line hits the inner cover and the old one hits the outer cover, so the traversal proceeds along the old one (one that hits outer cover).
- **R3:** If the point lies on either inner or outer cover, it follows the direction of the cover on which it lies and proceeds in such a manner that least number of changes in direction occur.

  As shown in Fig. 5(c), when the traversal reaches outer cover at v_1, it follows the direction of outer cover to reach v_2 and v_2 being on the inner cover, the traversal proceeds along the direction of inner cover to reach v_3. The traversal continues this way until it eventually finds a concavity/convexity line segment.

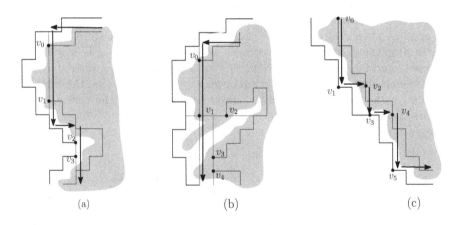

(a)   (b)   (c)

Fig. 5. Illustration of Rules (a) **R1**, (b) **R2**, and (c) **R3**

The traversal continues according to the rules till it reaches the start point where it concludes producing the sandwich cover. An outline of the algorithm in pseudo-code is given below.

Algorithm Outline
1. construct OIC (Sect. 2.1)
2. construct IIC (Sect. 2.2)
3. determine concavity and convexity line segments (Sect. 3.1)
4. v_s = start point (top left point of IIC)
5. $v_c = v_s$;
6. **do**
7.   check three immediate next vertices of v_c
8.   apply the appropriate rule (**R1**, **R2**, or **R3**)
9. v_d = next vertex in the direction according to rule
10.   update v_c i.e., $v_c = v_d$
11. **while**$(v_c \neq v_s)$;

In steps 1 and 2, the OIC and the IIC are constructed, thereof the corresponding lists L_{out}, L_{in}, L_x, and L_y are obtained. Using these lists, the concavity and convexity line segments are determined in Step 3. The start vertex, v_s, is assigned the top left point of IIC in Step 4 and the current vertex, v_c, is updated (Step 5). Step 7 determines which rule to be applied and the rule is applied in Step 8 and the next vertex v_d is determined in Step 9. The loop (Steps 6 through 11) continues until the traversal reaches v_s.

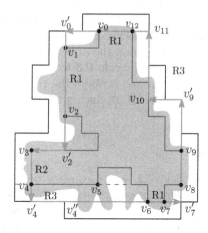

Fig. 6. Demonstration of the algorithm (Color figure online)

Figure 6 gives a brief demonstration of the algorithm. Outer isothetic cover (OIC), inner isothetic cover (IIC), and sandwich cover (SC) are shown in black, blue, and red respectively. Sandwich cover is obtained by starting an anticlockwise traversal at v_0 (top left point of IIC) and making progress from there on following the rules. Since vertex v_0 is on a convexity line segment defined by the convex edge $v_{12}v_0$, it proceeds along that line obeying the rules **R1**. At v_0' the traversal finds that it is on a convexity line segment defined by the convex edge v_1v_2. So, the traversal follows the direction of convexity line to reach v_2'

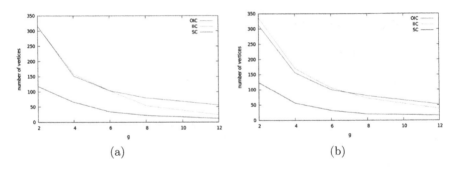

Fig. 7. Plots of grid size versus number of vertices on OIC, IIC and SC for the object (a) Plane and (b) Horse

as mentioned in rule **R1**. v_2' being on the inner cover it proceeds following the direction of v_2 to reach v_3. Point v_3 is on the convexity line defined by v_3v_4. So, it follows that line and reaches v_4 to meet another convexity line defined by v_4v_5. Since convexity line defined by v_4v_5 hits inner cover and the convexity line defined by v_3v_4 hits outer cover, the traversal proceeds along line defined by v_3v_4 obeying rule **R2** to reach v_4'. The point v_4' being on the outer cover proceeds along the direction associated with v_4' (rule **R3**) to reach v_4''. The point v_4'' lies on the convexity line of v_6v_7. So rule **R1** takes the traversal to v_7' and it being on the convexity line v_8v_9 rule **R1** again takes the traversal to v_9'. Now v_9' is on the outer cover so the traversal follows the direction of v_9' to reach v_{10} obeying **R3**. From v_{10} the traversal obeys the rule **R3** to reach v_{11}. The point v_{11} is on the convexity line defined by $v_{12}v_0$ so it obeys the rule **R1** and meets the start point v_0 and the algorithm concludes producing the sandwich cover as shown in Fig. 6 in red.

3.3   Time Complexity

The algorithm to derive sandwich cover consists of four steps. To estimate the running time of the algorithm, let n be the number pixels that constitute the boundary of the digital object and also let g be the grid size. $O(n/g)$ be the number of grid points on the contour of the digital object. This needs $O(g)O(n/g) = O(n)$ time, since the intersection of each grid edge with the digital object is checked in $O(g)$ time, and the number of grid points visited is bounded by $O(n/g)$. In Step 1 outer isothetic cover is constructed and it requires $O(n/g)$ time [2,3]. In Step 2 inner isothetic cover is constructed and its construction also takes $O(n/g)$ time. In Step 3, concavity and convexity lines are determined and the determination of these lines require searching in the lists L_x and L_y. The number of vertices in L_x and L_y each is $O(n/g)$ and lists are sorted so searching requires $O(\log(n/g))$ and the number of such search is bounded by $O(n/g)$. So, the total running time of Step 3 is bounded by $O(\frac{n}{g}\log(n/g))$. Steps 4 and 5 require constant time. Checking three vertices (Step 7) requires searching in L_x and/or L_y which consumes $O(3\log(n/g) \cong O(\log n/g))$ time. As the number of

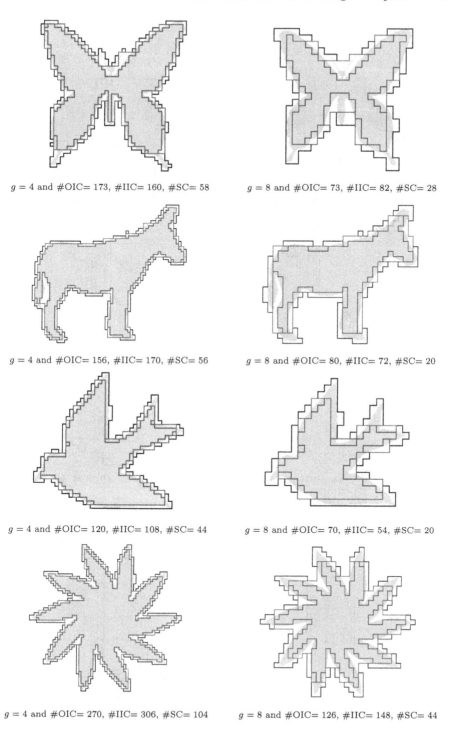

$g = 4$ and #OIC= 173, #IIC= 160, #SC= 58 $g = 8$ and #OIC= 73, #IIC= 82, #SC= 28

$g = 4$ and #OIC= 156, #IIC= 170, #SC= 56 $g = 8$ and #OIC= 80, #IIC= 72, #SC= 20

$g = 4$ and #OIC= 120, #IIC= 108, #SC= 44 $g = 8$ and #OIC= 70, #IIC= 54, #SC= 20

$g = 4$ and #OIC= 270, #IIC= 306, #SC= 104 $g = 8$ and #OIC= 126, #IIC= 148, #SC= 44

Fig. 8. Sandwich covers for four digital objects for two grid sizes, $g = 4$ and $g = 8$, along with the number of vertices of OIC, IIC, and SC

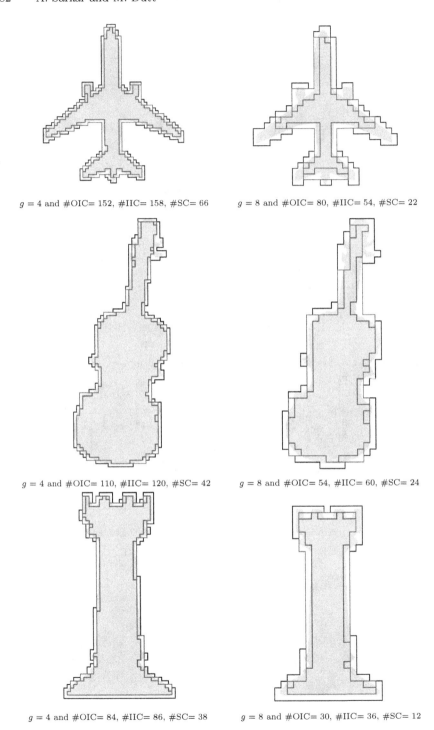

$g = 4$ and #OIC= 152, #IIC= 158, #SC= 66

$g = 8$ and #OIC= 80, #IIC= 54, #SC= 22

$g = 4$ and #OIC= 110, #IIC= 120, #SC= 42

$g = 8$ and #OIC= 54, #IIC= 60, #SC= 24

$g = 4$ and #OIC= 84, #IIC= 86, #SC= 38

$g = 8$ and #OIC= 30, #IIC= 36, #SC= 12

Fig. 9. Sandwich covers for three digital objects for two grid sizes, $g = 4$ and $g = 8$, along with the number of vertices of OIC, IIC, and SC

vertices on the sandwich cover is bounded by $O(n/g)$, the loop, Steps 6 to 11, requires $O(\frac{n}{g} \times \log(n/g))$ time. So, in effect the running time of the proposed algorithm to construct the sandwich cover of a digital object is bounded by $O(n/g) + O(n/g) + O(\frac{n}{g} \log(n/g)) + O(\frac{n}{g} \log(n/g)) \cong O(\frac{n}{g} \log(n/g))$.

4   Experimental Results and Analysis

The proposed algorithm is implemented in C in Ubuntu 12.04, 64-bit, kernel version 3.5.0-43-generic, the processor being Intel i5-3570, 3.4 GHz FSB. Four instances of sandwich covers for four different objects or two grid sizes, $g = 4$ and $g = 8$, along with the number of vertices of OIC, IIC, and SC, are shown in Fig. 8 and for three others in Fig. 9. As evident from the results, the number of vertices is significantly reduced at the same time they preserve the shape of the original object. It may be noted that the gain w.r.t. the reduction in the number of vertices for complex objects ('butterfly' in Fig. 8) is more than the simpler object ('rook' in Fig. 9). The plots in Fig. 7 show that the number of vertices for the sandwich cover for all the objects are noticeably smaller at lower grid sizes, however the difference decreases at higher grid sizes.

5   Conclusions

A combinatorial technique to construct sandwich cover is presented in this paper. The running time of the algorithm is $O(\frac{n}{g} \log(n/g))$, where n is the number of points on the perimeter of the digital object. Although the algorithm is used to construct sandwich cover from outer and inner isothetic cover in this paper, it can also be applied to construct minimum vertex isothetic polygon that lies inside the annular region defined by given two non intersecting simple isothetic polygon with one completely inside the other. The produced sandwich cover is not unique in the sense that there could be more than one sandwich cover with the same minimum number of vertices. In general the sandwich cover preserve the shape of the object but if the inner cover of the object has a curvature with a narrow neck (less than unit grid size) then the existence of the curvature in the produced sandwich cover may not be captured. This problem has applications in shape analysis of digital objects. Several metrics can be developed to use sandwich cover as shape descriptor of digital objects and can also be applied for shape classification.

References

1. Bemporad, A., Filippi, C., Torrisi, F.: Inner and outer approximations of polytopes using boxes. Comput. Geom. Theory Appl. **27**, 151–178 (2004)
2. Biswas, A., Bhowmick, P., Bhattacharya, B.B.: TIPS: on finding a tight isothetic polygonal shape covering a 2D object. In: Kalviainen, H., Parkkinen, J., Kaarna, A. (eds.) SCIA 2005. LNCS, vol. 3540, pp. 930–939. Springer, Heidelberg (2005)

3. Biswas, A., Bhowmick, P., Bhattacharya, B.B.: Construction of isothetic covers of a digital object: a combinatorial approach. J. Vis. Comun. Image Represent. **21**(4), 295–310 (2010)

4. Das, B., Dutt, M., Biswas, A., Bhowmick, P., Bhattacharya, B.B.: A combinatorial technique for construction of triangular covers of digital objects. In: Barneva, R.P., Brimkov, V.E., Šlapal, J. (eds.) IWCIA 2014. LNCS, vol. 8466, pp. 76–90. Springer, Heidelberg (2014)

5. Kamon, Y., Flash, T., Edelman, S.: Learning to grasp using visual information. In: Proceedings of the IEEE International Conference on Robotics and Automation, pp. 2470–2476 (1995)

6. Khan, J.F., Bhuiyan, S.M.A., Adhami, R.R.: Image segmentation and shape analysis for road-sign detection. IEEE Trans. Intell. Transp. Syst. **12**(1), 83–96 (2011)

7. Morales, A., Sanz, P., del Pobil, A.: Vision-based computation of three-finger grasps on unknown planar objects. In: IEEE International Conference on Intelligent Robots and Systems, pp. 1711–1716 (2002)

8. Nandy, S., Bhattacharya, B., Barrera, A.: Safety zone problem. J. Algorithms **37**, 538–569 (2000)

9. Naser, M.A., Hasnat, M., Latif, T., Nizamuddin, S., Islam, T.: Analysis and representation of character images for extracting shape based features towards building an OCR for bangla script. In: International Conference on Digital Image Processing, pp. 330–334 (2009)

10. Sklansky, J.: Minimum-perimeter polygons of digitized silhouettes. IEEE Trans. Comput. **21**(3), 1355–1364 (1972)

11. Sloboda, F., Zatko, B.: On boundary approximation. In: Proceedings of the Sixth International Conference on Computer Analysis of Images and Patterns, CAIP 1995, pp. 488–495 (1995)

Theoretical Foundations of Combinatorial Image Analysis – Grammars and Other Formal Tools

Picture Array Generation Using Pure 2D Context-Free Grammar Rules

K.G. Subramanian[1]($\boxtimes$), M. Geethalakshmi[2], N. Gnanamalar David[3], and Atulya K. Nagar[1]

[1] Department of Mathematics and Computer Science, Faculty of Science, Liverpool Hope University, Hope Park, Liverpool L16 9JD, UK
kgsmani1948@yahoo.com
[2] Department of Mathematics, Queen Mary's College, Chennai 600004, India
[3] Department of Mathematics, Madras Christian College, Chennai 600059, India

Abstract. Pure two-dimensional context-free grammar ($P2DCFG$) is a simple but effective non-isometric 2D grammar model to generate picture arrays. This 2D grammar uses only one kind of symbol as in a pure string grammar and rewrites in parallel all the symbols in a column or row by a set of context-free type rules. $P2DCFG$ and a variant called $(l/u)P2DCFG$, which was recently introduced motivated by the "leftmost" rewriting mode in string context-free grammars, have been investigated for different properties. In this paper we introduce another variant of $P2DCFG$ that corresponds to "rightmost" rewriting in string context-free grammars. The resulting grammar is called $(r/d)P2DCFG$ and rewrites in parallel all the symbols only in the rightmost column or the lowermost row of a picture array by a set of context-free type rules. Unlike the case of string context-free grammars, the picture language families of $P2DCFG$ and the two variants $(l/u)P2DCFG$ and $(r/d)P2DCFG$ are mutually incomparable, although they are not disjoint. We also examine the effect of regulating the rewriting in a $(r/d)P2DCFG$ by suitably adapting two well-known control mechanisms in string grammars, namely, control words and matrix control.

1 Introduction

A variety of two-dimensional (2D) picture array generating grammars [5,9–11,18,19] have been introduced and investigated by researchers, motivated by problems in different areas such as pattern generation and recognition, image description and analysis and others. These 2D grammars have been mainly developed based on the concepts and techniques of the well-investigated string grammar theory. They are basically of two varieties, namely, isometric array grammars in which geometric shape of the re-written portion of the array is preserved while non-isometric array grammars can alter the geometric shape. Here we consider a 2D picture grammar of the latter variety, called a pure 2D context-free grammar ($P2DCFG$) introduced in [17]. This 2D grammar represents a simple yet expressive non-isometric language generator of rectangular

© Springer International Publishing Switzerland 2015
R.P. Barneva et al. (Eds.): IWCIA 2015, LNCS 9448, pp. 187–201, 2015.
DOI: 10.1007/978-3-319-26145-4_14

picture arrays, involving only terminal symbols as in a pure string grammar [7] and using tables of context-free (CF) type rules. All the symbols in a column or a row of a rectangular array are re-written in parallel by CF type rules in a $P2DCFG$ and in order to maintain the rectangular form of the array, all the symbols rewritten are replaced by strings of equal length. A $P2DCFG$ allows rewriting any column or any row of a picture array by the rules of an applicable column rule table or row rule table respectively. In [1,2,16], further properties of this 2D grammar model are investigated.

Motivated by the notion of leftmost derivation [14] in a context-free string grammar in which only leftmost nonterminal in a sentential form of a derivation is rewritten, a variant of $P2DCFG$, referred to as $(l/u)P2DCFG$ was introduced in [6]. This variant considers a "leftmost rewriting mode" in terms of $P2DCFG$ in the sense that only the leftmost column or the uppermost row of a rectangular picture array is rewritten in a derivation step. In the case of context-free string grammars, the notion of a rightmost derivation is also known [14] in which only the rightmost nonterminal in a sentential form of a derivation is rewritten. We recall that for an ordinary derivation of a terminal word in a context-free string grammar, there is an equivalent leftmost as well as rightmost derivation of the word. On the other hand, it has been shown in [6] that the picture language classes of $P2DCFG$ (which correspond to "ordinary derivation mode") and $(l/u)P2DCFG$ (which correspond to "leftmost derivation mode") are incomparable. Here we investigate another variant of $P2DCFG$ by considering a "rightmost rewriting mode" in terms of $P2DCFG$ in the sense that only the rightmost column or the lowermost row of a rectangular picture array is rewritten in a derivation step, with all the symbols being rewritten by "equal length" rules. We denote the resulting class of 2D grammars by $(r/d)P2DCFG$ and the corresponding language class by $(r/d)P2DCFL$.

We show that this variant of $(r/d)P2DCFL$ is also incomparable with both the picture language classes of $P2DCFG$ and $(l/u)P2DCFG$. The effect of regulating the picture generation by controlling the application of rules with a regular control set on the labels of the tables of rules has been studied in [1,2,6,16,17] for the classes of $P2DCFL$ and $(l/u)P2DCFL$. Besides examining the effect of this kind of a control on $(r/d)P2DCFG$, we also consider matrix type of control on the tables of rules for all the three classes, namely, $P2DCFL$ and the two variants, $(l/u)P2DCFL$ and $(r/d)P2DCFL$.

2   Preliminaries

For notions related to formal language theory we refer to [13,14] and for array grammars and two-dimensional languages we refer to [5].

Given a finite alphabet Σ, a word or a string w is a sequence of symbols from Σ. The set of all words over Σ, including the empty word λ with no symbols, is denoted by Σ^*. The length of a word w is denoted by $|w|$. For any word $w = a_1 a_2 \ldots a_n$, we denote by $^t w$, the word $w = a_1 a_2 \ldots a_n$ $(n \geq 1)$

written vertically. For example, if $w = acbb$ over $\{a, b, c\}$, then ${}^t w = \begin{matrix} a \\ c \\ b \\ b \end{matrix}$. A two-dimensional $m \times n$ array (also called a picture array or a picture) p over an alphabet Σ is a rectangular array with m rows and n columns and is of the form

$$p = \begin{matrix} p(1,1) & \cdots & p(1,n) \\ \vdots & \ddots & \vdots \\ p(m,1) & \cdots & p(m,n) \end{matrix}$$

where each pixel $p(i,j) \in \Sigma, 1 \leq i \leq m, 1 \leq j \leq n$. The uppermost row of p is considered as the first row and the leftmost column as the first column of p. Likewise, The lowermost row of p is the last row and the rightmost column, the last column of p. We denote the number of rows and the number of columns of p, respectively, by $|p|_{row}$ and $|p|_{col}$ and call the pair $(|p|_{row}, |p|_{col})$ as the size of p. The set of all rectangular arrays over Σ is denoted by Σ^{**}, which includes the empty array λ. $\Sigma^{++} = \Sigma^{**} - \{\lambda\}$. A picture language is a subset of Σ^{**}.

We now recall a pure 2D context-free grammar introduced in [16,17].

Definition 1. *A pure 2D context-free grammar (P2DCFG) is a 4-tuple*

$$G = (\Sigma, P_1, P_2, \Gamma)$$

where

(i) Σ is a finite alphabet of symbols;

(ii) P_1 is a finite set of column rule tables c_i, $1 \leq i \leq m$, for some $m \geq 1$, where each c_i is a finite set of context-free rules of the form $a \rightarrow u, a \in \Sigma, u \in \Sigma^*$ having the property that for any two rules $a \rightarrow u, b \rightarrow v$ in c_i, we have $|u| = |v|$ i.e. the words u and v have equal length;

(iii) P_2 is a finite set of row rule tables r_j, $1 \leq j \leq n$, for some $n \geq 1$, where each r_j is a finite set of rules of the form $c \rightarrow {}^t x, c \in \Sigma, x \in \Sigma^*$ such that for any two rules $c \rightarrow {}^t x, d \rightarrow {}^t y$ in r_j, we have $|x| = |y|$;

(iv) $\Gamma \subseteq \Sigma^{**} - \{\lambda\}$ is a finite set of axiom arrays.

A derivation in a *P2DCFG* G is defined as follows: Let $p, q \in \Sigma^{**}$. A picture q is derived in G from a picture p, denoted by $p \Rightarrow q$, either *(i)* by rewriting in parallel all the symbols in a column of p, each symbol by a rule in some column rule table or *(ii)* rewriting in parallel all the symbols in a row of p, each symbol by a rule in some row rule table. All the rules used to rewrite a column (or row) are taken from the same table.

The picture language generated by G is the set of picture arrays $L(G) = \{M \in \Sigma^{**} | M_0 \Rightarrow^* M$ for some $M_0 \in \Gamma\}$. The family of picture languages generated by *P2DCFGs* is denoted by *P2DCFL*.

Example 1. Consider the $P2DCFG$ $G_1 = (\Sigma, P_1, P_2, \{M_0\})$ where $\Sigma = \{a, b, d\}$, $P_1 = \{c\}, P_2 = \{r\}$, where $c = \{a \rightarrow bab, d \rightarrow ada, e \rightarrow aea\}$, $r = \left\{ d \rightarrow \dfrac{d}{a}, a \rightarrow \dfrac{a}{b} \right\}$, and $M_0 = \begin{array}{c} a\,d\,a \\ b\,a\,b \\ a\,e\,a \end{array}$.

G_1 generates a picture language L_1 consisting of picture arrays p of size $(m, 2n + 1)$, $m \geq 3$, $n \geq 1$ with $p(1, j) = p(1, j + n + 1) = p(m, j) = p(m, j + n + 1) = a$, for $1 \leq j \leq n$; $p(1, n+1) = d$; $p(m, n+1) = e$; $p(i, n+1) = a$, for $2 \leq i \leq m - 1$; $p(i, j) = b$, otherwise. We note that a derivation in G_1, starting from the axiom array M_0, generates a picture array of the form

$$
\begin{array}{ccccc}
a \cdots a & d & a \cdots a \\
b \cdots b & a & b \cdots b \\
\cdots & & \cdots \\
\cdots & & \cdots \\
b \cdots b & a & b \cdots b \\
a \cdots a & e & a \cdots a
\end{array}
$$

since the column rule table is applicable to only the "middle" column $^t da \cdots ae$, rewriting in parallel all the symbols d and a in that column, thereby adding the symbol a to the left and right of d as well as e, while adding the symbol b to the left and right of every a in that column. Likewise, the row rule table r is applicable to only the uppermost row and adds a row of the form $b \cdots bab \cdots b$ below it. Also note that the application of the column rule table c can take place independent of the row rule table r and hence the number of rows and the number of columns in the generated picture arrays of L_1 need not be related by any proportion. On the otherhand, due to the nature of rules in the column table c, every generated picture array has an equal number of columns to the left and right of the middle column $^t (da \ldots ae)$. A member of L_1 is shown in Fig. 1(a) and on interpreting the symbol b as blank, the corresponding picture representing the letter I is shown in Fig. 1(b).

$$
\begin{array}{ll}
\begin{array}{ccccccc}
a & a & a & d & a & a & a \\
b & b & b & a & b & b & b \\
b & b & b & a & b & b & b \\
b & b & b & a & b & b & b \\
b & b & b & a & b & b & b \\
a & a & a & e & a & a & a
\end{array}
&
\begin{array}{ccccccc}
a & a & a & d & a & a & a \\
 & & & a & & & \\
 & & & a & & & \\
 & & & a & & & \\
 & & & a & & & \\
a & a & a & e & a & a & a
\end{array}
\end{array}
$$

Fig. 1. (a) (on the left) A picture array in the language L_1 (b) (on the right) The interpreted picture representing letter I.

We now recall (l/u) mode of derivation in a $P2DCFG$ introduced and investigated in [6].

Definition 2. *Let $G = (\Sigma, P_1, P_2, \Gamma)$ be a P2DCFG with the components as in Definition 1. A picture array M_2 is derived from M_1 in G with (l/u) mode of derivation, denoted by $M_1 \Rightarrow_{(l/u)} M_2$, by rewriting all the symbols and only* *these symbols, either in the leftmost column or in the uppermost row of M_1 using respectively, the column rule tables or the row rule tables. All the rules used to rewrite a column (or row) are taken from the same table.*

The picture language generated is defined as in the case of a P2DCFG but using $\Rightarrow_{(l/u)}$ derivations. The family of picture languages generated by P2DCFGs under $\Rightarrow_{(l/u)}$ derivations is denoted by $(l/u)P2DCFL$. For convenience, we write $(l/u)P2DCFG$ to refer to P2DCFG with $\Rightarrow_{(l/u)}$ derivations.

We illustrate with an example.

Example 2. Consider an $(l/u)P2DCFG$ $G_2 = (\Sigma, P_1, P_2, \{M_0\})$ where $\Sigma = \{a, b\}$, $P_1 = \{c\}$, $P_2 = \{r\}$ with $c = \{a \rightarrow ab, d \rightarrow da\}$, $r = \left\{a \rightarrow \begin{smallmatrix} a \\ a \end{smallmatrix}, b \rightarrow \begin{smallmatrix} b \\ b \end{smallmatrix}\right\}$, and $M_0 = \begin{smallmatrix} a & b \\ d & a \end{smallmatrix}$.

G_2 generates a picture language L_2 consisting of arrays p of size (m, n), $m \geq 2$, $n \geq 2$ with $p(m, 1) = d$; $p(m, j) = a$, for $2 \leq j \leq n$; $p(i, 1) = a$, for $1 \leq i \leq m-1$; $p(i, j) = b$, otherwise. A member of L_2 is shown in Fig. 2. A member of L_2 is shown in Fig. 2(a) and on interpreting the symbol b as blank, the corresponding picture representing the letter L is shown in Fig. 2(b).

$$
\begin{array}{llllll}
a & b & b & b & b & b \\
a & b & b & b & b & b \\
a & b & b & b & b & b \\
a & b & b & b & b & b \\
d & a & a & a & a & a
\end{array}
\qquad
\begin{array}{llllll}
a \\
a \\
a \\
a \\
d & a & a & a & a & a
\end{array}
$$

Fig. 2. (a) (on the left) A picture array in the language L_2 (b) (on the right) The interpreted picture representing letter L.

3   Pure 2D Context-Free Grammar with (r/d) Mode of Derivations

We now introduce another variant of a *P2DCFG*. The leftmost and rightmost derivation modes in a context-free grammar (CFG) in string language theory, are well-known [13,14], especially in the context of the study of parsers. It is also known that these derivation modes are equivalent to the "ordinary" derivations in a CFG in the sense of generating the same language class. Motivated to consider a corresponding notion of "leftmost kind" of derivation in pure 2D context-free grammars, the $(l/u)P2DCFG$ with an (l/u) mode of derivation was introduced in [6]. Here we consider the dual idea of rewriting the rightmost column of a picture array by a column rule table or the lowermost row by a row rule table. This corresponds to the "rightmost kind" of derivation mode in a string CFG. The interesting aspect is that this results in a picture language family which neither contains nor is contained in *P2DCFL* or $(l/u)P2DCFL$.

Definition 3. *Let $G = (\Sigma, P_1, P_2, \Gamma)$ be a P2DCFG with the components as in Definition 1. An (r/d) mode of derivation of a picture array M_2 from M_1 in G, denoted by $\Rightarrow_{(r/d)}$, is a derivation in G such that only the rightmost column or the lowermost row of M_1 is rewritten using respectively, the column rule tables or the row rule tables, to yield M_2. The generated picture language is defined as in the case of a P2DCFG but with $\Rightarrow_{(r/d)}$ derivations. The family of picture languages generated by P2DCFGs under $\Rightarrow_{(r/d)}$ derivations is denoted by $(r/d)P2DCFL$. For convenience, we write $(r/d)P2DCFG$ to refer to P2DCFG with $\Rightarrow_{(r/d)}$ derivations.*

We illustrate with an example.

$$
\begin{array}{cccccc}
a & b & b & b & b & a \\
b & b & b & b & b & a \\
b & b & b & b & b & a \\
b & b & b & b & b & a \\
a & a & a & a & a & b
\end{array}
$$

Fig. 3. A picture array in the language L_3

Example 3. Consider an $(r/d)P2DCFG$ $G_3 = (\Sigma, P_1, P_2, \{M_0\})$ where $\Sigma = \{a, b\}$, $P_1 = \{c\}$, $P_2 = \{r\}$ with $c = \{a \rightarrow ba, b \rightarrow ab\}$, $r = \left\{ a \rightarrow \dfrac{b}{a}, b \rightarrow \dfrac{a}{b} \right\}$, and $M_0 = \dfrac{a\ a}{a\ b}$.

G_3 generates a picture language L_3 consisting of arrays p of size (m, n), $m \geq 2$, $n \geq 2$ with $p(1, 1) = a$; $p(m, n) = b$; $p(m, j) = a$, for $1 \leq j \leq n-1$; $p(i, n) = a$, for $1 \leq i \leq m-1$; $p(i, j) = b$, otherwise. A member of L_3 is shown in Fig. 3. A sample derivation in $(r/d)P2DCFG$ G_3 starting from the axiom array M_0 and using the tables c, r, c, c in this order is shown in Fig. 4. The application of the column rule table c rewrites all symbols in the rightmost column in parallel and likewise, the application of the row rule table r rewrites all symbols in the lowermost row. We note that the application of a column table or a row table is independent of each other as in a *P2DCFG* and so cannot maintain any proportion between the number of columns and the number of rows in any generated picture array.

$$
M_0 = \begin{array}{cc} a & a \\ a & b \end{array} \Rightarrow_{(r/d)} \begin{array}{ccc} a & b & a \\ a & a & b \end{array} \Rightarrow_{(r/d)} \begin{array}{ccc} a & b & a \\ b & b & a \\ a & a & b \end{array} \Rightarrow_{(r/d)} \begin{array}{cccc} a & b & b & a \\ b & b & b & a \\ a & a & a & b \end{array} \Rightarrow_{(r/d)} \begin{array}{ccccc} a & b & b & b & a \\ b & b & b & b & a \\ a & a & a & a & b \end{array}
$$

Fig. 4. A sample derivation under (r/d) mode in Example 3

We now compare the generative power of $(r/d)P2DCFL$ with $(l/u)P2DCFL$ and $P2DCFL$.

Theorem 1. *Each pair of the three families of $P2DCFL$, $(l/u)P2DCFL$ and $(r/d)P2DCFL$ is incomparable but not disjoint, when the alphabet contains at least two symbols. All the three families coincide if we restrict to only a unary alphabet.*

Proof. The non-trivial picture language L_{RECT} of all rectangular picture arrays over $\{a, b\}$ belongs to all the three families of $P2DCFL$, $(l/u)P2DCFL$ and $(r/d)P2DCFL$. In fact the corresponding $P2DCFG$ needs to have only two tables

$$c = \{a \to aa, a \to ab, b \to ba, b \to bb\}, r = \left\{a \to \begin{matrix} a \\ a \end{matrix}, a \to \begin{matrix} a \\ b \end{matrix}, b \to \begin{matrix} b \\ a \end{matrix}, b \to \begin{matrix} b \\ b \end{matrix}\right\}$$

and axiom pictures a, b, for all the three modes of derivations in the $P2DCFG$. This shows that the three families are mutually not disjoint.

The incomparability of the families $P2DCFL$ and $(l/u)P2DCFL$ has been established in [6]. The picture language L_3 in Example 3 which belongs to $(r/d)P2DCFL$ cannot be generated by any $P2DCFG$ since every column in the picture arrays of L_3 involves the two symbols a, b only and so in order to generate the picture arrays of L_3 starting from an axiom array, we have to specify column rules for both a, b. In the (r/d) mode of derivation, the rightmost column will require a column rule that will rewrite b into $a \cdots ab$ and a into $b \cdots ba$. But then the table with these rules can be applied to any other column in a $P2DCFG$, resulting in picture arrays not in the language L_3.

On the other hand the picture language L_1 in Example 1 belongs to $P2DCFL$ but it cannot be generated by any $(r/d)P2DCFG$ as there is an unique middle column in every picture array of L_1 and to the left and right of this middle column there are an equal number of identical columns. Since only the rightmost column can be rewritten in an $(r/d)P2DCFG$, it is not possible to maintain this feature of "equal number of identical columns" if rightmost column rewriting is done. We also note that the picture generated in any intermediate step also belongs to the language which prevents the use of any other symbol other than a and b. This proves the incomparability of the families $P2DCFL$ and $(r/d)P2DCFL$.

Again, the picture language L_3 in Example 3, which belongs to (r/d) $P2DCFL$, cannot be generated by any $(l/u)P2DCFG$, since the leftmost column of every picture array in L_3 has the symbol a in the first row as well as the last row. The former will require a rule of the form $a \to ab \cdots b$ while the latter will require a rule of the form $a \to a \cdots ab$. But then the presence of these two rules in a column table gives a non-deterministic choice for rewriting a in the leftmost column in (l/u)mode of derivation, resulting in picture arrays not in the language L_3.

On the other hand, consider the picture language L'_3 consisting of picture arrays p of size (m, n), $m \geq 2$, $n \geq 2$, with $p(1, 1) = b$; $p(m, n) = a$; $p(1, j) = a$, for $2 \leq j \leq n$; $p(i, 1) = a$, for $2 \leq i \leq m$; $p(i, j) = b$, otherwise. L'_3 can be generated a $(l/u)P2DCFG$ with axiom array $M_0 = \begin{matrix} b & a \\ a & a \end{matrix}$ and column table c and row table r given by

$$c = \{a \to ab, b \to ba\}, r = \left\{a \to \begin{matrix} a \\ b \end{matrix}, b \to \begin{matrix} b \\ a \end{matrix}\right\}$$

A member of L_3' is shown in Fig. 5. It can be seen that L_3' cannot be generated by any $(r/d)P2DCFG$. This proves the incomparability of the families $(l/u)P2DCFL$ and $(r/d)P2DCFL$.

In the case of a unary alphabet with a single symbol, say, a, the column rules and the row rules can use only a and hence rewriting any column (or row) is equivalent to rewriting the leftmost or the rightmost column (or row) of a picture array. This shows that all the three families coincide in this case.   □

$$
\begin{matrix}
b & a & a & a & a & a \\
a & b & b & b & b & b \\
a & b & b & b & b & b \\
a & b & b & b & b & b \\
a & b & b & b & b & a
\end{matrix}
$$

Fig. 5. A picture array in the language L_3'.

Remark 1. In [2] (Proposition 3.14), the uniform membership problem (ump) for the class of $P2DCFG$ with unary alphabet, namely, "does the picture p belong to the language $L(G)$ generated by the $P2DCFG$ G, given p as well as G as input?", is shown to be in class P. As a consequence of the fact that in the case of an unary alphabet, rewriting any column (respectively row) of a picture array is equivalent to rewriting the leftmost or the rightmost column (respectively row), this result on ump holds for the classes of $(l/u)P2DCFG$ and $(r/d)P2DCFG$.

4   Regulating Rewriting in $(r/d)P2DCFG$ with Control Words

Based on the concept of regulating the rewriting in string and array grammars [3,4,14,15] with different control mechanisms, as for example, grammars with control languages and matrix grammars, a $P2DCFG$ with a control language on the labels of the column rule and row rule tables has been introduced in [17] and certain properties have been obtained. Further results on the effect of control have been established in [1,2,16]. This study on control language has been extended in [6] to the class $(l/u)P2DCFG$. Here we obtain corresponding results for the class $(r/d)P2DCFG$.

A $(r/d)P2DCFG$ with a control language is $G^c = (G, Lab, \mathcal{C})$ where $G = (\Sigma, P_1, P_2, \Gamma)$ is a $(r/d)P2DCFG$, Lab is a set of labels of the tables of G, with each table in $P_1 \cup P_2$ being assigned a distinct label and $\mathcal{C} \subseteq Lab^*$ is a string language. The words in Lab^* are called control words of G. Derivations $M_1 \Rightarrow_w M_2$ in G^c are done as in G in (r/d) mode except that if $w \in Lab^*$ and $w = l_1 l_2 \ldots l_m$, then the tables of rules with labels l_1, l_2, ..., and l_m are successively applied starting with the picture array M_1 to finally yield the picture array M_2, which is collected in the language if M_1 is an axiom array. If any of the labels in a control word is such that the corresponding table of rules cannot be applied to the picture array on hand, the derivation is discarded and it does not yield any array. The picture array language generated by G^c consists of all picture arrays obtained from axiom arrays of G with the derivations controlled as described above. We denote the family of picture languages generated by $(r/d)P2DCFGs$ with a regular control language by $(R)(r/d)P2DCFL$ and with a context-free control language by $(CF)(l/u)P2DCFG$.

Theorem 2. $(r/d)P2DCFL \subset (R)(r/d)P2DCFL \subset (CF)(r/d)P2DCFL$.

Proof. The inclusions follow by noting that a $(r/d)P2DCFG$ is a $(R)(r/d)$ $P2DCFG$ on setting the regular control language as Lab^* where Lab is the set of labels of the tables of the $(r/d)P2DCFG$ and the well-known fact [14] that the class of regular string languages is included in the class of context-free languages.

The proper inclusion in $(r/d)P2DCFL \subset (R)(r/d)P2DCFL$ can be seen by considering a picture language L_4 consisting of only square sized arrays p of the language L_3 given in the Example 3. This picture language can be generated by the $(r/d)P2DCFG$ G_3 in Example 3 with a regular control language $(cr)^*$. But by definition, the application of the column rule and row rule tables are independent in a $(r/d)P2DCFG$ and hence no $(r/d)P2DCFG$ can generate L_4 which consists of square sized arrays.

The proper inclusion of $(R)(r/d)P2DCFL$ in $(CF)(r/d)P2DCFL$ can be shown by considering a picture language L_5 consisting of picture arrays p as in Example 1 but of sizes $(n + 2, 2n + 1), n \geq 1$. The $(CF)(r/d)P2DCFG$ $G^c = (G_5, Lab, \mathcal{C})$ generates L_5, where $G_5 = (\Sigma, P_1, P_2, \{M_0\})$, $\Sigma = \{a, b, d\}$, $P_1 = \{c_1, c_2, c_3\}, P_2 = \{r\}$ with

$$c_1 = \{d \rightarrow ad, a \rightarrow ba\}, c_2 = \{d \rightarrow da, a \rightarrow ab\}, c_3 = \{a \rightarrow aa, b \rightarrow bb\},$$

$$r = \left\{ d \rightarrow \frac{a}{d}, a \rightarrow \frac{b}{a} \right\},$$

$M_0 = \begin{matrix} a\ d \\ b\ a \\ a\ d \end{matrix}$ and $Lab = \{c_1, c_2, c_3, r\}$ with c_1, c_2, c_3, r themselves being considered as the labels of the corresponding tables. The CF control language is $\mathcal{C} = \{(c_1r)^n c_2 c_3^n |\ n \geq 0\}$. The grammar G_5 generates the picture arrays of L_5, in the (r/d) derivation mode according to the control words of $\mathcal{C}$. Starting from the axiom array $M_0 = \begin{matrix} a\ d \\ b\ a \\ a\ d \end{matrix}$ the rightmost column of M_0 is rewritten using the column rule table c_1 and this is immediately followed by the row rule table r which rewrites once, all the symbols in the lowermost row. This can be repeated

n times (for some $n \geq 0$). Then the column rule table c_2 is applied once, followed by the application of the column rule table c_3, the same number of times as c_1 followed by r was done, thus yielding a picture array in L_5. But L_5 cannot be generated by any $(r/d)P2DCFG$ with regular control. In fact if a generation of a picture array $p \in L_5$ makes use of a regular control, then there will be no information available on the number of columns generated once the derivation "crosses" the middle column (made of one d as the first symbol and another d as the last symbol with all other symbols in the column being a's). Hence the columns to the left and right of this middle column cannot be generated in equal number. □

The notion of a control symbol or control character was considered in [2] while dealing with $(R)P2DCFG$. The idea is that In a $(R)P2DCFG$, the alphabet may contain some symbols called control symbols [2] which might not be ultimately involved in the picture arrays of the language generated. A picture language L_d was considered in [2] given by $L_d = \{p \in \{a,b\}^{++}| \ |p|_{col} = |p|_{row}, p(i,j) = a, \text{for } i = j, p(i,j) = b \text{ for } i \neq j\}$. It was shown in [2], that at least two control symbols are required to generate L_d using $(R)P2DCFG$. In [6], it was proved that L_d can be generated with a single control character using $(R)(l/u)P2DCFG$. We show here that an analogous result holds in the case of $(R)(r/d)P2DCFG$.

Lemma 1. *The language L_d can be generated by an $(R)(r/d)P2DCFG$ using a single control character. Also, L_d is not in $(r/d)P2DCFL$.*

Proof. Consider the $(R)(r/d)P2DCFG$ with the $(r/d)P2DCFG$ given by

$$(\{a,b,e\},\{c\},\{r\},\{\begin{smallmatrix} a & b \\ b & a \end{smallmatrix}\}) \text{ where}$$

$$c = \{a \to ea, b \to bb\}, r = \left\{ a \to \begin{smallmatrix} b \\ a \end{smallmatrix}, e \to \begin{smallmatrix} a \\ b \end{smallmatrix}, b \to \begin{smallmatrix} b \\ b \end{smallmatrix} \right\},$$

and control language $(cr)^*$ generates L_d. The idea in the generation of the picture arrays of L_d is that the symbol e in the alphabet acts as the control character. An application of the column table c produces e to the left of the only a in the last row and when this is followed by the application of the row table r (according to the control word), the symbol e "disappears", yielding an array in L_d. It can be seen that L_d cannot be generated by any $(r/d)P2DCFG$. □

In [2, p. 1730], a picture language L_{2d} given by $L_{2d} = \{p \in \{a,b\}^{++}| \ |p|_{col} = |p|_{row}, p(i,j) = b, \text{for } i = j \text{ and } i = j - 1, p(i,j) = a \text{ otherwise}\}$ was shown to be generated by a $(R)P2DCFG$ making use of four ([2] mentions three but the grammar actually involves four) control characters. We note that L_{2d} can be generated by a $(R)(r/d)P2DCFG$ G_{2d}^c with three control characters. In fact the grammar is essentially as given in [2] with a slight modification. For completeness we give this grammar here. The $P2DCFG$ in G_{2d}^c is $(\{a,b,b_1,b_2,e\},P_1,P_2,\{M_0\})$, $P_1 = \{c\}, P_2 = \{r_1,r_2\}$ with

$$c = \{a \to aa, b_2 \to b_2e\}, r_1 = \{a \to a, b_1 \to b, b_2 \to b\},$$

$$r_2 = \left\{ a \to \frac{a}{a}, b_1 \to \frac{b}{a}, b_2 \to \frac{b}{b_1}, e \to \frac{a}{b_2} \right\},$$

$M_0 = \frac{b\ a}{b_1\ b_2}$ and the regular control language $(cr_2)^*r_1$. We can likewise construct a $(R)(l/u)P2DCFG$ to generate L_{2d}. Also, in [2, p. 1728], it was shown that the family $(R)P2DCFL$ has nonempty intersection with the class LOC [5] of local picture languages whose pictures are defined by means of tiles i.e. square pictures of size $(2,2)$ and another class of picture languages, which we call here as PL, defined by a class of picture grammars referred to as Průša grammars [2]. This is done in [2] by showing that L_{2d} is in all the three families $P2DCFL$, LOC and PL. As a consequence of these remarks, we have the following Theorem 3.

Theorem 3. *All the three families $(R)P2DCFL$, $(R)(l/u)P2DCFL$ and $(R)(r/d)P2DCFL$ have non-empty intersection with the picture language families LOC and PL.*

5 Matrix Control on *P2DCFG*

In another type of regulating the use of rules in derivations, known as matrix control, a pre-specified finite sequence of rules is applied constituting a single step of derivation in the grammar [3,14]. Here we consider this kind of control in $P2DCFG$ and the two variants $(l/u)P2DCFG$ and $(r/d)P2DCFG$.

Definition 4. *A matrix $P2DCFG$ is a 3-tuple $G_m = (G, Lab, M)$ where $G = (\Sigma, P_1, P_2, \Gamma)$ is a $P2DCFG$, Lab is a set of labels of the tables of G, with each table in $P_1 \cup P_2$ being assigned a distinct label and M is a finite set of sequences, called matrices, of the form $m = [l_1, \cdots, l_n]$, $n \geq 1$, where $l_i \in Lab$, for all $1 \leq i \leq n$.*

Derivations in a matrix $P2DCFG$ are defined as in a $P2DCFG$ except that a single derivation step is done by the application of the tables of rules of a matrix m in M, in the order in which the labels of the tables are given in m. The family of picture languages generated by matrix $P2DCFG$ is denoted by $MP2DCFL$.

We illustrate with an example.

Example 4. Consider the picture language L_1' consisting of arrays belonging to the picture language L_1 in Example 1 but of size $(n + 2, 2n + 1)$, $n \geq 1$. Note that the arrays in L_1' maintain a proportion and hence L_1' cannot be generated by any $P2DCFG$ since the column tables and row tables can be independently applied in a $P2DCFG$. But L_1' is generated by the matrix $P2DCFG$ $G_{m1}' = (G_1, Lab, M)$ where G_1 is as in Example 1. In fact the column table c and the row table r in G_1 are $c = \{a \to bab, d \to ada, e \to aea\}$, $r = \left\{ d \to \frac{d}{a}, a \to \frac{a}{b} \right\}$,

$$a\ d\ a$$
and the axiom array is $M_0 = \begin{matrix} b\ a\ b \end{matrix}$. The label set $Lab = \{c, r\}$ and the set M
$$a\ e\ a$$
consists of the only matrix $m = [c, r]$.

Note that in a derivation in G'_m that starts with the axiom array, an application of the matrix m amounts to rewriting by the rules of the table c followed by the rules of r to constitute one step of derivation yielding an array of size $(4, 5)$

given by $\begin{matrix} a\ a\ d\ a\ a \\ b\ b\ a\ b\ b \\ b\ b\ a\ b\ b \\ a\ a\ d\ a\ a \end{matrix}$. The process can be repeated any number of times yielding

the arrays of L_1.

Analogous to matrix $P2DCFG$, we can define matrix $(l/u)P2DCFG$ and matrix $(r/d)P2DCFG$ as in Definition 4, except that we replace $P2DCFG$ by $(l/u)P2DCFG$ or $(r/d)P2DCFG$. We denote the resulting families of picture languages by $M(l/u)P2DCFL$ and $M(r/d)P2DCFL$.

We illustrate matrix $(r/d)P2DCFG$ by an example.

Example 5. Consider the picture language L'_3 consisting of arrays belonging to the picture language L_3 in Example 3 but of size (n, n), $n \geq 2$. The language L'_3 cannot be generated by any $(r/d)P2DCFG$ since the arrays in L'_3 maintain a proportion. But L'_3 is generated by the matrix $(r/d)P2DCFG$ $G'_{m3} = (G_3, Lab, M)$ where G_3 is as in Example 3. In fact the column table c and the row table r in G_3 are $c = \{a \to ba, b \to ab\}$, $r = \left\{a \to \begin{matrix} b \\ a \end{matrix}, b \to \begin{matrix} a \\ b \end{matrix}\right\}$,

and $M_0 = \begin{matrix} a\ a \\ a\ b \end{matrix}$. The label set $Lab = \{c, r\}$ and the set M consists of the only matrix $m = [c, r]$.

Starting from the axiom array, one step of derivation on applying the matrix
$$a\ b\ a$$
m yields the array $\begin{matrix} b\ b\ a \end{matrix}$. On repeating the process, we obtain the arrays of L'_3.
$$a\ a\ b$$

Lemma 2. *(i)* $P2DCFL \subset MP2DCFL$
(ii) $(l/u)P2DCFL \subset M(l/u)P2DCFL$
(iii) $(r/d)P2DCFL \subset M(r/d)P2DCFL$

Proof. The proper inclusions in (i) and (iii) are a consequence of the Examples 4 and 5, while the inclusions are straightforward, noting that every table t of rules in a $P2DCFG$ can be considered as a matrix $m = \{t\}$. The statement (ii) can be similarly shown.

Theorem 4. *(i)* $P2DCFL \subset MP2DCFL \subset (R)P2DCFL$
(ii) $(l/u)P2DCFL \subset M(l/u)P2DCFL \subset (R)(l/u)P2DCFL$
(iii) $(r/d)P2DCFL \subset M(r/d)P2DCFL \subset (R)(r/d)P2DCFL$

Proof. The inclusions and proper inclusions in the first half of the statements $(i), (ii)$ and (iii) follow from Lemma 2. We prove the second half of (i) and (iii). The second half of (ii) can be similarly shown. For every matrix of the form $m = [l_1, \cdots, l_n]$ in a given matrix $P2DCFG$ G_m, where l_i, $1 \le i \le n$, are the labels of the tables of rules, we associate a word $w = l_1 \cdots l_n$. If $w_1, \cdots, w_p$ are the words obtained in this way from the matrix $P2DCFG$ G_m, then we form a $(R)P2DCFG$ where the $P2DCFG$ is the same as the one in the matrix $P2DCFG$ but the set of control words is $\{w_1, \cdots, w_p\}^*$. It can be seen that the $(R)P2DCFG$ constructed generates the picture language generated by G_m. This proves the inclusion $MP2DCFL \subseteq (R)P2DCFL$. The inclusion $M(r/d)P2DCFL \subseteq (R)(r/d)P2DCFL$ is similar.

$$\begin{array}{ccccccccccc} a & a & a & b & b & d & b & b & a & a & a \\ a & a & a & b & b & d & b & b & a & a & a \\ a & a & a & b & b & d & b & b & a & a & a \\ a & a & a & b & b & d & b & b & a & a & a \\ a & a & a & b & b & d & b & b & a & a & a \end{array}$$

Fig. 6. A picture array in the language $L_{a,b}$

For the proper inclusion $MP2DCFL \subset (R)P2DCFL$, we consider the picture language $L_{a,b}$ generated by the $P2DCFG$ $(\{a, b, d\}, \{c_1, c_2\}, \{r\}, \{d\})$ where

$$c_1 = \{d \to adb\}, c_2 = \{d \to bda\}, r = \left\{d \to \begin{matrix} d \\ d \end{matrix}, a \to \begin{matrix} a \\ a \end{matrix}, b \to \begin{matrix} b \\ b \end{matrix}\right\}$$

with a regular control language $\{(c_1r)^m(c_2r)^n | m, n \ge 1\}$. We note that this regular language can be expressed as $(c_1r)^+(c_2r)^+$ in terms of the Kleene-plus operation [13] in string languages. The picture arrays generated are of size $(m+n, 2m + 2n + 1)$, $m, n \ge 1$. A member of this language is shown in Fig. 6. We note that the first m columns are made of only a, the next n columns are made of only b and there is a middle column made of only d while to the right of this middle column, there are n columns made of only b followed by m columns only made of a. On the other hand no matrix $P2DCFG$ can handle this feature. In fact, if there is a matrix with a column table that produces the columns of a and also a column table that produces the columns of b, then picture arrays which are not in the desired form will be generated. Likewise, if there are two independent matrices with one having a column table that produces the columns of a and another matrix having a column table that produces the columns of b, any of them can be applied again yielding pictures not in the language.

For the proper inclusion $M(r/d)P2DCFL \subset (R)P2DCFL$, we consider a similar picture language generated in the (r/d) mode, by the $P2DCFG$ $(\{a, b, d\}, \{c_1, c_2\}, \{r\}, \{d\})$ where

$$c_1 = \{d \to ad\}, c_2 = \{d \to bd\}, r = \left\{d \to \begin{matrix} d \\ d \end{matrix}, a \to \begin{matrix} a \\ a \end{matrix}, b \to \begin{matrix} b \\ b \end{matrix}\right\}$$

with a regular control language $\{(c_1 r)^m (c_2 r)^n | m, n \geq 1$. It can be seen that matrix $P2DCFG$ in (r/d) mode cannot generate this language.

6 Concluding Remarks

Another variant of $P2DCFG$ [16,17] rewriting only the rightmost column or the lowermost row of a picture array is considered and properties of the resulting family $(r/d)P2DCFL$ of picture languages are obtained. We can also consider and examine other variants of $P2DCFG$ having a mixed mode of derivation, as for example, rewriting the leftmost column along with the lowermost row or the rightmost column along with the uppermost row. In [2], membership problem and the effect of substitution rules of the form $a \to b$ have been elaborately explored for the class $P2DCFL$. These questions and other properties remain to be investigated in the $(l/u)P2DCFG$ as well as $(r/d)P2DCFG$. Also we note that grammars are relevant to the problem of generation of fractals, as certain kind of Lindenmayer system (also called, L system) [8,12], which uses context-free grammar (CFG) type of rules, has been used to generate fractals but the application of the CFG type of rules is done in parallel. The question of generation of fractals by the Pure 2D CF grammar models is a possible problem of investigation and we believe that this might require a different approach of applying the tables of rules of this 2D grammar.

Acknowledgements. The authors are grateful to the referees for their very useful comments which helped improve the presentation of the paper.

References

1. Bersani, M.M., Frigeri, A., Cherubini, A.: On some classes of 2D languages and their relations. In: Aggarwal, J.K., Barneva, R.P., Brimkov, V.E., Koroutchev, K.N., Korutcheva, E.R. (eds.) IWCIA 2011. LNCS, vol. 6636, pp. 222–234. Springer, Heidelberg (2011)
2. Bersani, M.M., Frigeri, A., Cherubini, A.: Expressiveness and complexity of regular pure two-dimensional context-free languages. Int. J. Comput. Math. **90**, 1708–1733 (2013)
3. Dassow, J., Păun, G.: Regulated Rewriting in Formal Language Theory. Springer, Heidelberg (1989)
4. Freund, R.: Control mechanisms on #-context-free array grammars. In: Păun, Gh. (ed.) Mathematical Aspects of Natural and Formal Languages, pp. 97–137. World Scientific Publishing, Singapore (1994)
5. Giammarresi, D., Restivo, A.: Two-dimensional languages. In: Rozenberg, G., Salomaa, A. (eds.) Handbook of Formal Languages, vol. 3, pp. 215–267. Springer, Heidelberg (1997)
6. Křivka, Z., Martín-Vide, C., Meduna, A., Subramanian, K.G.: A variant of pure two-dimensional context-free grammars generating picture languages. In: Barneva, R.P., Brimkov, V.E., Šlapal, J. (eds.) IWCIA 2014. LNCS, vol. 8466, pp. 123–133. Springer, Heidelberg (2014)

7. Maurer, H.A., Salomaa, A., Wood, D.: Pure grammars. Inform. Control **44**, 47–72 (1980)
8. Mishra, J.: Classification of linear fractals through L-system. In: Proceedings of the International Conferences Emerging Trends in Engineering and Technology, pp. 1–5 (2008)
9. Nakamura, A.: Picture languages. In: Davis, L.S. (ed.) Foundations of Image Understanding. International Series in Engineering and Computer Science, pp. 127–155. Kluwer Academic Publishers, Norwell (2001)
10. Rosenfeld, A.: Picture Languages. Academic Press, Reading (1979)
11. Rosenfeld, A., Siromoney, R.: Picture languages - a survey. Lang. Des. **1**, 229–245 (1993)
12. Rozenberg, G., Salomaa, A.: The Mathematical Theory of L-Systems. Academic Press, New York (1980)
13. Rozenberg, G., Salomaa, A. (eds.): Handbook of Formal Languages (3 Volumes). Springer, Berlin (1997)
14. Salomaa, A.: Formal Languages. Academic Press, Reading (1973)
15. Siromoney, R., Subramanian, K.G., Rangarajan, K.: Control on kolam arrays. Inform. Control **32**, 272–275 (1976)
16. Subramanian, K.G., Ali, R.M., Geethalakshmi, M., Nagar, A.K.: Pure 2D picture grammars and languages. Discrete Appl. Math. **157**(16), 3401–3411 (2009)
17. Subramanian, K.G., Nagar, A.K., Geethalakshmi, M.: Pure 2D picture grammars (P2DPG) and P2DPG with regular control. In: Brimkov, V.E., Barneva, R.P., Hauptman, H.A. (eds.) IWCIA 2008. LNCS, vol. 4958, pp. 330–341. Springer, Heidelberg (2008)
18. Subramanian, K.G., Rangarajan, K., Mukund, M. (eds.): Formal Models, Languages and Applications. Series in Machine Perception and Artificial Intelligence, vol. 66. World Scientific, Singapore (2006)
19. Wang, P.S.-P. (ed.): Array Grammars, Patterns and Recognizers. Series in Computer Science, vol. 18. World Scientific, Singapore (1989)

Scanning Pictures the Boustrophedon Way

Henning Fernau[1], Meenakshi Paramasivan[1]($\boxtimes$), Markus L. Schmid[1], and D. Gnanaraj Thomas[2]

[1] Fachbereich 4 – Abteilung Informatik, Universität Trier, 54286 Trier, Germany
{Fernau,Paramasivan,MSchmid}@uni-trier.de
[2] Department of Mathematics, Madras Christian College, Chennai 600059, India
dgthomasmcc@yahoo.com

Abstract. We are introducing and discussing finite automata working on rectangular-shaped arrays (i.e., pictures) in a boustrophedon reading mode. We prove close relationships with the well-established class of regular matrix (picture) languages. We derive several combinatorial, algebraic and decidability results for the corresponding class of picture languages. For instance, we show pumping and interchange lemmas for our picture language class. We also explain similarities and differences to the status of decidability questions for classical finite string automata. For instance, the non-emptiness problem for our picture-processing automaton model(s) turns out to be NP-complete. Finally, we sketch possible applications to character recognition.

1 Introduction

Syntactic considerations of digital images have a tradition of about five decades. They should (somehow) reflect methods applied to picture processing. However, one of the basic methods of scanning pictures in practice have not been thoroughly investigated from a more theoretical point of view: that of using space-filling curves. Here, we start such an investigation with what can be considered as the most simple way of defining space-filling curves: scanning line after line of an image, alternating the direction of movement every time when the image boundary is encountered (more information on the use of space-filling curves in connection with image processing or picture languages can be found in [12,14,20,23]).

We consider finite automata that work this way. We show that they are (essentially) equivalent to regular matrix languages (RMLs) as introduced in a sequence of papers of Rani Siromoney and her co-authors already in the early 1970s. These two-dimensional picture languages have connection with generation of kolam patterns [17,24]. Possibly surprisingly enough, we also present quite a number of new results for this class of picture languages, including a discussion of natural decidability questions lacking so far.

2 Our Model and Some Examples

2.1 General Definitions

In this section, we briefly recall the standard definitions and notations regarding one- and two-dimensional words and languages.

© Springer International Publishing Switzerland 2015
R.P. Barneva et al. (Eds.): IWCIA 2015, LNCS 9448, pp. 202–216, 2015.
DOI: 10.1007/978-3-319-26145-4_15

Let $\mathbb{N} := \{1, 2, 3, \ldots\}$ be the set of natural numbers. For a finite alphabet Σ, a *string* or *word* (*over* Σ) is a finite sequence of symbols from Σ, and ε stands for the *empty string*. The notation Σ^+ denotes the set of all nonempty strings over Σ, and $\Sigma^* := \Sigma^+ \cup \{\varepsilon\}$. For the *concatenation* of two strings w_1, w_2 we write $w_1 \cdot w_2$ or simply $w_1 w_2$. We say that a string $v \in \Sigma^*$ is a *factor* of a string $w \in \Sigma^*$ if there are $u_1, u_2 \in \Sigma^*$ such that $w = u_1 \cdot v \cdot u_2$. If u_1 or u_2 is the empty string, then v is a *prefix* (or a *suffix*, respectively) of w. The notation $|w|$ stands for the length of a string w.

A *two-dimensional word* (also called a *picture*, a *matrix* or an *array*) *over* Σ is a tuple

$$W := ((a_{1,1}, a_{1,2}, \ldots, a_{1,n}), (a_{2,1}, a_{2,2}, \ldots, a_{2,n}), \ldots, (a_{m,1}, a_{m,2}, \ldots, a_{m,n})),$$

where $m, n \in \mathbb{N}$ and, for every i, $1 \le i \le m$, and j, $1 \le j \le n$, $a_{i,j} \in \Sigma$. We define the *number of columns* (or *width*) and *number of rows* (or *height*) of W by $|W|_c := n$ and $|W|_r := m$, respectively. For the sake of convenience, we also denote W by $[a_{i,j}]_{m,n}$ or by a matrix in a more pictorial form. If we want to refer to the jth symbol in row i of the picture W, then we use $W[i, j] = a_{i,j}$. By Σ^{++}, we denote the set of all (non-empty) pictures over Σ. Every subset $L \subseteq \Sigma^{++}$ is a *picture language*. $L' = \Sigma^{++} - L$ is the *complement* of the picture language L.

Let $W := [a_{i,j}]_{m,n}$ and $W' := [b_{i,j}]_{m',n'}$ be two non-empty pictures over Σ. The *column concatenation* of W and W', denoted by $W \oplus W'$, is undefined if $m \ne m'$ and is the picture

$$
\begin{array}{ccccccc}
a_{1,1} & a_{1,2} & \cdots & a_{1,n} & b_{1,1} & b_{1,2} & \cdots & b_{1,n'} \\
a_{2,1} & a_{2,2} & \cdots & a_{2,n} & b_{2,1} & b_{2,2} & \cdots & b_{2,n'} \\
\vdots & \vdots & \ddots & \vdots & \vdots & \vdots & \ddots & \vdots \\
a_{m,1} & a_{m,2} & \cdots & a_{m,n} & b_{m',1} & b_{m',2} & \cdots & b_{m',n'}
\end{array}
$$

otherwise. The *row concatenation* of W and W', denoted by $W \ominus W'$, is undefined if $n \ne n'$ and is the picture

$$
\begin{array}{cccc}
a_{1,1} & a_{1,2} & \cdots & a_{1,n} \\
a_{2,1} & a_{2,2} & \cdots & a_{2,n} \\
\vdots & \vdots & \ddots & \vdots \\
a_{m,1} & a_{m,2} & \cdots & a_{m,n} \\
b_{1,1} & b_{1,2} & \cdots & b_{1,n'} \\
b_{2,1} & b_{2,2} & \cdots & b_{2,n'} \\
\vdots & \vdots & \ddots & \vdots \\
b_{m',1} & b_{m',2} & \cdots & b_{m',n'}
\end{array}
$$

otherwise. In order to denote that, e. g., $U \ominus V$ is undefined, we also write $U \ominus V = $ undef.

Example 1. Let

$$W_1 := \begin{smallmatrix} a & b & a \\ b & c & a \\ a & b & b \end{smallmatrix}, \quad W_2 := \begin{smallmatrix} b & c \\ b & a \\ c & a \end{smallmatrix} \quad \text{and} \quad W_3 := \begin{smallmatrix} a & b & c \\ c & b & b \end{smallmatrix}.$$

Then $W_1 \ominus W_2 = W_1 \oplus W_3 = $ undef, but

$$W_1 \oplus W_2 = \begin{smallmatrix} a & b & a & b & c \\ b & c & a & b & a \\ a & b & b & c & a \end{smallmatrix} \quad \text{and} \quad W_1 \ominus W_3 = \begin{smallmatrix} a & b & a \\ b & c & a \\ a & b & b \\ a & b & c \\ c & b & b \end{smallmatrix}.$$

For a picture W and $k, k' \in \mathbb{N}$, by W^k we denote the k-fold column-concatenation of W, by W_k we denote the k-fold row-concatenation of W, and $W_{k'}^k = (W^k)_{k'}$.

2.2 Boustrophedon Finite Automata

We now give the main definition of this paper, introducing a new automaton model for picture processing.

Definition 1. *A boustrophedon finite automaton, or BFA for short, can be specified as a 7-tuple $M = (Q, \Sigma, R, s, F, \#, \square)$, where Q is a finite set of states, Σ is an input alphabet, $R \subseteq Q \times (\Sigma \cup \{\#\}) \times Q$ is a finite set of rules. A rule $(q, a, p) \in R$ is usually written as $qa \to p$. The special symbol $\# \notin \Sigma$ indicates the border of the rectangular picture that is processed, $s \in Q$ is the initial state, F is the set of final states.*

We are now going to discuss the notions of configurations, valid configurations and an according configuration transition to formalize the work of BFAs, based on snapshots of their work.

Let $\square$ be a new symbol indicating an erased *position and let $\Sigma_+ := \Sigma \cup \{\#, \square\}$. Then $C_M := Q \times \Sigma_+^{++} \times \mathbb{N}$ is the set of configurations of M.*

A configuration $(p, A, \mu) \in C_M$ is valid if $1 \le \mu \le |A|_r$ and, for every i, $1 \le i \le \mu - 1$, the ith row equals $\# \square^{|A|_c - 2} \#$, for every j, $\mu + 1 \le j \le |A|_r$, the jth row equals $\# w \#$, $w \in \Sigma^{|A|_c - 2}$, and, for some ν, $0 \le \nu \le |A|_c - 2$, $w \in \Sigma^{|A|_c - \nu - 2}$, the μth row equals $\# \square^\nu w \#$, if μ is odd and $\# w \square^\nu \#$, if μ is even. Notice that valid configurations model the idea of observable snapshots of the work of the BFA.

If (p, A, μ) and (q, A', μ) are two valid configurations such that A and A' are identical but for one position (i, j), where $A'[i, j] = \square$ while $A[i, j] \in \Sigma$, then $(p, A, \mu) \vdash_M (q, A', \mu)$ if $pA[i, j] \to q \in R$. If (p, A, μ) and $(q, A, \mu + 1)$ are two valid configurations, then $(p, A, \mu) \vdash_M (q, A, \mu + 1)$ if the μth row contains only $\#$ and $\square$ symbols, and if $p\# \to q \in R$. The reflexive transitive closure of the relation $\vdash_M$ is denoted by $\vdash_M^*$.

The BFA M is deterministic, or a BDFA for short, if for all $p \in Q$ and $a \in \Sigma \cup \{\#\}$, there is at most one $q \in Q$ with $pa \to q \in R$.

The language $L(M)$ accepted by M is then the set of all $m \times n$ pictures A over Σ such that

$$(s, \#_m \oplus A \oplus \#_m, 1) \vdash_M^* (f, \#_m \oplus \square_m^n \oplus \#_m, m)$$

for some $f \in F$.

Note that the automaton works on a picture with a first and last column of only $\#$ symbols, but only the part in between these border columns is accepted. In other words, the computation starts with scanning the left uppermost corner of the picture and then working through the picture row-by-row, as the ox turns, i.e., the boustrophedon way, until the last entry of the last row is scanned. The following illustrates how a BFA scans some input picture and how a picture of a

valid configuration looks like; it can be seen that the sequence of □ only indicates
how far the input has been processed:

```
# a  b  a  b  a  #
→ → → → → → →  ↓
# b  c  a  c  b  #  ↓
↓ ← ← ← ← ← ← ← ←
↓ # a  b  b  b  b  #
→ → → → → → → → ↓
# a  b  c  b  b  #  ↓
↓ ← ← ← ← ← ← ← ←
↓ # c  b  b  a  a  #
→ → → → → → → →
```

```
#  □ □ □ □ □  #
#  □ □ □ □ □  #
#  □ □ □ b  b  #
#  a  b  c  b  b  #
#  c  b  b  a  a  #
```

Notice that since rules of the form $p\# \to q$ need not be present in R, so that
in some natural sense the classical regular string languages are a special case of
BFA languages.

Example 2. The *set of tokens* L of all sizes and of all proportions is accepted
by the BFA $M = (Q, \Sigma, R, s, F, \#, \Box)$, where $Q = \{s, s_1, s_2, s_3, s_4\}$, $\Sigma = \{\mathrm{x}, \cdot\}$,
$R = \{s\mathrm{x} \to s_1, s_1\cdot \to s_1, s_1\# \to s_2, s_1\# \to s_4, s_2\cdot \to s_2, s_2\mathrm{x} \to s_3, s_3\# \to s, s_3\# \to s_4, s_4\mathrm{x} \to s_4\}$, and $F = \{s_4\}$.

A sample token of L accepted by M is

```
x · · ·
x · · ·
x · · ·
x · · ·
x x x x
```

We now recall the notion of two-dimensional right-linear grammars (2RLG)
as given in [7]. The original definition of a 2RLG (under the name of a regular
matrix grammar (RMG)) and the properties of the corresponding class of picture
languages called RML can be found in [8,15,16,21].

Definition 2. *A two-dimensional right-linear grammar (2RLG) is defined by a
7-tuple* $G = (V_h, V_v, \Sigma_I, \Sigma, S, R_h, R_v)$, *where:*

- V_h *is a finite set of horizontal nonterminals;*
- V_v *is a finite set of vertical nonterminals;*
- $\Sigma_I \subseteq V_v$ *is a finite set of intermediates;*
- Σ *is a finite set of terminals;*
- $S \in V_h$ *is a starting symbol;*
- R_h *is a finite set of horizontal rules of the form* $V \to AV'$ *or* $V \to A$, *where* $V, V' \in V_h$ *and* $A \in \Sigma_I$;
- R_v *is a finite set of vertical rules of the form* $W \to aW'$ *or* $W \to a$, *where* $W, W' \in V_v$ *and* $a \in \Sigma$.

There are two phases of derivation of the 2RLG. In the first phase, a horizontal string of intermediate symbols is generated by means of the string grammar
$G_h = (V_h, \Sigma_I, S, R_h)$, denoted by $H(G)$. In the second phase, treating each intermediate as a start symbol, the vertical generation of the actual picture is done
in parallel, by applying a finite set of right-linear rules R_v. In order to produce
a rectangular-shaped picture, the rules of R_v must be applied in parallel; also
this means that the rules of the form $V_i \to a_i$ are all applied in every column

simultaneously to finish the picture with the generation of its last row. These grammars make sure that the columns can grow only in downward direction.

We note that our model is closely connected with 2RLG, as we will show more precisely in the following. The formalization of 2RLG that we chose is closer to our model than the original one due to Siromoney and her co-authors.

3   Characterization Results

Clearly, BFAs are a special form of 4-NFA (4-way nondeterministic finite automata, see [7]), and it is known that for these 2-dimensional automata, the deterministic variant is weaker regarding its descriptive capacity compared to the nondeterministic one. Hence, and also because of the practical relevance of the deterministic model, the following result is interesting.

Theorem 1. *BDFAs and BFAs describe the same class of picture languages.*

Proof. Apply the well-known subset construction. This works out as our BFAs are syntactically the same as classical finite automata, only the interpretation of their processing is different.   □

Next, we examine the question whether the boustrophedon processing mode of our automata is essential. To this end, let us consider yet another interpretation of finite automata, this time termed *returning finite automata*, or RFA for short. Syntactically, they are identical to BFA, so they can be again described by a 7-tuple $M = (Q, \Sigma, R, s, F, \#, \square)$. However, they always process rows from left to right. Formally, this means that we can carry over all parts of the definition of BFA apart from the notion of a valid configuration, which needs to be slightly modified. Now, a configuration $(p, A, \mu) \in C_M$ is *valid* if $1 \leq \mu \leq |A|_r$ and, for every i, $1 \leq i \leq \mu - 1$, the ith row equals $\# \square^{|A|_c-2} \#$, for every j, $\mu + 1 \leq j \leq |A|_r$, the jth row equals $\#w\#$, $w \in \Sigma^{|A|_c-2}$, and, for some ν, $0 \leq \nu \leq |A|_c - 2$, $w \in \Sigma^{|A|_c-\nu-2}$, the μth row equals $\# \square^\nu w\#$.

Theorem 2. *BFAs and RFAs describe the same class of picture languages.*

Proof. We first show how an RFA can simulate a BFA. The basic idea can be summarised as follows. On the first row, which is scanned from left to right by both automata, the RFA simulates the BFA one to one. Assume that the BFA, while moving on to the second row, changes into a state q, scans the row from right to left and enters a state p when the beginning of this row is reached. In order to simulate this behaviour, the RFA stores its current state q in the finite state control and guesses the state p. It then scans the second row from left to right (starting in state p) by applying the transitions of the BFA in reverse direction. When the end of the row is reached, the computation only proceeds if the *RFA* is in state q. This procedure is then repeated.

More formally, the states of the RFA are like BFA states or triples thereof. These triples simulate the processing of even rows and are like (ℓ, q, r), where q is the actual state, r is the state that the RFA should reach after finishing the current even row at the right border and ℓ is the state in which the RFA starts simulating the current row (left border). The formal definition is as follows.

Let $M = (Q, \Sigma, R, s, F, \#, \square)$ be some BFA. Then, the equivalent RFA $M' = (Q', \Sigma, R', s, F', \#, \square)$ is defined by $Q' = Q \cup (Q \times Q \times Q)$,

$$R' = \{pa \to q \mid pa \to q \in R, a \in \Sigma\}$$
$$\cup \{(\ell, q, r)a \to (\ell, p, r) \mid pa \to q \in R, \ell, r \in Q, a \in \Sigma\}$$
$$\cup \{p\# \to (\ell, \ell, q) \mid p\# \to q \in R, \ell \in Q\}$$
$$\cup \{(\ell, r, r)\# \to q \mid \ell\# \to q \in R, r \in Q\},$$

and $F' = F \cup \{(\ell, q, q) \mid \ell \in F, q \in Q\}$. The formal (induction) proof of the correctness of the construction is left to the reader.

The converse direction, simulating RFAs with BFAs, can be seen in a similar way. $\qquad\square$

We can likewise define finite automata that read all rows in a right-to-left fashion. A similar construction as in the previous theorem shows (again) that this model is equivalent to BFAs. All conversions between the different FA models for picture processing are at worst quadratic. This also shows that the direction of the rotation mentioned in the next theorem does not matter.

Theorem 3. *A picture language can be described by a BFA if and only if its image, rotated by 90 degrees, is in RML.*

Proof. We provide two simulations to show the claim.

Let $G = (V_h, V_v, \Sigma_I, \Sigma, S, R_h, R_v)$, be a 2-dimensional right-linear grammar. The rotation can be interpreted as $H(G)$ describing the leftmost column of the picture, while the second phase of G then means to generate all rows, starting from the intermediate string from $H(G)$. We are going to construct an equivalent RFA $M = (Q, \Sigma, R, s, F, \#, \square)$, which is sufficient for giving a BFA thanks to Theorem 2. Let $Q = (V_h \cup \{f\}) \times (V_v \cup \{f\}) \cup \{s\}$, where $f \notin V_h \cup V_v$, and $F = \{(f, f)\}$. Let R contain the following rules:

- $sa \to (S', A')$, if $S \to AS' \in R_h$ and $A \to aA' \in R_v$,
- $sa \to (f, A')$, if $S \to A \in R_h$ and $A \to aA' \in R_v$,
- $(X, A)a \to (X, A')$, if $X \in V_h \cup \{f\}$ and $A \to aA' \in R_v$,
- $(X, A)a \to (X, f)$, if $A \to a \in R_v$, $X \in V_h$,
- $(X, f)\# \to (X', A)$, if $X \to AX' \in R_h$,
- $(X, f)\# \to (f, A)$, if $X \to A \in R_h$,
- $(f, A)a \to (f, f)$, if $A \to a \in R_v$.

The idea of the construction is that the generation of columns of G is performed in the second component of the state pairs, whereas the first component corresponds to the generation of the axiom (i.e., the first row of the pictures generated by G). The crucial difference is that the first symbol of the axiom (which in the case of RFA is the first column instead of the first row) is generated and then the first row is generated before the second letter of the axiom is generated in the next row. Hence, the two phases of the picture construction of G is dovetailed.

The converse is seen as follows. Let $M = (Q, \Sigma, R, s, F, \#, \square)$ be some RFA. We construct an equivalent 2-dimensional right-linear grammar $G = (V_h, V_v, \Sigma_I, \Sigma, S, R_h, R_v)$ (generating the rotated picture) with $V_h = Q \cup \{S\}$, $\Sigma_I = Q \times Q$, and rules

- $S \rightarrow (s, r)r \in R_h$ for all $r \in Q$,
- $q \rightarrow (q, r)r \in R_h$ for all $q, r \in Q$,
- $q \rightarrow \varepsilon$ for all $q \in Q$,
- $(p, r) \rightarrow (q, r)a \in R_v$ for all $pa \rightarrow q \in R$, $r \in Q$,
- $(p, q) \rightarrow \varepsilon \in R_v$ for all $p\# \rightarrow q \in R$.

The astute reader will have noticed that we took the freedom to incorporate erasing productions for convenience, but these can be avoided by using standard formal language constructions. This concludes the proof. $\square$

Due to Theorem 3, we can inherit several properties for the class of picture languages described by BFAs. For instance, the class is not closed under rotation by 90 degrees, also known as quarter turns, see [16]. On the positive side, RML (and hence BFA picture languages) are closed under Boolean operations. More precisely, it was shown in [16] that RML (and hence BFA picture languages) are closed under union. We supplement this by the following two results.

Theorem 4. *BFA picture languages are closed under complementation.*

Proof. First, let us recall from Theorem 1 that BDFA and BFA describe the same class of picture languages. Let $M = (Q, \Sigma, R, s, F, \#, \square)$ be some BDFA. Without loss of generality, we assume that M is complete. Then we can construct a BDFA $\overline{M}$ by state complementation, i.e., $\overline{M} = (Q, \Sigma, R, s, Q - F, \#, \square)$. On some input picture $A \in \Sigma^{++}$, M reaches the same state as $\overline{M}$ and, furthermore, since both M and $\overline{M}$ are deterministic, there exists exactly one state $q \in Q$ that can be reached by M and $\overline{M}$ on input A. This directly implies that $A \in L(M)$ if and only if $A \notin L(\overline{M})$; thus, $L(\overline{M}) = \overline{L(M)}$. Hence BFA picture languages are closed under complementation. $\square$

Notice that the previous theorem has become easy because we have a deterministic model for BFAs, in contrast to what has been established for RML before. De Morgan's law now immediately yields:

Corollary 1. *BFA picture languages are closed under intersection.*

Conversely, the results we derive in the following for BFAs can be immediately read as results for RML, as well.

4   Pumping and Interchange Lemmas

Since in the pictures of an RML, the first row as well as the columns are generated by regular grammars, there are two ways to apply the pumping lemma for regular languages: we can pump the first row, which results in repetitions of a column-factor of the picture, or we can pump each column individually, which will only lead to a rectangular shaped picture if the pumping exponents are, in a sense, well-chosen. Hence, we can conclude a horizontal and a vertical pumping lemma for RML (see [9]) and, due to Theorem 3, these pumping lemmas carry over to BFA languages:

Lemma 1. *Let M be a BFA. Then there exists an $n \in \mathbb{N}$, such that, for every $W \in L(M)$ with $|W|_r \geq n$, $W = X \ominus Y \ominus Z$, $|X \ominus Y|_r \leq n$, $|Y|_r \geq 1$ and, for every $k \geq 0$, $X \ominus Y_k \ominus Z \in L(M)$.*

Lemma 2. *Let M be a BFA and let $W \in L(M)$ with $|W|_r = m$. Then there exist $n, r_1, r_2, \ldots, r_m \in \mathbb{N}$, such that, for every $W \in L(M)$ with $|W|_c \geq n$,*

$$W = (x_1 \oplus y_1 \oplus z_1) \ominus (x_2 \oplus y_2 \oplus z_2) \ominus \ldots \ominus (x_m \oplus y_m \oplus z_m),$$

$|x_i \ominus y_i|_c \leq n$, $|y_i|_c \geq 1$, $1 \leq i \leq m$, and, for every $k \geq 1$,

$$W = (x_1 \oplus y_1^{(k\, t_1)} \oplus z_1) \ominus (x_2 \oplus y_2^{(k\, t_2)} \oplus z_2) \ominus \ldots \ominus (x_m \oplus y_m^{(k\, t_m)} \oplus z_m) \in L(M),$$

where $t_i = \frac{lcm(r_1, r_2, \ldots, r_m)}{r_i}$, $1 \leq i \leq m$.

Lemma 1 is straightforward and in order to see that Lemma 2 holds, it is sufficient to note that n is the maximum of all the pumping lemma constants for the individual rows (recall that each row is generated by an individual regular grammar) and the r_i are the lengths of the factors that are pumped. Obviously, not every way of pumping the rows results in a rectangular shaped picture, so we can only pump by multiples of the t_i.

While the vertical pumping lemma has the nice property that a whole row-factor can be pumped, in the horizontal pumping lemma we can only pump factors of each individual row, that are independent from each other. As a result, this lemma does not guarantee the possibility of pumping by 0, i.e., removing a factor, which, for classical regular languages, often constitutes a particularly elegant way of showing the non-regularity of a language.

However, it can be shown that also for BFA there exists a horizontal pumping lemma that pumps whole column-factors (which then also translates into a vertical pumping lemma for RML that pumps whole row-factors).

Lemma 3. *Let M be a BFA and let $m \in \mathbb{N}$. Then there exists an $n \in \mathbb{N}$, such that, for every $W \in L(M)$ with $|W|_r \leq m$ and $|W|_c \geq n$, $W = X \oplus Y \oplus Z$, $|X \oplus Y|_c \leq n$, $|Y|_c \geq 1$ and, for every $k \geq 0$, $X \oplus Y^k \oplus Z \in L(M)$.*

Proof. Let Q be the set of states of M, let q_0 be the start state and let $n = |Q|^m + 1$. Furthermore, let $W \in L(M)$ with $|W|_r = m$ (the case $|W|_r < m$ can be handled analogously) and $|W|_c = n' \geq n$. Since M accepts W, there is an accepting computation $(p_1, W_1, m_1) \vdash_M^* (p_k, W_k, m_k)$ for W, i.e., $(p_1, W_1, m_1) = (s, \#_m \oplus A \oplus \#_m, 1)$ and $(p_k, W_k, m_k) = (f, \#_m \oplus \square_m^{n'} \oplus \#_m, m)$. We can now consider the extended configurations (p_i, W_i', m_i), where W_i' is like W_i with the only difference that each $\square$ symbol is replaced by the state that has been entered by producing this occurrence of $\square$. Since W has m rows, the maximum number of different columns in W'_k is $|Q|^m$ and since W has at least $n = |Q|^m + 1$ columns, we can conclude that $W'_k = X' \oplus \alpha' \oplus Y' \oplus \alpha' \oplus Z'$, where $|\alpha'|_c = 1$. Furthermore, $|X' \oplus \alpha' \oplus Y'|_c \leq n$ and $|\alpha' \oplus Y'|_c \geq 1$. Now let $W = X \oplus \alpha \oplus Y \oplus \alpha \oplus Z$, where $|X|_c = |X'|_c$, $|Y|_c = |Y'|_c$ and $|Z|_c = |Z'|_c$. By definition of BFA, for every $i \geq 0$, M accepts $X \oplus (\alpha \oplus Y)^i \oplus \alpha \oplus Z$. $\square$

We wish to point out that in a similar way, we can also prove a row and a column interchange lemma (the only difference is that the number n has to be chosen large enough to enforce repeating pairs of states):

Lemma 4. *Let M be a BFA. Then there exists an $n \in \mathbb{N}$, such that, for every $W \in L(M)$ with $|W|_r \geq n$, there exists a factorisation $W = V_1 \ominus X \ominus V_2 \ominus Y \ominus V_3$, $|X|_c \geq 1$, $|Y|_c \geq 1$, such that $V_1 \ominus Y \ominus V_2 \ominus X \ominus V_3 \in L(M)$.*

Lemma 5. *Let M be a BFA and let $m \in \mathbb{N}$. Then there exists an $n \in \mathbb{N}$, such that, for every $W \in L(M)$ with $|W|_r \leq m$ and $|W|_c \geq n$, there exists a factorisation $W = V_1 \oplus X \oplus V_2 \oplus Y \oplus V_3$, $|X|_c \geq 1$, $|Y|_c \geq 1$, such that $V_1 \oplus Y \oplus V_2 \oplus X \oplus V_3 \in L(M)$.*

5   Complexity Results

Only few complexity results have been obtained so far for RML. The only reference that we could find was an unpublished manuscript of Dassow [2] that also merely classified the decidability versus undecidability status of several decision problems. Here, we give the exact complexity status of the basic decidability questions for RML, formulated in terms of BFA.

We will only look into classical formal language questions, which are:

- Universal membership: Given a B(D)FA M and a picture A (as input of some algorithm), is A accepted by M?
- Non-emptiness: Given a B(D)FA M, is there some picture A accepted by M?
- Inequivalence: Given two B(D)FAs M_1 and M_2, do both automata accept the same set of pictures?

Also, we shortly discuss the issue of minimization in the context of BDFAs. Our complexity considerations will be concerned with standard complexity classes, like (N)L, i.e., (non-)deterministic logarithmic space, (N)P, i.e., (non-)deterministic polynomial time, and PSPACE, i.e., polynomial space. All hardness reductions that we sketch are implementable in deterministic logarithmic space.

Theorem 5. *The universal membership problem for BFAs is NL-complete.*

Proof. As universal membership is NL-hard for NFAs (working on strings), NL-hardness is clear. As a configuration of a BFA can be specified (basically) by the state and a pointer into the input picture, membership in NL is obvious for this problem, as well. $\square$

By a similar argument applied to deterministic devices, we conclude:

Corollary 2. *The universal membership problem for BDFAs is L-complete.*

We now turn to the problem of deciding whether or not a given B(D)FA accepts a non-empty language. Interestingly, membership in NP is not that easy to derive as it is usually the case.

Theorem 6. *The non-emptiness problem for B(D)FAs is NP-complete, even for unary input alphabets.*

Proof. We reduce from the well-known NP-complete intersection emptiness problem for finite automata with a one-letter input alphabet, see [10]. The idea is to run each of the input automata $A_1, \ldots, A_m$ in one line. We can assume that all state alphabets are disjoint. All transition rules of each A_i are also transition rules of the BFA M we are going to construct. For each final state f_i of A_i, we introduce a rule $f_i\# \rightarrow s_{i+1}$, where s_{i+1} is the initial state of A_{i+1}. The set F_m of final states of A_m is the set of final states of M. The intersection $L(A_1) \cap \cdots \cap L(A_m)$ is non-empty if and only if some word a^k is accepted by all these automata, which means that the picture a_m^k is accepted by M.

Conversely, let $M = (Q, \Sigma, R, s, F, \#, \square)$ be some B(D)FA. Given $q_0, q_f \in Q$, we first associate to M the (string-processing) NFA $A[q_0, q_f]$ with state set Q, initial state q_0, the (only) final state q_f and (unary) input alphabet $\{a\}$, as well as with a transition $pa \rightarrow q$ if and only if there is some $x \in \Sigma$ (notice that $x \neq \#$) such that $px \rightarrow q \in R$. Our nondeterministic algorithm for deciding whether or not $L(M) \neq \emptyset$ now works as follows:

- First, it guesses $2r \leq 2|Q|$ states $q_0^1, q_f^1, q_0^2, q_f^2, \ldots, q_0^r, q_f^r$ and verifies the following properties:
  - $q_0^1 = s$ and $q_f^r \in F$;
  - for each $j = 1, \ldots, r-1$: $q_f^j\# \rightarrow q_0^{j+1} \in R$.
  If these tests are passed, the computation continues (otherwise, it simply aborts).
- Call (as a black box) the NP algorithm that decides the non-emptiness of intersection for finite automata with unary input alphabet with the r automata $A[q_0^j, q_f^j]$, $1 \leq j \leq r$, as input.
- Output that $L(M) \neq \emptyset$ if and only if $\bigcap_{j=1}^r L(A[q_0^j, q_f^j]) \neq \emptyset$.

The construction is correct, as by the pumping lemmas we know that any n-state BFA M either accepts a picture with at most n rows and at most n^n columns, or it does not accept any picture at all. Hence, it is sufficient to guess at most $2n$ states as indicated. □

Theorem 7. *The inequivalence problem for BDFAs is NP-complete.*

Proof. The hardness immediately transfers from Theorem 6. As the BDFA picture languages are closed under Boolean operations (and the constructions can be carried out in polynomial time), given two BDFAs M_1 and M_2, we can construct a BDFA M such that the picture language accepted by M is the symmetric difference of the picture languages of M_1 and M_2; hence, M_1 and M_2 are equivalent if and only if the picture language of M is empty.   □

From what is known about the string language case, we immediately infer:

Corollary 3. *The inequivalence problem for BFAs is PSPACE-complete.*

Let us finally comment shortly about minimization. Here, the question is, given a BDFA A, to find a BDFA A^* that has as few states as possible but describes the same pictures as A does. Notice that this problem can be solved in polynomial time for DFAs accepting words. It might be tempting to use this well-known algorithm and apply it to a given BDFA A. In fact, this would result, in general, in a smaller automaton A' that is also picture-equivalent to A. However, in general A^* and A' can differ significantly.

Let us explain the problems with this approach with a simple example. Consider the (string) language

$$L = \{a^{385}\}^+ \# \{a^{385}\}^+ \# \{a^{385}\}^+ .$$

The smallest DFA accepting this language has 1159 states. The pictures that are accepted by this automaton (as a BDFA) are pictures of three rows, where each row has a number of a's that is a multiple of 385. However, as the length of rows have to synchronize, it is sufficient to make sure that at least one row has the right length, so that $L' = \{a^{385}\}^+ \# \{a\}^+ \# \{a\}^+$ would be another string language, whose minimal state deterministic finite automaton has only 388 states, that accepts the same picture language. As $385 = 5*7*11$, we can even see that the minimal DFA for

$$L'' = \{a^5\}^+ \# \{a^7\}^+ \# \{a^{11}\}^+ ,$$

which has 26 states, again accepts the same picture language when interpreted as a BDFA. Still, it remains unclear to us if this is the minimal deterministic automaton for the picture language in question. Even more, we do not see a general efficient methodology how to obtain minimal-state BDFAs. The only method that we can propose is brute-force, cycling through all smaller BDFAs and then testing for equivalence. This can be easily implemented in polynomial space, so that we can conclude (in terms of a decision problem).

Proposition 1. *The question to determine whether a given BFA is minimal-state can be solved in polynomial space.*

Notice that we could state this (even) for nondeterministic devices, but as indicated above, we do not know anything better for deterministic ones, either. Basically the same results could be stated for RFAs, as well (also for this type of automata, we could consider a deterministic variant). Especially, our (bad) example carries over, as the input alphabet is a singleton set.

6   Possible Applications to Character Recognition

Character recognition has always been the testbed application for picture processing methods. We refer to [4,13] and the literature quoted therein. In this regard, we are now going to discuss the recognition of some classes of characters, also (sometimes) showing the limitations of our approach, making use of the pumping lemmas that we have shown above.

For example, consider the set K of all L tokens of all sizes with fixed proportion i.e., the ratio between the two arms of L being 1. The three members of K are as follows:

$$
\begin{array}{ccc}
& & \texttt{x . . .} \\
& \texttt{x . .} & \texttt{x . . . .} \\
\texttt{x .} & \texttt{x . . } & \texttt{x . . . .} \\
\texttt{x x} \ , & \texttt{x x x} \ , & \texttt{x x x x}
\end{array}
$$

We claim that K is not accepted by any BFA. Suppose there exists a BFA to accept K. Then by Lemma 1 there exists an $n \in \mathbb{N}$, such that, for every $W \in K$ with $|W|_r \geq n$, $W = X \ominus Y \ominus Z$, $|X \ominus Y|_r \leq n$, $|Y|_r \geq 1$ and, for every $k \geq 0$, $X \ominus Y_k \ominus Z \in K$. But, unfortunately, for many values of k we get L tokens with unequal arms which are not members of K which gives a contradiction to our assumption.

On the other hand, as pointed out by Example 2, if we do not require the ratio between the two arms to be fixed, then the corresponding set of pictures can be recognised by a BFA. Similarly, the characters A (if given in the form ꓥ), E, F, H, I, P (if given in the form Ꮲ), T, U (if given in the form ⊔) can be recognised by BFA, if we do not require fixed proportions. In particular, this means that ꓲ, ꬲ, Ꮇ, Ꮋ, ꓑ, Ꭲ are valid characters as well. Note that the character I plays a special role: this set of characters can only be recognised by a BFA if it is given in the form $\{ \cdot_n^{k_1} \oplus \mathbf{x}_n \oplus \cdot_n^{k_2} \mid k_1 \leq k, k_2 \leq k, \text{ and } n \in \mathbb{N} \}$ for some fixed constant $k \in \mathbb{N}$ (i.e., a BFA is not able to recognise the set of all vertical lines).

However, if we insist on fixed proportions, then it can be easily shown that the character classes mentioned above cannot be recognised by BFAs. For example, if the length of an arm of a character (or the distance between two parallel arms) is only allowed to grow in proportion to the length of another arm, then the vertical or horizontal pumping lemma shows that this class of characters cannot be recognised by a BFA.

More generally, even single diagonal lines cannot be detected by BFA, which excludes several classes of characters from the class of BFA languages, e.g., A, K, M, N, X. We shall prove this claim more formally, by applying the vertical interchange lemma.

Let L be the set of pictures of diagonal lines from the upper-left corner to the lower-right corner, i.e.,

$$
L = \left\{ \boxtimes, \boxed{\diagdown}, \boxed{\diagdown}, \boxed{\diagdown}, \boxed{\diagdown}, \ldots \right\} .
$$

If L can be recognised by a BFA, then, according to Lemma 4, there is a picture $W \in L$ with $W = V_1 \ominus X \ominus V_2 \ominus Y \ominus V_3$, $|X|_c \geq 1$, $|Y|_c \geq 1$, and $W' = V_1 \ominus Y \ominus V_2 \ominus X \ominus V_3 \in L(M)$. The following illustrates how this leads to a contradiction:

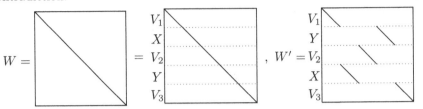

We chose L to only contain square pictures with a diagonal connecting the upper-left corner with the lower-right one for presentational reasons. In the same way, it can be shown that the set of single continuous diagonal lines cannot be recognised by BFA.

7　Discussions

Scanning pictures line by line 'as the ox turns' is for sure not a new invention in image processing. We have tried to derive a formal model that does mirror this strategy. On the one hand, we have shown that this formal model is pretty stable, as it has various characterizations, and it is even linked to RML, one of the earliest formal models of picture processing. On the other hand, there are quite some natural operations under which we would hope such a model to be closed, as, for example, quarter turns.

There are more powerful models than ours that have been proposed for picture processing, like 4-way NFAs or OTAs, see [7]. These have better closure properties, but also much weaker decidability results; for instance, the emptiness problem for such devices is undecidable.

However, OTAs are related to our model in the sense that they process a picture diagonal by diagonal, whereas our model process it row by row. The additional power seems to come from the fact that during a computation, OTAs label positions of the pictures by states and this labelling depends not only on the current symbol, but also on the state labels of the upper and left neighbours (i.e., OTAs are special versions of cellular automata). This means that information can be passed from top to bottom in every single column, whereas BFA can only accumulate information of a whole row. Notice that, when we remove this option from the way OTAs work, we arrive at a model that is possibly even closer to ours, the only difference being the way images are scanned. Clearly, diagonal scans can (now) detect diagonal lines, but now there is no way to detect vertical or horizontal ones, as would be the case for RML or BFA.

Conversely, we have seen that diagonals cannot be detected by neither RML nor BFA. Possibly, a more thorough study of different scanning schemes from the point of view of the (typical) classes of images that can be accepted would lead to new insights telling how images should be scanned by computers in practice.

The complexity status of variants of the membership problem for NFAs and DFAs is well understood. As we have seen, by the very definition of the work of BFAs and BDFAs, these results immediately transfer to this type of picture processing automata, as well. We have made this explicit above for the uniform membership problem, but the same observation also holds for the fixed membership problem. This should also enable practical implementations in future works. However, several other interesting problems appear to be much harder for picture automata, compared to their string-processing counterparts. Whether dimensionality is the core problem is still a matter of investigation.

It might be interesting to enhance the power of the automata that scan images. As we have seen, many interesting decidability questions are already pretty hard for finite automata; however, if we now extend our basic models, for example, from finite automata to weak forms of counter automata, we might be able to stay within the same complexity classes for these decision problems, while significantly increasing the usefulness of these models. For instance, we can use counters or linear-type grammars to recognize X-shaped letters (or diagonals).

The easiest way to deal with this might be to think about processing pictures with automata using two or more heads in a synchronized fashion, as already proposed in [5]. From a formal language point of view, using two heads, scanning row by row from left to right and right to left in parallel corresponds to even-linear languages as introduced in [1] and further generalized in [3,6,18,22]. We also mention bio-inspired models of computation that are closely linked to these language classes, as discussed in [11].

We are currently investigating the relationships of BFA languages with other types of picture languages introduced in the literature. We are determining the exact complexity status of the state minimization problem, as well. Also, learnability issues might be of interest, as only very few results are known about learning picture languages, see [19] as one example.

References

1. Amar, V., Putzolu, G.: On a family of linear grammars. Inf. Cont. **7**, 283–291 (1964). (Now Information and Computation)
2. Dassow, J.: Grammatical picture generation (2007). http://theo.cs.uni-magdeburg. de/lehre06w/picgen/grampicgen-text1.pdf
3. Fernau, H.: Even linear simple matrix languages: formal language properties and grammatical inference. Theor. Comput. Sci. **289**, 425–489 (2002)
4. Fernau, H., Freund, R.: Bounded parallelism in array grammars used for character recognition. In: Perner, P., Wang, P., Rosenfeld, A. (eds.) SSPR 1996. LNCS, vol. 1121, pp. 40–49. Springer, Heidelberg (1996)
5. Fernau, H., Freund, R., Holzer, M.: Character recognition with k-head finite array automata. In: Amin, A., Dori, D., Pudil, P., Freeman, H. (eds.) SPR 1998 and SSPR 1998. LNCS, vol. 1451, pp. 282–291. Springer, Heidelberg (1998)
6. Fernau, H., Sempere, J.M.: Permutations and control sets for learning non-regular language families. In: Oliveira, A.L. (ed.) ICGI 2000. LNCS (LNAI), vol. 1891, pp. 75–88. Springer, Heidelberg (2000)

7. Giammarresi, D., Restivo, A.: Two-dimensional languages. In: Rozenberg, G., Salomaa, A. (eds.) Handbook of Formal Languages, vol. III, pp. 215–267. Springer, Berlin (1997)
8. Krithivasan, K., Siromoney, R.: Array automata and operations on array languages. Int. J. Comput. Math. 4(A), 3–30 (1974)
9. Krithivasan, K., Siromoney, R.: Characterizations of regular and context-free matrices. Int. J. Comput. Math. 4(A), 229–245 (1974)
10. Lange, K.J., Rossmanith, P.: The emptiness problem for intersections of regular languages. In: Havel, I.M., Koubek, V. (eds.) MFCS 1992. LNCS, vol. 629, pp. 346–354. Springer, Heidelberg (1992)
11. Nagy, B.: On a hierarchy of $5' \rightarrow 3'$ sensing Watson-Crick finite automata languages. J. Logic Comput. 23(4), 855–872 (2013)
12. Niedermeier, R., Reinhardt, K., Sanders, P.: Towards optimal locality in mesh-indexings. Discrete Appl. Math. 117, 211–237 (2002)
13. Fernau, H., Freund, R., Holzer, M.: Regulated array grammars of finite index. In: Păun, G., Salomaa, A. (eds.) Grammatical Models of Multi-Agent Systems, pp. 157–181 (Part I) and 284–296 (Part II). Gordon and Breach, London (1999)
14. Sagan, H.: Space-Filling Curves. Springer, Heidelberg (1994)
15. Siromoney, G., Siromoney, R., Krithivasan, K.: Abstract families of matrices and picture languages. Comput. Graph. Image Process. 1, 284–307 (1972)
16. Siromoney, G., Siromoney, R., Krithivasan, K.: Picture languages with array rewriting rules. Inf. Control 22(5), 447–470 (1973). (Now Information and Computation)
17. Siromoney, G., Siromoney, R., Krithivasan, K.: Array grammars and kolam. Comput. Graph. Image Process. 3, 63–82 (1974)
18. Siromoney, R.: On equal matrix languages. Inf. Control 14, 133–151 (1969). (Now Information and Computation)
19. Siromoney, R., Mathew, L., Subramanian, K.G., Dare, V.R.: Learning of recognizable picture languages. In: Nakamura, A., Nivat, M., Saoudi, A., Wang, P.S.P., Inoue, K. (eds.) ICPIA 1992. LNCS, vol. 654, pp. 247–259. Springer, Heidelberg (1992)
20. Siromoney, R., Subramanian, K.G.: Space-filling curves and infinite graphs. In: Ehrig, H., Nagl, M., Rozenberg, G. (eds.) Graph Grammars 1982. LNCS, vol. 153, pp. 380–391. Springer, Heidelberg (1983)
21. Subramanian, K.G., Revathi, L., Siromoney, R.: Siromoney array grammars and applications. Int. J. Pattern Recogn. Artif. Intell. 3, 333–351 (1989)
22. Takada, Y.: Learning even equal matrix languages based on control sets. In: Nakamura, A., Nivat, M., Saoudi, A., Wang, P.S.P., Inoue, K. (eds.) ICPIA 1992. LNCS, vol. 654, pp. 274–289. Springer, Heidelberg (1992)
23. Witten, I.H., Wyvill, B.: On the generation and use of space-filling curves. Softw. Pract. Experience 13, 519–525 (1983)
24. Yanagisawa, K., Nagata, S.: Fundamental study on design system of kolam pattern. Forma 22, 31–46 (2007)

Accepting H Iso-Array System

V. Masilamani[1], D.K. Sheena Christy[2(✉)], D.G. Thomas[3], A.K. Nagar[4],
and T. Robinson[3]

[1] Department of Computer Science and Engineering,
IIITD&M Kanchipuram, Chennai 600 036, India
`masila@iiitdm.ac.in`
[2] Department of Mathematics, SRM University, Kattankulathur 603 203, India
`sheena.lesley@gmail.com`
[3] Department of Mathematics, Madras Christian College, Chennai 600 059, India
`dgthomasmcc@yahoo.com, robin.mcc@gmail.com`
[4] Department of Mathematics and Computer Science,
Liverpool Hope University, Liverpool, UK
`nagara@hope.ac.uk`

Abstract. In this paper, we introduce the notion of accepting H iso-array system which is defined by a finite set of permitting iso-arrays as a counter part of the well investigated accepting splicing systems defined by Mitrana et al. (2010). We compare the generative power of this system with that of some existing models such as controlled sequential pasting system, regular iso-array grammar, context-free iso-array grammar and study some properties of accepting H iso-array systems.

Keywords: DNA computing · Splicing system · iso-picture languages · Kolam patterns

1 Introduction

In the syntactic approach of pattern recognition and image analysis, there have been several studies on theoretical models in the last few decades for generating or recognizing two dimensional objects, pictures and picture languages [2].

Motivated by the study of recognizable rectangular picture languages using rectangular tiles [2], a new concept of recognizability has been introduced and studied for a class of picture languages called iso-picture languages through iso-triangular tiling systems (ITS) [5].

Iso-arrays are made up of isosceles right angled triangles and an iso-picture is a picture, formed by catenating iso-arrays of same size [5,6]. By making use of iso-picture languages, one can generate variety of picture languages that cannot be generated by earlier models available in the literature [2]. The hexagonal picture languages, rectangular picture languages, languages of rhombuses and triangles are some of the examples of iso-picture languages. One application of the study of iso-picture languages is its use in the generation of interesting kolam patterns [7]. Another application of this study is the use of tiling patterns to decorate

© Springer International Publishing Switzerland 2015
R.P. Barneva et al. (Eds.): IWCIA 2015, LNCS 9448, pp. 217–231, 2015.
DOI: 10.1007/978-3-319-26145-4_16

and cover floors and walls. Iso-pictures generation has application in computer graphics. The isosceles triangles used in iso-picture generation are the primitive elements in image synthesis in computer graphics. Hence it is appropriate to think of image synthesis (generation) model using triangle primitives. On the other hand, if the image synthesized has the triangles as primitives then to recognize objects that are present in the image, a recognizing system is required. In this paper, image accepting / recognizing models using isosceles triangles as primitives are proposed. To enable parallel processing for image accepting or recognition bio-inspired parallel computing models have been studied in this paper.

Bio-inspired computing is a field devoted to tackling complex problems using computational methods modeled after designed principles encountered in nature. One of the topics covered in bio-inspired computing is DNA computing [11,12]. There has been a lot of interest in the study of formal language theory applied to DNA computing. A specific model of DNA recombination is the splicing operation which consists of "cutting" DNA sequences and then "pasting" the fragments, under the action of restriction enzymes and ligases. The H system was introduced by Head [3] for generating string languages. V. Mitrana et al. [10] proposed a novel view of splicing systems and defined the concept of an accepting splicing system for words. Two ways of iterating the splicing operation in the generating case and the computational power of generating splicing systems defined by these operations were investigated. Arroyo et al. [1] introduced the generalization of accepting splicing system with permitting and forbidding words and studied the computational power of the variants of accepting splicing systems. In [8] the notion of accepting H-array splicing system and the computational power of this system has been proposed and studied.

A simple and effective parallel splicing on iso-arrays has been introduced [9] to generate iso-array languages. Splicing rules that involve tiles of isosceles right angled triangles are considered. Two iso-arrays are spliced in four directions namely horizontal, vertical, right and left. Some closure properties of H-iso-array systems under language theoretic operations – union, concatenation; geometric operation – reflection about the base line and rotation by $90°$, $180°$ and $270°$ are considered. The class of array languages accepted by these systems have been compared with the classes of array languages generated by existing models and considered a variant of accepting H-array system with permitting and forbidding arrays.

The present paper introduces the notion of accepting H iso-array system which is defined by a finite set of permitting iso-arrays as a counter part of the well investigated accepting splicing systems for words defined by Mitrana et al. [10]. We compare the generative powers of the accepting H-iso-array system with that of a sequential generating model called controlled sequential pasting system and generative powers of other existing models of two dimensional iso-array grammars, namely regular iso-array grammars, context-free iso-array grammars and iso-array local systems.

2   Basic Definitions

In this section we recall the basic notions of iso-picture language [5].

Definition 1. Let $\Sigma_I = \left\{ {}_{S_1}\triangle{}^{S_3}_{S_2\ A}, \; {}_{S_3}\triangledown{}^{S_2}_{B\ S_1}, \; {}_{S_3}\triangleleft{}^{S_1}_{C}|S_2, \; S_2|{}^{S_3}_{D}\triangleright{}^{S_1} \right\}$

The sides of each tile in Σ_I are of length $\frac{1}{\sqrt{2}}, 1, \frac{1}{\sqrt{2}}$. An iso-array is an arrangement of isosceles right angled triangles of tiles from the set Σ_I.

A U-iso-array of size m is formed exclusively by m number of $\triangle_A$ tiles on side S_2 and it is denoted by U_m. It will have m_2 tiles in total (including the m number of A tiles on S_2). Similarly D-iso-array, L-iso-array and R-iso-array are formed exclusively by B-tile, C-tile and D-tile on side S_2 respectively.

Example 1. The following are the iso-arrays of size three.

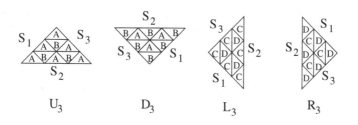

$$U_3 \qquad D_3 \qquad L_3 \qquad R_3$$

Fig. 1. iso-arrays of size three.

Definition 2. Iso-arrays of same-size can be catenated using the following four types of catenations of iso-arrays.
(i) Horizontal Catenation ($\ominus$):
The only possible catenation is $U \ominus D$.
(ii) Vertical Catenation ($\oslash$):
$L \oslash R$ is the only possible catenation of iso-arrays.
(iii) Right Catenation ($\oslash$):
The following catenations are possible under right catenation:
$D \oslash U, \; R \oslash U, \; D \oslash L, \; R \oslash L$.
(iv) Left Catenation ($\obslash$):
The following catenations are possible under left catenation:
$U \obslash D, \; U \obslash L, \; L \obslash R, \; R \obslash D$.
The catenation can be defined between any two gluable iso-arrays of same size.

Example 2. The horizontal catenation of U and D-iso-arrays of size 3 is shown in Fig. 2.

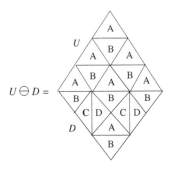

$$U \ominus D =$$

Fig. 2. Horizontal catenation of U and D-iso-arrays of size 3.

Definition 3. Let Σ_I be the finite alphabet of tiles A, B, C and D. An iso-picture of size (n, m), $(n, m \geq 1)$ over Σ_I is a picture formed by catenating n-iso-arrays of size m. The number of tiles in any iso-picture of size (n, m) is nm^2.

Any two iso-pictures of sizes (n_1, m) and $(n_2, m), n_1, n_2, m \geq 1$ can be cate-nated using the rules of catenation of iso-arrays, provided the sides of iso-pictures are gluable. The set of all iso-pictures over Σ_I is denoted by Σ_I^{**}. An iso-picture language L over Σ_I is a subset of Σ_I^{**}.

The following definitions enumerate the basic concepts of iso-array splicing rules and the main notion of H iso-array systems [13].

Definition 4. Let $\#$ and $\$$ be two special symbols not in Σ_I.

The splicing rule $\alpha_1 \# \alpha_2 \$ \alpha_3 \# \alpha_4$ over Σ_I is called the

(i) horizontal splicing rule if $\alpha_1 = U_m$ or Λ, $\alpha_2 = D_m$ or Λ, $\alpha_3 = U_m$ or Λ and $\alpha_4 = D_m$ or Λ

(ii) vertical splicing rule if $\alpha_1 = L_m$ or Λ, $\alpha_2 = R_m$ or Λ, $\alpha_3 = L_m$ or Λ and $\alpha_4 = R_m$ or Λ

(iii) right splicing rule

(a) $\alpha_1 = D_m$ or Λ, $\alpha_2 = U_m$ or Λ, $\alpha_3 = D_m$ or Λ and $\alpha_4 = U_m$ or Λ (or)

(b) $\alpha_1 = R_m$ or Λ, $\alpha_2 = U_m$ or Λ, $\alpha_3 = R_m$ or Λ and $\alpha_4 = U_m$ or Λ (or)

(c) $\alpha_1 = D_m$ or Λ, $\alpha_2 = L_m$ or Λ, $\alpha_3 = D_m$ or Λ and $\alpha_4 = L_m$ or Λ (or)

(d) $\alpha_1 = R_m$ or Λ, $\alpha_2 = L_m$ or Λ, $\alpha_3 = R_m$ or Λ and $\alpha_4 = L_m$ or Λ

(iv) left splicing rule if

(a) $\alpha_1 = U_m$ or Λ, $\alpha_2 = D_m$ or Λ, $\alpha_3 = U_m$ or Λ and $\alpha_4 = D_m$ or Λ (or)

(b) $\alpha_1 = U_m$ or Λ, $\alpha_2 = L_m$ or Λ, $\alpha_3 = U_m$ or Λ and $\alpha_4 = L_m$ or Λ (or)

(c) $\alpha_1 = L_m$ or Λ, $\alpha_2 = R_m$ or Λ, $\alpha_3 = L_m$ or Λ and $\alpha_4 = R_m$ or Λ (or)

(d) $\alpha_1 = R_m$ or Λ, $\alpha_2 = D_m$ or Λ, $\alpha_3 = R_m$ or Λ and $\alpha_4 = D_m$ or Λ.

The set of all horizontal, vertical, left and right splicing rules over Σ_I are denoted by $R_\ominus$, $R_\oplus$, $R_\oslash$, $R_\oslash$ respectively.

Definition 5. An H iso-array scheme is a tuple $\sigma = (\Sigma_I, R_\ominus, R_\oplus, R_\oslash, R_\oslash)$ where Σ_I is an alphabet, $R_\ominus$ is a finite set of horizontal splicing rules. Similarly, $R_\oplus, R_\oslash, R_\oslash$ are finite sets of vertical, right and left splicing rules.

An H iso-array system is defined by $S = (\sigma, E)$ where E is finite subset of Σ_I^{**}. We define $\sigma(L)$ as

$$
\sigma(L) = \left\{ p \in \Sigma_I^{**} \middle/ \begin{array}{c} (p_1, p_2) \overset{\ominus}{\to} p \text{ (or)} \\ (p_1, p_2) \overset{\oplus}{\to} p \text{ (or)} \\ (p_1, p_2) \overset{\oslash}{\to} p \text{ (or)} \\ (p_1, p_2) \overset{\oslash}{\to} p \end{array} \right\}
$$

for some $p_1, p_2 \in L$.
$\sigma^*(L)$ is defined iteratively as follows:
$\sigma^0(L) = L$, $\sigma^{i+1}(L) = \sigma^i(L) \cup \sigma(\sigma^i(L))$, for $i \geq 0$
$$\sigma^*(L) = \bigcup_{i=0}^{\infty} \sigma^i(L).$$

The family of iso-picture languages generated by these splicing systems is denoted by FHIA.

Example 3. Horizontal and Vertical Splicing Rule:

Let
$$\Sigma_I = \{ \text{} \}$$

and $E = \{w\}$ where $w = $

Let $p_1 = w$ and $p_2 = w$ and the splicing rules are

$$R_{\oplus} = \{ \text{} \}$$

$$R_{\ominus} = \{ \text{} \}$$

Then $(p_1, p_2) \overset{\ominus}{\vdash} p_3$ and $(p_1, p_2) \overset{\oplus}{\vdash} p_4$ where a member generated by the system is given in Fig. 3.

3   Accepting H Iso-Array System

In this section we introduce the notion of accepting H Iso-array system with two methods of iterated splicing namely uniform and non-uniform methods. We study the generative power of this system and compare with that of well-known systems such as controlled sequential pasting system, regular iso-array grammar (RIAG) and context free iso-array grammar (CFIAG).

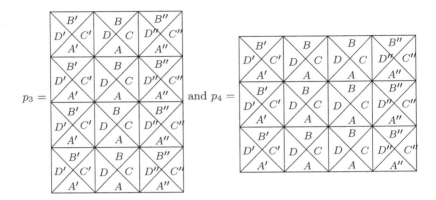

Fig. 3. A generated member of FHIA by using vertical (p_3) and horizontal (p_4) splicing rule.

Definition 6. An accepting H iso-array system (AHIAS) is a pair $\Gamma = (S, P)$ where $S = (\sigma, E)$ is a H iso-array system and P is a finite subset of Σ_I^{**}. Let $w \in \Sigma_I^{**}$. We define the uniform method (usual way) of the iterated array splicing of Γ as follows:

$\sigma^0(E, w) = \{w\}$,
$\sigma^{i+1}(E, w) = \sigma^i(E, w) \cup \sigma(\sigma^i(E, w) \cup E)$, $i \geq 0$
$\sigma^*(E, w) = \bigcup_{i=0}^{\infty} \sigma^i(E, w)$.

The language accepted by an accepting H iso-array system Γ is $L(\Gamma) = \{w \in \Sigma_I^{**} | \sigma^*(E, w) \cap P \neq \phi\}$.

The class of languages accepted by accepting H iso-array system is denoted by $\mathcal{L}(AHIAS)$.

Let $\Gamma = (S, P)$ be an accepting H iso-array system and $w \in \Sigma_I^{**}$. The non-uniform variant of iterated splicing of Γ is defined as follows.

$\tau^0(E, w) = \{w\}$;
$\tau^{i+1}(E, w) = \tau^i(E, w) \cup \sigma(\tau^i(E, w), E)$, $i \geq 0$,
$\tau^*(E, w) = \bigcup_{i=0}^{\infty} \tau^i(E, w)$.

An iso-array $w \in \Sigma_I^{**}$ is said to be accepted by Γ in non-uniform way if $\tau^*(E, w) \cap P \neq \phi$.

The language accepted by Γ in non-uniform way is $L_n(\Gamma) = \{w \in \Sigma_I^{**} | \tau^*(E, w) \cap P \neq \phi\}$.

The class of all languages accepted by AHIAS in non-uniform way is denoted by $\mathcal{L}_n(AHIAS)$.

Example 4. Let $\Gamma = (S, P)$ be an uniform variant of AHIAS with $S = (\sigma, E)$ where $\sigma = (\Sigma_I, R_\ominus, R_\oplus, R_\oslash, R_\oslash)$

Here $\Sigma_I = \{ \overset{\triangle}{{}_a}, \overset{b}{\triangledown}, \triangleleft_c, {}_d\triangleright, \overset{\triangle}{{}_{a'}}, \overset{b'}{\triangledown}, \triangleleft_{c'}, {}_{d'}\triangleright \}$, $E = \{\Lambda\}$ and

$$P = \{p\} \text{ where } p = \begin{array}{|c|c|} \hline d\,\backslash c & d'\,\backslash c' \\ \text{b} & \text{b}' \\ \hline a & a' \\ d'\,\backslash c & d\,\backslash c \\ \text{b}' & \text{b} \\ a' & a \\ \hline \end{array}$$

$R_{\mathbb{O}} = \{P_1 : \langle_{c}\!\!\!\triangleleft | \#\Lambda\$\Lambda\# | \triangleright_{d'} \rangle \quad P_2 : \langle_{c'}\!\!\!\triangleleft | \#\Lambda\$\Lambda\# | \triangleright_{d} \rangle\}, \quad R_{\ominus} = R_{\oslash} = R_{\obslash} = \phi.$

$$w = \begin{array}{|c|c|c|} \hline d\,\backslash c & d'\,\backslash c' & d'\,\backslash c' \\ \text{b} & \text{b}' & \text{b}' \\ \hline a & a' & a' \\ d'\,\backslash c & d\,\backslash c & d\,\backslash c \\ \text{b}' & \text{b} & \text{b} \\ a' & a & a \\ \hline \end{array} = q$$

$\sigma^0(E, w) = \{w\}$

$\sigma^1(E, w) = \sigma^0(E, w) \cup \sigma(\sigma^0(E, w) \cup E)$

$$= \{w\} \cup \left\{ \begin{array}{|c|c|} \hline d\,\backslash c & d'\,\backslash c' \\ \text{b} & \text{b}' \\ \hline a & a' \\ d'\,\backslash c & d\,\backslash c \\ \text{b}' & \text{b} \\ a' & a \\ \hline \end{array} \right\}$$

Clearly $\sigma^1(E, w) \cap P \neq \phi$. Hence w is accepted by Γ. The language accepted by this AHIAS is $L_1 = \{u \in \Sigma_I^{**}/u = p^n \mathbb{O} q^n, r \mathbb{O} s^n, r^n \mathbb{O} s, p^n \mathbb{O} s, n \geq 1\}$ where

$$r = \begin{array}{|c|} \hline d\,\backslash c \\ \text{b} \\ \hline a \\ d'\,\backslash c' \\ \text{b}' \\ a' \\ \hline \end{array} \quad \text{and} \quad s = \begin{array}{|c|} \hline d'\,\backslash c' \\ \text{b} \\ \hline a' \\ d\,\backslash c \\ \text{b} \\ a \\ \hline \end{array} \quad (\, p^n = p\mathbb{O}p\mathbb{O}\ldots\mathbb{O}p, \; n \text{ times})$$

We now recall the definition of controlled sequential pasting systems (CSPS) [4] to compare the generative power of AHIAS with that of CSPS.

Definition 7. A controlled sequential pasting system $(CSPS)$ is a 5-tuple (Σ_I, I, P, C, t_0) where Σ_I is a finite nonempty set of isosceles right angled triangular tiles, I is the finite set of edge labels of tiles in Σ_I. P is a finite set of tables $\{T_1, T_2, \ldots, T_k\}$, and each $T_i, i = 1, 2, \ldots, k$ is one of left, right, up, down, leftup, leftdown, rightup or rightdown table consisting, respectively, of a finite set of left, right, up, down, leftup, leftdown, rightup or rightdown rules only. The rules of the tables are applied in parallel to the respective edges of the pattern derived. C is a control language over P and $t_0 \in \Sigma_I^{++}$ is the axiom.

Suppose $t_i \underset{w}{\Rightarrow} t_{i+1}$ denotes the derivation of the pattern t_{i+1} being generated from t_i in one step by applying the control word $w \in C$. A pattern t_j is generated from t_0 by applying the control word $w = w_1, w_2 \ldots w_j \in C$ where $w_i \in C(1 \leq i \leq j)$ if there exists a sequence of derivations $t_0 \underset{w_1}{\Rightarrow} t_1 \underset{w_2}{\Rightarrow} t_2 \underset{w_3}{\Rightarrow} \cdots \underset{w_j}{\Rightarrow} t_j$ and is denoted by $t_0 \overset{*}{\underset{w}{\Rightarrow}} t_j$.

The set of all patterns generated by the system is denoted by

$$T(S) = \left\{ t_j \in \Sigma_I^{**} : t_0 \overset{*}{\underset{c}{\Rightarrow}} t_j / j \geq 0, c \in C \right\}.$$

The control language C may be either regular, context-free, or context-sensitive, in which case we attach the corresponding name to the control. The class of all languages generated by CSPS is denoted by $\mathcal{L}(CSPS)$.

Example 5. Let us consider the language L_2 to be the set of all staircases of fixed proposition with tiles A, B, C and D. The following sequential pasting system $S = (\Sigma_I, I, P, C, t_0)$ generates the set of all staircases of fixed proposition using a regular control.

Here $\Sigma_I = \left\{ \begin{array}{c} a_1 \triangle a_3 \\ a_2 \end{array}, \begin{array}{c} b_2 \\ b_3 \nabla b_1 \end{array}, \begin{array}{c} c_3 \\ c_1 \end{array} c_2, \begin{array}{c} d_1 \\ d_2 \triangleright d_3 \end{array} \right\}$

$I = \{a_1, a_2, a_3, \ldots, d_1, d_2, d_3\}, \quad t_0 = \triangleright D$

$P = \{R_1, R_2, R_{u_1}, R_{u_2}, U\}$
$R_1 = \{(b_1, a_1)\}, \; R_2 = \{(a_3, b_3)\}$
$R_{u_1} = \{(d_1, b_3)\}, \; R_{u_2} = \{(a_3, c_1)\}$
$U = \{(c_3, d_3)\}$
$C = \{R_{u_1}(R_1 R_2 R_1 R_{u_2} U R_{u_1})^n / n \geq 1\}$
A sample derivation is shown in Fig. 4.

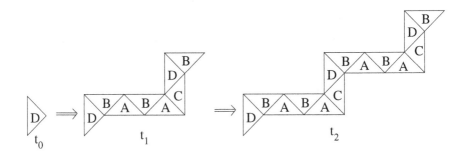

Fig. 4. Staircase

Theorem 1. *The classes $\mathcal{L}(AHIAS)$ and $\mathcal{L}(CSPS)$ are incomparable.*

Proof. To prove that $\mathcal{L}(CSPS)$-$\mathcal{L}(AHIAS) \neq \phi$, let us consider the language L_2 given in Example 5, L_2 cannot be accepted by any AHIAS. In fact, considering various possibilities of splicing rules in an AHIAS to obtain members of L_2 alone we observe that the staircases of unequal proportions are also accepted by the system eventhough P may be finite and E is finite. Hence we cannot find any finite $P \subseteq \Sigma_I^{**}$ such that $\sigma^*(A, w) \cap P \neq \phi$. Hence $L_2 \notin \mathcal{L}(AHIAS)$.

To Prove $\mathcal{L}(AHIAS)$-$\mathcal{L}(CSPS) \neq \phi$, consider the language L_1 accepted by the AHIAS, Γ as shown in Example 4. But it cannot be generated by any

CSPS since we cannot find any regular control pattern of pasting system for the language L_1. Hence the classes $\mathcal{L}(AHIAS)$ and $\mathcal{L}(CSPS)$ are incomparable. $\square$

We now review the notions of regular iso-array grammar and context free iso-array grammar [6].

Definition 8. A regular iso-array grammar (RIAG) is a structure $G = (N, T,$
$P, S)$ where $N = \left\{ \triangle_A, \nabla^B, \triangleleft^C, \triangleright_D \right\}$, $T = \left\{ \triangle_a, \nabla^b, \triangleleft^c, \triangleright_d \right\}$ are finite
sets of symbols called nonterminals and terminals; $N \cap T = \phi$. $S \in N$ is the start symbol or the axiom.
P consists of rules of the following forms:

(1) (2)

(3) (4)

(5) (6)

Similar rules can be given for the other tiles ∇_B, $\triangleleft^C$, $\triangleright_D$.

The regular iso-array language (RIAL) generated by G is defined by $\{W \mid S \Rightarrow^*_G W, W$ is a finite connected array over $T\}$ and is denoted by $L(G)$.

Definition 9. A context-free iso-array Grammar (CFIAG) is a structure $G = (N, T, P, S)$ where

$$N = \left\{ \triangle_A, \nabla^B, \triangleleft^C, \triangleright_D \right\}, T = \left\{ \triangle_a, \nabla^b, \triangleleft^c, \triangleright_d \right\}$$

are finite nonempty sets of symbols called nonterminals and terminals, $N \cap T = \phi$.
$S \in N$ is the start symbol or the axiom. P consists of rules of the form $\alpha \to \beta$,
where α and β are finite connected arrays of one or more triangular tiles over

$N \cup T \cup \left\{ \triangle_{\#A}, \nabla^{\#B}, \triangleleft^{\#C}, \triangleright_{\#D} \right\}$ and satisfy the following conditions:

1. The shapes of α and β are identical.
2. α contains exactly one nonterminal and possibly one or more #'s.
3. Terminals in α are not rewritten.
4. The application of the rule $\alpha \to \beta$ preserves the connectedness of the host array (that is, the application of the rule to a connected array results in a connected array).

The rule $\alpha \rightarrow \beta$ is applicable to a finite connected array γ over $N \cup T \cup$
$\left\{ \text{\includegraphics{#A}}, \text{\includegraphics{#B}}, \text{\includegraphics{#C}}, \text{\includegraphics{#D}} \right\}$, if α is a subarray of γ and in a direct derivation step,
one of the occurrences of α is replaced by β, yielding a finite connected array δ.
We write $\gamma \Rightarrow_G \delta$. The reflexive transitive closure of $\Rightarrow_G$ is denoted by $\Rightarrow_G^*$.

The context free iso-array language (CFIAL) generated by G is defined by
$\{\delta : S \Rightarrow_G^* \delta, \delta \text{ is a finite connected array over } T\}$ and is denoted by $L(G)$.

The class of all languages generated by RIAG and CFIAG are denoted by
$\mathcal{L}(RIAG)$ and $\mathcal{L}(CFIAG)$.

Theorem 2. *The classes $\mathcal{L}(AHIAS)$ and $\mathcal{L}(RIAG)$ are incomparable.*

Proof. To prove that $\mathcal{L}(RIAG)$-$\mathcal{L}(AHIAS) \neq \phi$, consider the language L_3, the
set of all hexagons of all sizes. It is proved that L_3 is generated by RIAG [6]. But
it cannot be accepted by any AHIAS, as we cannot find any finite P, a subset
of Σ_I^{**} such that $\sigma^*(E, w) \cap P \neq \phi$ with w being of very large size.

To prove that $\mathcal{L}(AHIAS)$-$\mathcal{L}(RIAG) \neq \phi$, let us consider L_4, the set of
all triangles of all sizes with tiles $\triangle_a$ and ∇^b. It cannot be generated by
any regular iso-array grammar as the junction in (triangle figure with a, a, b, a) cannot be
handled by the regular iso-array grammar [5]. The language L_4 can be accepted
by the following uniform variant of AHIAS $\Gamma = (S, P)$ with $S = (\sigma, E)$ where

$$\sigma = (\Sigma_I, R_\ominus, R_\oplus, R_\oslash, R_\obackslash), E = \{\Lambda\} \text{ and } P = \left\{ \text{(triangle figure with a, a, b, a)} \right\}$$

Here $\Sigma_I = \left\{ \triangle_a, \nabla^b \right\}$
$R_\ominus = \{ \triangle_a \text{ \#} \nabla_b \text{\$} \triangle_a \text{\#} \Lambda\}$ and $R_\oplus = R_\oslash = R_\obackslash = \phi$.   □

Theorem 3. *The class $\mathcal{L}(AHIAS)$ and the $\mathcal{L}(CFIAG)$ are incomparable.*

Proof. To prove that $\mathcal{L}(CFIAG)$-$\mathcal{L}(AHIAS) \neq \phi$, consider the language L_5 is
the set of all digitized right angled triangles of all sizes. L_5, generated by the
following CFIAG:
$G = (N, T, P, S)$ where

$$N = \left\{ \triangle_A, \nabla_B, \triangleleft_C, \triangleright_D, \nabla_B', \triangleright_D' \right\},$$

$$T = \left\{ \triangle_a, \nabla_b, \triangleleft_c, \triangleright_d, \triangle_{a'}, \nabla_{b'}, \triangleleft_{c'}, \triangleright_{d'} \right\}$$

P consists of the following rules:

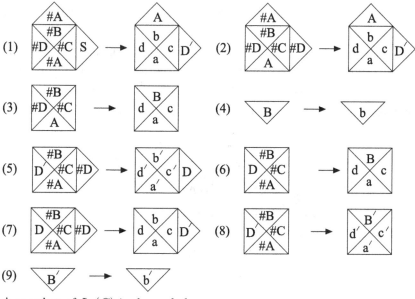

A member of $L_5(G)$ is shown below

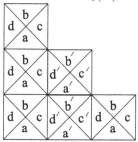

But $L_5(G)$ cannot be accepted by any AHIAS, as there is no finite P, a subset of Σ_I^{**} such that $\sigma^*(E, w) \cap P \neq \phi$, for $w \in L_5(G)$.

To prove that $\mathcal{L}(AHIAS)$-$\mathcal{L}(CFIAG) \neq \phi$, let us consider the language L_1 considered in Example 4. It cannot be generated by any CFIAG but it can accepted by a AHIAS. Hence the classes $\mathcal{L}(AHIAS)$ and $\mathcal{L}(CFIAG)$ are incomparable.   □

Definition 10. Let p be an iso-picture of size (n, m). We denote by $B_{n',m'}(p)$, the set of all subiso-pictures of p of size (n', m'), where $n' \leq n, m' \leq m$.

Let p be an iso-picture over Σ_I. Then $\hat{p}$ is an iso-picture obtained by surrounding p with special boundary symbols ⟨symbols⟩ $\notin \Sigma_I$.

An iso-picture language $L \subseteq \Sigma_I^{**}$ is called local if there exists a finite set θ of iso-arrays of size 2 over $\Sigma_I \cup \{$ ⟨symbols⟩ $\}$ such that $L = \{p \in \Sigma_I^{**}/B_{1,2}(\hat{p}) \subseteq \theta\}$ and is denoted by $L(\theta)$.

The family of local iso-picture languages will be denoted by $ILOC$.

Theorem 4. *The classes $\mathcal{L}(AHIAS)$ and $ILOC$ are incomparable but not disjoint.*

Proof. To prove $ILOC$-$\mathcal{L}(AHIAS) \neq \phi$. let us consider the language L_6 as the set of all iso-picture language of rhombuses, where the elements (tiles) along the two diagonals are represented by the tiles $\overset{A_2}{\triangle}$ and $\overset{B_2}{\triangledown}$ and elements in the remaining positions are represented by tiles $\overset{A_1}{\triangle}$ and $\overset{B_1}{\triangledown}$, a member of which is shown below.

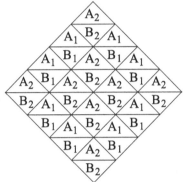

The language L_6 is local [5]. But L_6 cannot be accepted by any AHIAS in uniform way, as it is not possible to fix a finite P such that $\sigma^*(E, w) \cap P \neq \phi$, where $w \in L_6$.

To prove $\mathcal{L}(AHIAS)$-$ILOC \neq \phi$, let L_7 be the iso-picture language of rhombuses of size $(2, m)$, $m \geq 1$ over $\Sigma_I = \{\overset{A}{\triangle}, \overset{B}{\triangledown}\}$. It is shown that L_7 is not local [5]. But the language L_7 can be accepted by the uniform variant of AHIAS, $\Gamma = (S, P)$ where $S = (\sigma, E)$, $\sigma = (\Sigma_I, R_\ominus, R_\oplus, R_\oslash, R_\oslash)$ with $E = \{\Lambda\}$,

$$P = \left\langle \begin{array}{c} \overset{A}{\triangle} \\ \overset{A}{\triangle}\,\overset{B}{\triangle}\,\overset{A}{\triangle} \\ \overset{A}{\triangle}\,\overset{B}{\triangle}\,\overset{A}{\triangle}\,\overset{B}{\triangle}\,\overset{A}{\triangle} \\ \overset{B}{\triangle}\,\overset{A}{\triangle}\,\overset{B}{\triangle}\,\overset{A}{\triangle}\,\overset{B}{\triangle} \\ \overset{B}{\triangle}\,\overset{A}{\triangle}\,\overset{B}{\triangle} \\ \overset{B}{\triangle} \end{array} \right\rangle ;$$

$\Sigma_I = \{\overset{A}{\triangle}, \overset{B}{\triangledown}\}$, $R_\oplus = \phi$, $R_\ominus = \phi$,

$R_\oslash = \{\overset{A}{\triangle}\#\Lambda\#\overset{A}{\triangle}\#\Lambda\}$,

$R_\oslash = \{\overset{B}{\triangledown}\#\Lambda\#\overset{B}{\triangledown}\#\Lambda\}$.

To show that $\mathcal{L}(AHIAS) \cap ILOC \neq \phi$, let L_8 be the iso-picture language of parallelograms of sizes $(2, m)$, $m \geq 1$ over the tiles $\overset{A}{\triangle}$ and $\overset{B}{\triangledown}$. One member of L_8 is shown below.

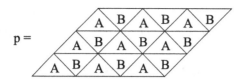

$p =$

L_8 is in ILOC [5]. L_8 can be accepted by the AHIAS $\Gamma = (S, P)$ where
$S = (\sigma, E)$ with $\sigma = (\Sigma_I, R_\ominus, R_\oplus, R_\oslash, R_\oslash)$ where $\triangle_A$, $\triangledown^B$, $E = \{\Lambda\}$.
$P = \{p\}$
$R_\oplus = R_\ominus = \phi$
$R_\oslash = \{\triangledown^B \#\Lambda\# \triangledown^B \#\Lambda\}$,
$R_\oslash = \{\triangle_A \#\Lambda\# \triangle_A \#\Lambda\}$.   □

Notation:

$(x)^n = x \oplus x \oplus \ldots \oplus x = \boxed{X}\boxed{X}\boxed{\cdots}\boxed{X}$ with n times x along the vertical
(column) direction.

$(x)_n = x \ominus x \ominus \ldots \ominus x = \begin{array}{c}\boxed{X}\\\boxed{X}\\\vdots\\\boxed{X}\end{array}$ with n times x along the horizontal (row)

direction.

$_n(x) = x \oslash x \oslash \ldots \oslash x = $ with n times x along the right direction.

$^n(x) = x \oslash x \oslash \ldots \oslash x = $ with n times x along the left direction.

If $X = p =$, then $|X|_\ominus$ is the number of

A tiles catenated along a horizontal direction to be 3, $|X|_\oplus$ is the number of B
tiles catenated along a vertical direction to be 3.

If $Y =$ 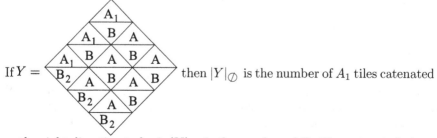 then $|Y|_\oslash$ is the number of A_1 tiles catenated

along the right direction to be 3, $|Y|_\oslash$ is the number of B_2 tiles catenated along
the left direction to be 3.

Theorem 5. *Let Γ be an AHIAS with $R_\ominus = R_\odot = R_\oslash = \phi$. There exists a positive integer ℓ_1 such that if $p \in L(\Gamma)$ with $|p|_{⏀} > \ell_1$ then $p \,⏀\, y \,⏀\, p \in L(\Gamma)$ for any $y \in \Sigma_I^{**}$ with $|y|_\ominus = |p|_\ominus$.*

Proof. Let $\Gamma = (S, P)$ be an AHIAS with $S = (\sigma, E)$ where $\sigma = (\Sigma_I, R_{⏀}, R_\ominus, R_\odot, R_\oslash)$, E an initial language, P is a finite subset of Σ_I^{**} and $R_{⏀} \neq \phi$, $R_\ominus = R_\odot = R_\oslash = \phi$. Let $\ell_1 = \max\{|x|_{⏀}/x \in P\}$. Let us assume that $p \in L(\Gamma)$ and $|p|_{⏀} > \ell_1$. Then $p \in L(\Gamma) - P$.

We prove that for any iso-array $y \in \Sigma_I^{**}$ with $|y|_\ominus = |p|_\ominus$, $p \,⏀\, y \,⏀\, p \in L(\Gamma)$. In order to prove the statement, we show that $(\sigma_{R_{⏀}}^i(E, p) - \{p\}) \subseteq \sigma_{R_{⏀}}^i(E, p \,⏀\, y \,⏀\, p)$ for $i \geq 1$, by induction hypothesis on i. The result is true for $i = 1$. Assume that an iso-array $s \in (\sigma_{R_{⏀}}^{i+1}(E, p) - \{p\})$ for $i \geq 1$, then by induction hypothesis, $s \in \sigma_{R_{⏀}}^{i+1}(E, p \,⏀\, y \,⏀\, p)$, as if s is obtained from a pair of iso-arrays (p, x) or (x, p) for $x \neq p$, then s can also be obtained from a distinct pair of iso-arrays $(p \,⏀\, y \,⏀\, p, x)$ or $(x, p \,⏀\, y \,⏀\, p)$ by applying the same splicing rules which have been applied for the iso-array (p, x) or (x, p).

Similarly, if z is obtained from a pair of iso-arrays (p, p) then s an also be obtained from a pair of iso-arrays $(p \,⏀\, y \,⏀\, p, p \,⏀\, y \,⏀\, p)$. Hence it completes the proof of the statement. □

Similar to the Theorem 5, we have

Theorem 6. *Let Γ be an AHIAS with*

(i) $R_{⏀} = R_\odot = R_\oslash = \phi$, *there exists a positive integer ℓ_2 such that if $p \in L(\Gamma)$ with $|p|_\ominus > \ell_2$, then $p \ominus y \ominus p \in L(\Gamma)$ for any $y \in \Sigma_I^{**}$ with $|y|_{⏀} = |p|_{⏀}$.*

(ii) $R_{⏀} = R_\ominus = R_\odot = \phi$, *there exists a positive integer ℓ_3 such that if $p \in L(\Gamma)$ with $|p|_\oslash > \ell_3$, then $p \oslash y \oslash p \in L(\Gamma)$ for any $y \in \Sigma_I^{**}$ with $|y|_\oslash = |p|_\oslash$.*

Similar results hold good for the languages accepted by AHIAS in non-uniform way. In otherwords we have:

Theorem 7. *Let Γ be an AHIAS with*

(i) $R_\ominus = R_\odot = R_\oslash = \phi$, *there exists an integer $k_1 > 0$ such that if $w \in L_n(\Gamma)$ with $|w|_\ominus \geq k_1$ then either $w \,⏀\, y \in L_n(\Gamma)$ (or) $y \,⏀\, w \in L_n(\Gamma)$ for $y \in \Sigma_I^{**}$ with $|y|_\ominus = |w|_\ominus$.*

(ii) $R_{⏀} = R_\odot = R_\oslash = \phi$, *there exists an integer $k_2 > 0$ such that if $w \in L_n(\Gamma)$ with $|w|_{⏀} \geq k_2$ then either $w \ominus y \in L_n(\Gamma)$ (or) $y \ominus w \in L_n(\Gamma)$ for $y \in \Sigma_I^{**}$ with $|y|_{⏀} = |w|_{⏀}$.*

(iii) $R_{⏀} = R_\ominus = R_\odot = \phi$, *there exists an integer $k_3 > 0$ such that if $w \in L_n(\Gamma)$ with $|w| \geq k_3$ then either $w \oslash y \in L_n(\Gamma)$ (or) $y \oslash w \in L_n(\Gamma)$ for $y \in \Sigma_I^{**}$ with $|y|_\oslash = |w|_\oslash$.*

Theorem 8. $\mathcal{L}_n(AHIAS) \subset \mathcal{L}(AHIAS)$.

Proof. Proof of the inclusion is omitted.

For the strict inclusion, we consider the language L_1 from Example 4. But the language L_1 cannot be accepted by any non-uniform variant of AHIAS, since we cannot find any finite P, a subset of Σ_I^{**} and also by Theorem 7. □

References

1. Arroyo, F., Castellanos, J., Dassow, J., Mitrana, V., Sanchez-Couso, J.R.: Accepting splicing systems with permitting and forbidding words. Acta Inform. **50**, 1–14 (2013)
2. Gimmarresi, D., Restivo, A.: Two-dimensional languages. In: Salomaa, A., Rozenberg, G. (eds.) Handbook of Formal Languages, vol. 3, pp. 215–267. Springer, Heidelberg (1997)
3. Head, T.: Formal language theory and DNA: an analysis of the generative capacity of specific recombinant behaviours. Bull. Math. Biol. **49**, 735–759 (1987)
4. Kalyani, T., Sasikala, K., Dare, V.R., Robinson, T.: Controlled sequential pasting systems. In: Thangavel, P. (ed.) Algorithms and Artificial Systems, pp. 137–151. Allied Publisher (2003)
5. Kalyani, T., Dare, V.R., Thomas, D.G.: Local and recognizable iso picture languages. In: Pal, N.R., Kasabov, N., Mudi, R.K., Pal, S., Parui, S.K. (eds.) ICONIP 2004. LNCS, vol. 3316, pp. 738–743. Springer, Heidelberg (2004)
6. Kalyani, T., Dare, V.R., Thomas, D.G., Robinson, T.: Iso-array acceptors and learning. In: Sakakibara, Y., Kobayashi, S., Sato, K., Nishino, T., Tomita, E. (eds.) ICGI 2006. LNCS (LNAI), vol. 4201, pp. 327–339. Springer, Heidelberg (2006)
7. Kalyani, T., Dare, V.R., Thomas, D.G.: Iso-array grammars and picture languages. Int. J. Math. Sci. **6**, 369–384 (2007)
8. Masilamani, V., Sheena Christy, D.K., Thomas, D.G., Nagar, A.K., Thamburaj, R.: Accepting H-array splicing systems. In: Proceedings of Ninth International Conference on Bio-Inspired Computing : Theories and Applications, Communications in Computer and Information Science, vol. 472, pp. 313–317 (2014)
9. Masilamani, V., Sheena Christy, D.K., Thomas, D.G., Kalyani, T.: Parallel splicing on iso-arrays. In: Proceedings of Fifth International Conference on Bio-Inspired Computing: Theories and Applications, pp. 1535–1542 (2010)
10. Mitrana, V., Petre, I., Rogojin, V.: Accepting splicing systems. Theor. Comput. Sci. **411**, 2414–2422 (2010)
11. Paun, G., Rozenberg, G., Salomaa, A.: Computing by splicing. Theor. Comput. Sci. **168**, 321–336 (1996)
12. Paun, G., Rozenberg, G., Salomaa, A.: DNA Computing: New Computing Paradigms. Springer, Berlin (1998)
13. Sheena Christy, D.K., Masilamani, V., Thomas, D.G.: Iso-array splicing grammar system. In: Bansal, J.C., Singh, P.K., Deep, K., Pant, M., Nagar, A.K. (eds.) Proceedings of Seventh International Conference on Bio-Inspired Computing: Theories and Applications (BIC-TA 2012). AISC, vol. 201, pp. 157–167. Springer, Heidelberg (2012)

Construction of Perfect Auto-correlation Arrays and Zero Cross-correlation Arrays from Discrete Projections

Benjamin Cavy[1] and Imants Svalbe[2]($\boxtimes$)

[1] Polytech Nantes, University of Nantes, Nantes, France
[2] Department of Physics and Astronomy, Monash University, Melbourne, Australia
imants.svalbe@monash.edu

Abstract. This paper presents a new method by which sets of discrete 1D projections can be used to construct large families of 2D discrete arrays. These compact arrays have targeted, specific periodic correlation values that span the full range between perfect auto-correlation to zero cross-correlation. The array size is variable and the array elements can be binary or contain grey integer values. Arrays with these properties are useful for digital signal synchronisation, communications and water-marking. We show that multiple copies of zero cross-correlation arrays can be co-located without interference and that the presence of individual arrays is able to be determined independently. The arrays with perfect periodic auto-correlation also have high aperiodic auto-correlation and optimally low cross-correlation, making them well-suited for use as digital watermarks.

Keywords: Discrete radon projection · Perfect auto-correlation · Zero-cross-correlation · Ghost images

1 Introduction

Considerable theoretical and practical effort is invested in constructing large families of discrete arrays that have strong auto-correlation and low cross-correlation properties. These arrays have important applications in signal processing for precise data synchronisation, in computer science for digital watermarking and secure, robust communication over encrypted channels and in physics to minimise the energy of finite discrete lattice structures [1].

Ideally, these arrays are composed of signed, unitary elements, usually as entries of ± 1, or the roots of unity over complex numbers or quaternions. The optimum periodic auto-correlation of these arrays is perfect, yielding a peak value of N for an N element array when aligned and zero at each of all other $N-1$ alignments. The cross-correlation between members of a family of arrays should also be as close to zero as possible. We show here, for the first time, that it is possible to create families of 2D functions A, B, C, ... whose periodic

© Springer International Publishing Switzerland 2015
R.P. Barneva et al. (Eds.): IWCIA 2015, LNCS 9448, pp. 232–243, 2015.
DOI: 10.1007/978-3-319-26145-4_17

cross-correlations are perfect, i.e. $A * B$, $A * C$, $B * C$, $\ldots = 0$ across all elements of a 2D array.

Construction of these arrays is based on ghost functions [2,6,8]. Ghost functions are signed elements placed on an array in such a way that their intensity values sum to zero under periodic or aperiodic projection for a predetermined number of discrete view angles.

Section 2 presents the theoretical basis to construct, by discrete projection, non-zero 2D functions A and B where the cross-correlation $A * B = 0$. Section 3 uses a result from prior work on ghost functions, [13], that showed how multiple copies of A, as A', A'', $A'''\ldots$, and B as B', B'', $B'''\ldots$, can be nested within the same 2D array and still retain the property that $(A' + A'' + A''') * (B' + B'' + B''') = 0$. Section 4 extends this approach to construct multiple functions A, B, $C\ldots$, such that $A * B * C \cdots = 0$. The construction of $p \times p$ arrays with perfect auto-correlation is a corollary of this result. Section 5 presents some perfect auto-correlation results and gives an example application of these arrays as digital watermarks. Section 6 presents a summary of our findings and outlines areas for future work.

2   Theory

Our approach is motivated by the result found in the 1998 PhD work of Olivier Phillipé, who proved that the process of discrete projection [5] preserves aperiodic auto-correlation: the projection at discrete angle $p : q$ of the 2D auto-correlation of any data is the same as the auto-correlation of the 1D projection of that data at the angle $p : q$.

The 1D projection of a 2D image or function is defined in (1). The 1D array $proj_{p_i q_i}(b)$ stores the summed image values that are projected into discrete bins, where the bins are labelled by the integer b. For each b, the sum of $f(k, l)$ is taken along the line inside $f(k, l)$ where $b = -kq_i + lp_i$.

The cross correlation of two functions is defined in (2). There $C_{fg(r,s)}$ is the sum of the inner product of f and g displaced by relative integer coordinates (r, s). When $f = g$, the cross-correlation defined in (2) becomes the auto-correlation.

$$proj_{pi,qi}(b) = \sum_{k=-\infty}^{+\infty} \sum_{l=-\infty}^{+\infty} f(k, l)\Delta(b - kq_i + lp_i) \tag{1}$$

$$C_{f,g}(r, s) = \sum_{k=-\infty}^{+\infty} \sum_{l=-\infty}^{+\infty} f(k, l).g(k - r, l - s) \tag{2}$$

The property of preserving correlations under projection is in turn a consequence of the central slice theorem which underlies all direct (non-iterative) methods of tomographic reconstruction.

The Fourier transform of each 1D projected view of an object forms a 1D slice of the 2D Fourier transform of the object itself. This applies for real sampled projection data [7] as well as for the large variety of discrete projection methods: examples of the latter are given in [4,9].

The correlation (and convolution) property shown by Phillipé then follows directly, as $g * h = Fg \cdot Fh$, where '$*$' denotes the convolution or correlation operation, F is the discrete Fourier transform, '$\cdot$' is the scalar product and g and f are 1D, 2D or nD discrete functions.

Here we choose to use the Finite Radon Transform (FRT) [9] to compute discrete projections. The FRT sums array elements that lie on parallel straight lines, under the assumption of periodic (i.e. wrapped) boundary conditions, over a fixed set of discrete angles. The array size is restricted to be square in 2D (or a cube in 3D) with sides of prime length.

Any $p \times p$ block of discrete data, $I(x,y)$, can be represented by its FRT, $R(t,m)$. $R(t,m)$ is comprised of $p + 1$ 1D projections at discrete angle $m, 0 \leq m \leq p$, with each projected ray having an offset index, t, $0 \leq t < p$. Angle m corresponds to the discrete vector $m : 1$, i.e. m steps across for each step down on a 2D grid. There is a unique mapping of the $p + 1$ angles $m : 1$ onto the set of $p + 1$ directed vectors $x_m : y_m$ through the Farey/Haros series fractions [12], when the vectors x_m, y_m are ordered by increasing vector length $(x_m^2 + y_m^2)$.

The set of FRT projections can be inverted to recover, exactly, $I(x,y)$ from $R(t,m)$. The FRT has the useful property that the p elements of each 1D projection tile all elements of the 2D $p \times p$ image space exactly once, for each of the $p + 1$ projected views. The Fourier transform (FFT) of each FRT projection is thus a periodic 1D projection of the 2D Fourier transform of the image. This exact tiling property means that the FFT can also be replaced by other mappings such as the number theoretic transform [3].

2.1   Construction of Zero Cross-correlation Arrays

Consider two $p \times p$ images, A and B. Let $C = A * B$ be the $p \times p$ periodic 2D cross-correlation of A and B; we want to construct 2D arrays, A and B, so that $C = 0$.

Proof. Let $A(m)$, $B(m)$ and $C(m)$ be the FRT projection of image A, B and C (respectively) at discrete angle m. Given $C = 0$, the 2D/1D correlation property [5] requires that

$$C(m) = A(m) * B(m) = 0 \tag{3}$$

for all $m, 0 \leq m \leq p$.

This condition is satisfied if:

A has zero projected sums at half the $(p + 1)$ angles, e.g., $A(m) = 0$ for $0 \leq m < (p + 1)/2$

and

B has zero projected sums at the other half of the angles, e.g. $B(m) = 0$ for $(p + 1)/2 \leq m \leq p$.   □

As correlation operations are linear, then for $D = A + B$, with $A \times B = 0$
$D * A = (A + B) * A = A * A + A * B = A * A$ and similarly $D * B = B * B$. We
will use this property in Sect. 3.

A ghost image is comprised of $+1$ and -1 valued pixels on an otherwise zero
background. Each $+1$ pixel is positioned to line up with one -1 pixel to give a
zero projected sum along pre-selected directions. A minimal ghost is a ghost that
uses $N(+1)$ and $N(-1)$ pixels to produce zero-projected sums for N directions.
The auto-correlation for a minimal ghost has a peak value of $2N$.

The peak-to side-lobe ratio (PSL) and merit factor (M) are used to measure
the quality of correlations. PSL quantifies the correlation peak relative to the
largest off-peak magnitude. M is defined as the square of the peak correlation
value divided by the sum of the squared off-peak values. The values for PSL and
M should be as high as possible for strong auto-correlation functions and be as
low as possible for cross-correlations between distinct or orthogonal functions.

Methods to construct periodic minimal ghosts that have N zero -sum projected views were presented in [10,11]. The cross-correlation between $p \times p$ minimal ghosts in $N = (p+1)/2$ directions shown in [13] have values ranging between
-2 to $+2$, giving a $PSL = (p+1)/2$. For $1 < N \le (p+1)/2$, we can show that
$M = 2N/(4N - 3)$, so that for these cases, $M > 0.5$.

In Sect. 5 we construct examples of 'perfect' zero-mean pxp 'binary' or grey
arrays where the periodic auto-correlation has $PSL = M = p^2 - 1$ and where
the periodic cross-correlation between arrays can be optimally low (0 or $\pm p$).

2.2 Example of Zero-correlation Arrays

Figure 1 shows two 19×19 arrays, A and B, each being a ghost designed to
have zero-sums at 10 distinct angles; their auto-correlations, $A * A$ and $B * B$
are also shown. Their cross-correlation $C = A * B$, as expected, is exactly zero.
The auto-correlations have peaks of 20 and off-peak absolute values ≤ 2, with
$M = 0.541$.

The symmetry of A (here even) and B (here odd) always appears as oppo-
sites ($A^t = A$, $B^t = -B$, where t denotes the transpose). Each disjoint angle
set includes as members the four-fold symmetric angles, $m, -m, m', -m'$, where
$-m = p - m$ and m' is the inverse of m, $mm' = 1 \ mod(p)$. By choice, the arrays
A and B are generated using different starting locations, to ensure there is no
spatial overlap of any of the $+1$ or -1 points between A and B (i.e. $A.B = 0$).

3   Multiple Copies of A and B

Copies of N zero-sum direction ghosts can be constructed through the Fourier
transform of a minimal periodic ghost array with N zero-sums [13]. The FFT of
the FRT maps each discrete angle m to its complementary angle $-1/m = -m'$
(the 1D slice in Fourier space is perpendicular to the 1D slice in real space). The
location of each matching positive and negative value in the Fourier transformed
array must form zero-sums at angles $-m'$ from each original zero-sum angle m.

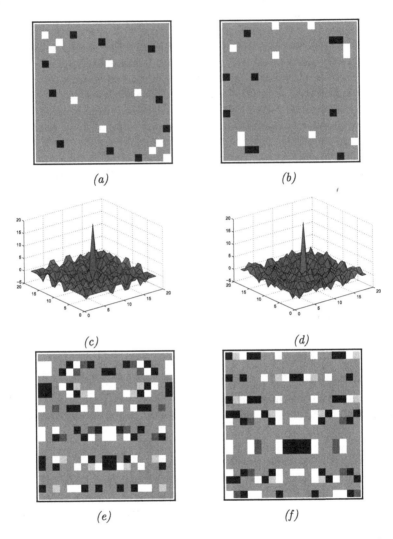

Fig. 1. (a) A is a 19×19 array $I(x, y)$ with 10 entries of $+1$ and 10 entries of -1 that has projections with zero-sums along 10 discrete directions. (b) Array B has the same properties as array A, but with zero sums in the 10 FRT directions that are disjoint from those for A, making $A * B = 0$. For images (a) and (b), black $= -1$, grey $= 0$, white $= +1$ values. (c) and (d) are the auto-correlations for A and B. (e) and (f) are the FRT $R(t, m)$ of A and B. Grey rows in (e) and (f) are discrete angles with zero-sum projected views. Note that the 10 zero-sum angles (shown as grey) for (e) and the 10 zero sum angles for (f) are complementary.

Assigning a $+1$ and -1 to the location at any distinct matching signed values in the FFT array constructs copies of a new ghost, each with N zero-sums. The number of ghost copies that are made varies with the number of distinct values in the 2D FFT of the original ghost (more copies are possible for ghosts with a smaller number of zero-sum angles).

For the case $N = (p + 1)/2$, [13] showed that the distribution of the values of the Fourier transform of these ghosts corresponds to the $(p - 1)/2$ unique positive and $(p - 1)/2$ negative values of $\sin(2\pi i/p)$ (with zero being the sole remaining distinct value). That means $(p - 1)/2$ new distinct ghost copies can be generated from the original ghost. All $(p - 1)/2$ ghost copies can be inserted into the same $p \times p$ array without overlap of any of their $+1$ or -1 points.

3.1   Example of Multiple Co-residents Ghosts

Multiple, non-overlapping copies of each minimal ghost can be made using the method outlined in Sect. 3.

These copies can be added into the same image space without changing any of their zero-sum projection properties.

Figure 2a shows the array AP, formed by pooling the $(19 - 1)/2 = 9$ ghosts built from ghost A of Fig. 1a, into a 19×19 array. The array AP retains zero-sums in 10 directions. Figure 2b shows the compound ghost BP, formed by adding together the 9 minimal ghost copies of the ghost B (Fig. 1b). The ghost BP also has 10 zero sum projections, but along those directions in which ghost AP is not zero. The auto-correlation peak for $AP * AP$ (and $BP * BP$) then becomes $(p + 1)(p - 1)/2 = (p^2 - 1)/2$. The off-peak values do however increase, because of the larger density of non-zero points means that more intersections will occur for more off-peak translations.

For the example shown in Fig. 2, auto-correlations $AP * AP$ and $BP * BP$ have a relatively strong peak correlation value of $180 = 9 \times 20$, but a relatively poor merit factor of 0.082. However the cross-correlation of the pooled ghosts, $AP * BP$, is exactly zero everywhere, as these pooled ghosts still have zero-sums at disjoint sets of angles (as did A and B).

Because the initial ghosts A and B are spatially distinct, we can also pool the copies of ghosts made from AP and BP. The array $DP_{19} = AP + BP$, shown in Fig. 3a, contains 9 copies from A and 9 copies made from B. Each individual ghost copy contained in array AP or BP can be retrieved by periodic correlation with DP_{19}. Using array DP_{19} to test the cross-correlation of any ghost member from AP with any ghost member from BP still results in zero; $DP_{19} * AP = AP * AP$, $DP_{19} * A = A * A$; the same applies when correlating DP_{19} with BP and B. The auto-correlation $DP_{19} * DP_{19}$ has peak value $360 = 19^2 - 1$, with side lobes between $+87$ and -89 and a merit factor $M = 0.178$, being the sum of the auto-correlation M values for $AP * AP$ and $BP * BP$.

Figure 3b shows the equivalent of the array DP_{19} in Fig. 3a, but now for a 373×373 array, which has a total of 374 distinct FRT projection angles. Here the zero-sums for each individual ghost A and B sum to zero under projection for two sets of 187 ($= 374/2$) disjoint angles.

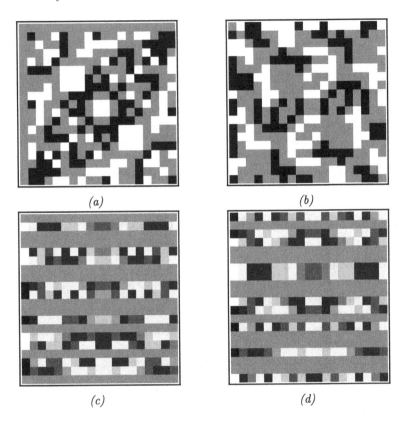

Fig. 2. Compound 19×19 ghost arrays (a) AP and (b) BP contain 9 copies of A and 9 copies of B respectively. Each array retains zero-sum projections at 10 angles. Black $= -1$, grey $= 0$, white $= +1$ values. (c) and (d) are the FRT $R(t, m)$ of AP and BP respectively. Grey rows in (c) and (d) are discrete angles with zero-sum projected views. Note the zero-sum angles for (c) and (d) are complementary.

Now each single ghost A and B has a peak auto-correlation value of 374, but with the same side lobe values as before, ± 2, ± 1 or 0. Each of the 186 members of the family AP copied from ghost A has zero cross-correlation with each of the 186 members of the family BP copied from ghost B. The equivalent array DP_{373} in Fig. 3b contains 186 copies of A and 186 copies of B. The auto-correlation $DP_{373} * DP_{373}$ has peak value $139, 128 = 373^2 - 1$, with sides lobes between $+10, 713$ and $-10, 623$ and a merit factor of 0.008.

4   Zero Cross-correlation Between Multiple Functions

Splitting the fixed set of $p + 1$ view angles for $p \times p$ arrays into more than two disjoint sets is also possible, for example to construct arrays where $A * B * C * D = 0$, as shown in Fig. 4.

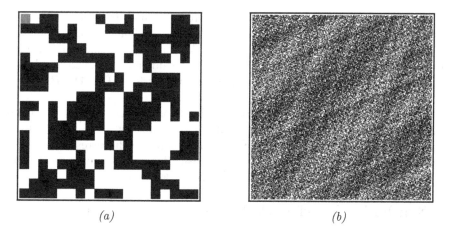

(a)   (b)

Fig. 3. (a) 19×19 array DP_{19} contains 9 copies of ghost A and 9 copies of ghost B. The single remaining zero-valued element is at the top left corner. (b) The equivalent array DP_{373} of size 373×373 contains 186 copies of an even ghost and 186 copies of an odd ghost, both ghosts have zero-sums over 187 disjoint discrete directions. Black $= -1$, grey $= 0$, white $= +1$ values.

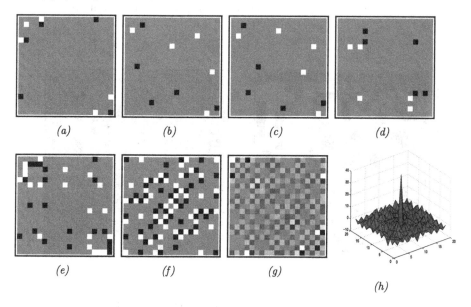

(a)   (b)   (c)   (d)

(e)   (f)   (g)

(h)

Fig. 4. (a)–(d) Four 19×19 arrays A, B, C and D. Each array has zero-sum discrete projections at 5 disjoint angles, black $= -1$, grey $= 0$, white $= +1$. (e) Array $E = A + B + C + D$, black $= -1$, grey $= 0$, white $= +1$. (f) Cross correlation $AB = A * B$, black $= -1$, grey $= 0$, white $= +1$. (g) Cross-correlation $ABC = A * B * C$ has grey values that range from -7 (shown as black) through zero (shown as mid-grey) to $+7$ (shown as white). Periodic cross-correlation $ABCD = ABC * D$, by design, is exactly everywhere zero (image not shown). (h) The periodic auto-correlation $E * E$, $peak = 40$, side lobes $+6$ to -5, $M = 0.755$.

Using the FFT to generate copies of a zero-sum ghost provides a choice in fixing the sign of each ghost copy, as +1 and −1 can equally well be assigned as −1 and +1 at each FFT threshold value. We have found that random or a more strategic shuffling of the sign assignments for different threshold levels can significantly improve the auto-correlation PSL values for $AP * AP$, $BP * BP$ and $DP_{19} * DP_{19}$.

For example, shuffling the signs of the ghost copies that make up the composite array DP_{19} in Fig. 3a increased the merit factor for $DP_{19} * DP_{19}$ from 0.178 to $360 = 19^2 - 1$, as shown in Fig. 5a. The latter result corresponds to the auto-correlation from a perfect array. This result leads to a simple method to construct perfect arrays for arbitrary $p \times p$ grids.

5 Synthesis of Perfect Auto-correlation Arrays

A perfect $p \times p$ array has, by [5], $p + 1$ 1D projections, each of which must have a perfect 1D auto-correlation. Making $p - 1$ copies of an $N = 1$ ghost for each of the $p + 1$ directions (with half the projections having reversed signs) gives a peak of $(p + 1).(p - 1) = p^2 - 1$, with a constant value of -1 at all other grid locations. A 373×373 perfect array generated using this method is shown in Fig. 5b. These perfect arrays have the same origins as those built from Gauss sums and quadratic residues [1].

(a)   (b)

Fig. 5. Changing the signs chosen for individual elements of a compound ghost can improve the auto-correlation of the compound array. Shuffling the signs of the ghost elements of Fig. 3a gives the 19×19 array (a), which has perfect auto-correlation. Using the same construction methods, the array shown in Fig. 3b can be converted to produce (b) a 373×373 perfect array.

Synthesizing a 2D function from its 1D projected views means we can choose (at random) which $(p + 1)/2$ of the $p + 1$ angles have their signs reversed.

3	-19	3	-8	3	3	3	3	3	3	3
3	3	3	14	3	-8	3	3	3	-8	-19
3	3	3	-8	-30	3	25	3	3	3	-8
-8	-8	-8	-8	25	3	3	3	-8	3	3
3	3	25	-19	3	3	-8	-19	3	-8	14
14	3	-8	-19	25	14	-19	3	-8	3	-8
14	3	-8	-41	3	-8	14	14	3	14	-8
-19	-8	-8	-19	-30	-8	-8	-19	3	-8	3
-8	3	3	-19	3	3	3	3	-8	3	14
3	14	-8	3	-8	3	-19	14	3	-8	3
-8	3	3	3	3	-8	3	-8	3	3	3

Fig. 6. An 11×11 grey array with perfect periodic auto-correlation.

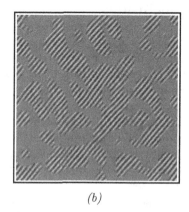

(a)   (b)

Fig. 7. (a) The 256×256 'cameraman' test image, with the array of Fig. 6 encoded as an additive watermark. The array was inserted in an 11×11 patch of the Fourier domain phase of the original data. (b) Image of the differences between the integer-quantised encoded and original data.

This means the family of these $p \times p$ perfect arrays has of order $(p+1)!/((p+1)/2)!$ distinct members. Further details on the methods used to construct binary and grey perfect arrays can be found in [14].

Example perfect arrays A and B made by this method have a cross-correlation that is minimised when the angle sets for A and B and maximally different. Under these conditions, A and B exhibit a zero cross-correlation peak and a symmetric distribution of off-peak values equal to either $+p$ or $-p$ which is the lowest possible cross-correlation value for arrays that each have maximal or perfect auto-correlation. The periodic auto-correlation of these $p \times p$ arrays has peak value $p^2 - 1$, with a uniform off-peak value of -1. At optimal shift, the aperiodic auto-correlation of these arrays has a merit factor M of around 3.

Figure 6 shows an example 11×11 array of grey integer values that has perfect periodic auto-correlation and has a merit factor of 3.859 for aperiodic auto-correlation. The compact size and strong correlation property of this array makes it highly suited for digital watermarking applications.

A scaled version of this array was used to perturb the phase angles over an 11×11 patch of the DFT of a 256×256 8-bit test image (cameraman). The resulting encoded image is visually indistinct from the original (psnr of 50.811). The quantised spatial domain image differences with respect to the original data range from -2 to $+2$, as shown in Fig. 7. The watermark is easily recovered by cross-correlation of the encoding array with the Fourier phase domain of the encoded data. The correlation peak to background ratio is robust to scaling of the image size $>40\%$ and to rotational misalignment of the image up to 1 degree.

6 Conclusions

Construction of arrays with zero cross-correlation have previously been contrived by summing pairs of correlations with complementary off-peak signs, i.e. arrays A and B made so that $A*A + B*B = 0$.

Here we have, we believe for the first time, designed arrays with strong correlation properties where, directly, $A*B = 0$.

Multiple copies of these arrays can be interlaced without overlap or loss of their strong correlation properties. This may make them suitable for use as 'selective switches' in digital watermarking and communication applications. A single compound watermark could be used to control access to shared data for multiple authorised users.

Using discrete projected views to design arrays with strong correlation properties is a new approach. It complements the current technique of folding 1D rows or columns to construct or 'inflate' larger 2D arrays from known 1D examples [1]. Our discrete projection approach can be applied to the construction of 3D arrays, based on ghosts that have zero-sums onto directed planes.

Extending our result to arrays with non-prime dimensions and unequal lengths would be less restrictive. We have shown it is possible to use our 1D projection method to construct arrays of general size $N \times M$ with aperiodic auto-correlation merit factors above 3. At present, each family of $p \times p$ perfect arrays, for prime p where $modulus(p, 4) = 3$, is comprised of exactly p distinct members. Each of the $p \times (p-1)/2$ cross-correlations between these family members is optimal, with periodic cross-correlation values being $\leq p$. It would be desirable to be able to build large families of $M \times N$ arrays with strong auto-correlation and where the cross-correlation between each family member is optimally or near-optimally low.

Acknowledgements. The authors thank Andrew Tirkel for sharing his expertise on sequences and the generation of perfect auto-correlation arrays. BC acknowledges support from Polytech Nantes and the School of Physics and Astronomy, Monash University as hosts for his internship.

References

1. Blake, S., Hall, T., Tirkel, A.: Arrays over roots of unity with perfect auto-correlation and good ZCZ cross correlation. Adv. Math. Commun. **7**(3), 231–242 (2013)
2. Brunetti, S., Dulio, P., Peri, C.: Discrete tomography determination of bounded sets in $\mathbb{Z}^n$. Discrete Appl. Math. **83**, 20–30 (2015)
3. Chandra, S., Svalbe, I.: Exact image representation via a number-theoretic radon transform. IET Comput. Vision **8**(4), 338–346 (2014)
4. Grigoryan, A.: Method of paired transforms for reconstruction of images from projections: discrete model. IEEE Trans. Image Process. **12**(9), 985–994 (2003)
5. Guédon, J.: The Mojette Transform: Theory and Applications. ISTE, Wiley (2009, Chapter 3, section 322)
6. Herman, G.T., Kuba, A.: Advances in Discrete Tomography and Its Applications. Birkhauser, Boston (2007)
7. Kak, A., Slaney, M.: Principles of Computerized Tomographic Imaging. SIAM, Philadelphia (2001)
8. Louis, A.K.: Ghosts in tomography: the null space of the radon transform. Math. Meth. Appl. Sci. **3**, 1–10 (1981)
9. Matúš, F., Flusser, J.: Image representation via a finite radon transform. IEEE T-PAMI **15**(10), 996–1006 (1993)
10. Svalbe, I., Normand, N.: Properties of minimal ghosts. In: Debled-Rennesson, I., Domenjoud, E., Kerautret, B., Even, P. (eds.) DGCI 2011. LNCS, vol. 6607, pp. 417–428. Springer, Heidelberg (2011)
11. Svalbe, I., Normand, N., Nazareth, N., Chandra, S.: On constructing minimal ghosts. In: DICTA 2010, Sydney, Australia, December 2010. http://dx.doi.org/10.1109/DICTA.2010.56
12. Svalbe, I.: Sampling properties of the discrete radon transform. Discrete Appl. Math. **139**, 265–281 (2004)
13. Svalbe, I.: Near-perfect correlation functions based on zero-sum digital projections. In: DICTA 2011, Noosa, Queensland, Australia, pp. 627–632, December 2011. http://dx.doi.org/10.1109/DICTA.2011.111
14. Tirkel, A., Cavy, B., Svalbe, I.: Families of multi-dimensional arrays with optimal correlations between all members. Electron. Lett. **51**, 1167–1168 (2015)

From Theory to Applications

Character Segmentation of Hindi Unconstrained Handwritten Words

Soumen Bag[1]($\boxtimes$) and Ankit Krishna[2]

[1] Department of Computer Science and Engineering, ISM Dhanbad, Dhanbad, India
bagsoumen@gmail.com
[2] Department of Computer Science and Engineering, IIIT Bhubaneswar,
Bhubaneshwar, India
ankitkrishna.id@gmail.com

Abstract. The proper character level segmentation of printed or handwritten text is an important preprocessing step for optical character recognition (OCR). It is noticed that the languages having cursive nature in writing make the segmentation problem much more complicated. Hindi is one of the well known language in India having this cursive nature in writing style. The main challenge in handwritten character segmentation is to handle the inherent variability in the writing style of different individuals. In this paper, we present an efficient character segmentation method for handwritten Hindi words. Segmentation is performed on the basis of some structural patterns observed in the writing style of this language. The proposed method can cope with high variations in writing style and skewed header lines as input. The method has been tested on our own database for both printed and handwritten words. The average success rate is 96.93 %. The method yields fairly good results for this database comparing with other existing methods. We foresee that the proposed character segmenattion technique can be used as a part of an OCR system for cursive handwritten Hindi language.

Keywords: Character segmentation · Handwritten word · Header line detection · Hindi language · Lower modifier · Upper modifier · Structural approach · OCR

1 Introduction

In last two decades several works have been done in the computer recognition of handwritten words. But few of us believe that a computer will ever be able to read humans' handwriting as good as human. Even so, it does not hurt to try to develop technology which can approach the recognition ability of humans. Solving the character segmentation problem is one of the keys to putting character recognition technology to practical use.

Character segmentation is an operation that seeks to decompose an image of a sequence of characters into subimages of individual symbols [7] as shown in

© Springer International Publishing Switzerland 2015
R.P. Barneva et al. (Eds.): IWCIA 2015, LNCS 9448, pp. 247–260, 2015.
DOI: 10.1007/978-3-319-26145-4_18

(Fig. 1). It is one of the decision processes in a system for optical character recognition (OCR). The performance of character segmentation techniques depend on the quality of the scanned document because due to poor quality scanning and ink bleeding, it generally happens that the neighboring characters in the scanned image touch each other. Character segmentation is a major challenge for such degraded documents [14].

A wide variety of line, word, and character segmentation methods for printed documents of Indian languages are reported in the literature [12]. But segmentation of cursive handwriting still remains one of the most challenging problems in the area of handwritten character recognition. Bangla and Hindi are the two very popular Indian languages having this cursive property in writing style. Some of the prior works on Bangla handwritten character segmentation include in [4,6,8,13,16]. But evidence of works on Hindi character segmentation are just a few in number [5,9]. In this paper, we present a novel technique for segmentation of unconstrained handwritten Hindi words which is highly efficient over the other existing methodologies in literature. The key features of our proposed method are summarized as follows:

- Extensive use of the structural properties of characters in the segmentation process.
- Efficient to handle inputs with highly skewed header lines.
- Covers many different handwriting styles written by different individuals and gives correct output for them.

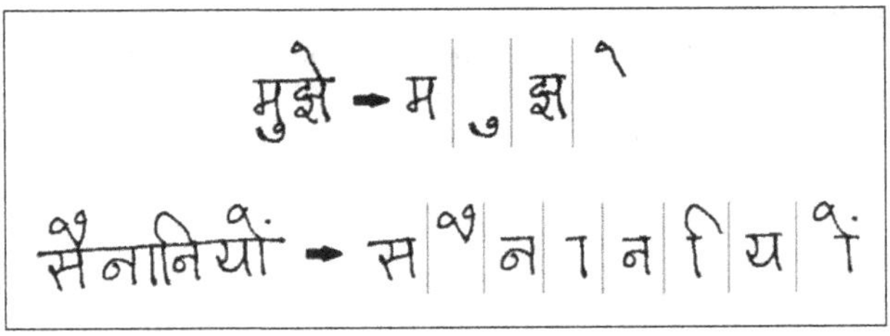

Fig. 1. Segmentation of handwritten Hindi words.

The rest of the paper is organized as follows. Section 2 describes the characteristics of Hindi language. Section 3 presents the proposed methodology for accurate segmentation of handwritten Hindi words. Experimental results and related discussions are reported in Sect. 4. The concluding notes are given in Sect. 5.

2   Properties of Hindi Language

Devanagari is the script for writing Hindi language. It consists of 14 vowels and 33 consonants. A sample set of basic Hindi alphabets is shown in Fig. 2(a).

The writing style is from left to right. There is no concept of lower or upper case alphabets as in English language [2]. Half form characters in Hindi increases the language complexity for recognition. The half characters may touch with full characters to make the characters called conjuncts or compound (see Fig.2(b)). In each conjunct character, the right part is a full consonant, and the left part is always a half consonant. When two or more characters are combined to form a word, the horizontal lines touch each other and generate a header line called *shirorekha*. The vowels modifiers can be placed at the left, right (or both), top, or bottom of the consonant. The vowels above the header line are called *ascenders* or upper modifiers and vowels below the consonants are called *descenders* or lower modifiers.

अ	आ	इ	ई	उ	ऊ	ऋ	ए	ऐ	ओ
औ	क	ख	ग	घ	ङ	च	छ	ज	झ
ञ	ट	ठ	ड	ढ	ण	त	थ	द	ध
न	प	फ	ब	भ	म	य	र	ऴ	व
श	ष	स	ह						

(a) Sample images of Hindi Basic characters.

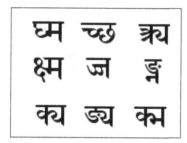

(b) Sample images of Hindi conjunct characters.

Fig. 2. Hindi alphabet set.

3 Proposed Methodology

The proposed method has three phases. A preliminary segmentation process extracts the header line and delineates the upper-strip from the rest in phase 1. This yields vertically separated middle zone and bottom zone components that may be conjuncts, touching characters, characters with lower modifier attached to it, shadow characters, or a combination of these. In phase 2, statistical information about these intermediate individual components is collected and the segmentation of upper modifier is performed. In phase 3, this statistical information is used again to select the components on which further segmentation needs to be attempted. This separates the lower modifiers from the middle zone components. This segmentation methodology is performed in the following hierarchical order as shown in Fig. 3.

1. Scan the handwritten Hindi words needed to be segmented and perform the binarization using an adaptive-cum-interpolative method as described in [3]. It is noticed that binarization plays an important role in character segmentation. In this method, a multiscale framework is added to adaptive version of Otsu's method [11] to handle noises in different scales. To convert Otsus method to an adaptive model, instead of computing the global threshold value for the whole image, it computes the local threshold value for each pixel by observing the intensity behavior of its neighbor pixels.
2. Detect the header line and remove it completely.
3. Segment the upper modifiers left in the upper zone, if any, and make appropriate joining if required.
4. Identify the middle zone components containing any lower modifiers and segment these lower modifiers if required.
5. Finally, the segmented result is presented for further recognition process.

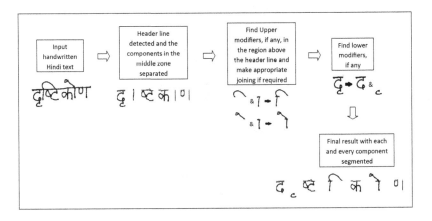

Fig. 3. System architecture of the proposed method.

3.1   Detection and Removal of Header Lines

Figure 4 outlines the proposed method for detecting and removing the header lines even if they are skewed in nature. The following steps discuss the method in detail.

1. Perform thinning of binarized handwritten Hindi words to get single pixel thin skeletons using Huang's method [10].
2. Find the start row, end row, start column, and end column for the span of a word.

3. Get the horizontal density of number of object pixels for each row in the upper half of the word height, i.e., from 'start row' to '(start row+end row)/2'.
4. Find the highest density row (marked as 'record') from the above list and consider to be the approximate header line row.
5. Divide the entire word width into stripes. The number of stripes is equal to ((2×width)/lower height) of the input word, where width = (end column-start column) and lower height = (end row-record).
6. Find the row having highest density of object pixel for each stripe by scanning from 'start row' to 'record'+7. This threshold value is set as per the experimental analysis.
7. Find the difference between 'record' and the local maximum row (from step 6) for each stripe and accordingly shift the entire stripe upwards or downwards based on the sign of the difference.

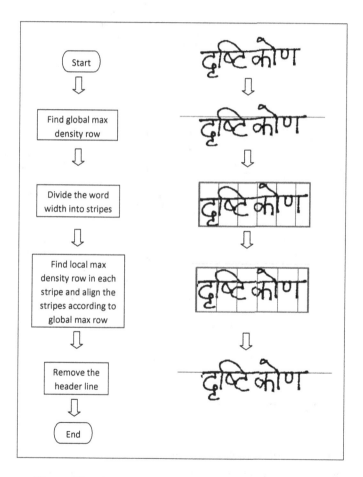

Fig. 4. Flowchart for header line detection and removal.

8. Finally, remove the row 'record' and appropriate number of rows above and below of it (according to the pen width of the input word) to get rid of the header line.

After removal of header line, we calculate the vertical density of object pixels for each column. We identify the columns having zero count for the above vertical density and use them as breakdown columns to separate the components at these positions.

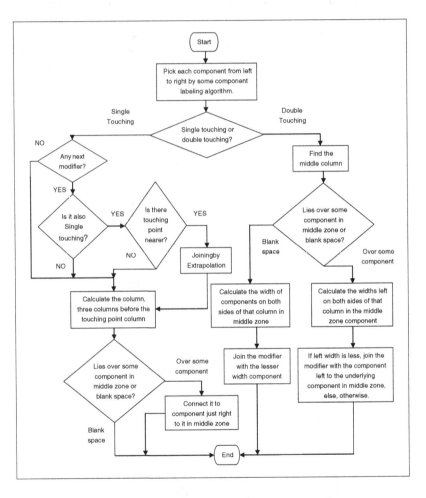

Fig. 5. Flowchart for upper modifier segmentation.

3.2 Segmentation of Upper Modifiers

To deal with upper modifiers separately, we extract it out one by one using Rosenfeld and Kak component labeling algorithm [15]. Figure 5 outlines the proposed technique for upper modifier segmentation. The steps are as follows:

1. The upper modifiers in Hindi language can be classified into two classes. This classification is done based on the number of times they touch the header line at different positions as shown in Fig. 6. They touch either once (↑↑) or twice (↑ ↑) as per the characteristics of writing style.

2. If it is a single touch, then check the class of the next modifier (if there is any) in the sequence from left to right. If there is no next modifier or the next modifier is not a single touching modifier then go to step 3. But if it is a single touching modifier, then check whether the touching points of present and next modifiers are nearer to each other or not. If they are not, then go to step 3. Otherwise, join them by applying the extrapolation method used in [1] and then move to the next step. All different possibilities of upper modifier are shown in Fig. 7.

3. Find the column which is four columns preceding from the touching point (see Fig. 8). If this column lies over blank space in middle zone then connect it to the component in the middle zone which is just right to the above calculated column. If it lies over some component in the middle zone then simply represent it separately.

4. If it is double touching, then find the middle column for the column span of the upper modifier. In Fig. 9, the middle column is marked by red line and the column span of the upper modifier is marked by green lines. If the middle column lies over some component in the middle zone (see Fig. 9(a)), then calculate the width on left (W_l) and right (W_r) sides of that column in the middle zone of that component. If W_l is less than W_r then join the modifier with the component in the middle zone which is just left to the underlying component in the middle zone. Otherwise, join it with the component in the middle zone which is just right to the underlying component in the middle zone.

5. If the middle column lies over blank space in the middle zone as shown in Fig. 9(b), then calculate the width of the components on both sides (C_l and C_r) of that column in the middle zone. If C_l is less than C_r then join the modifier with the component of width C_l; otherwise, join the modifier with the component of width C_r.

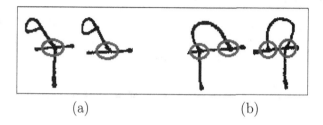

(a)   (b)

Fig. 6. (a) Single touching upper modifiers. (b) Double touching upper modifiers. Red circle indicates the touching point (Color figure online).

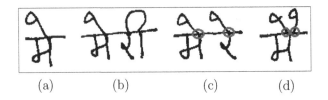

(a)   (b)   (c)   (d)

Fig. 7. (a) No next modifier. (b) Next modifier is not a single touching modifier.
(c) Next modifier is single touching but touching points are not nearer to each other.
(d) Next modifier is single touching and touching points are nearer to each other. Red
circle indicates the touching point (Color figure online).

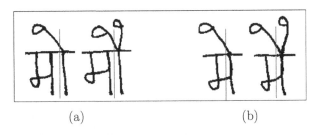

(a)   (b)

Fig. 8. Calculated column (marked in red) lies over (a) blank space in middle zone;
(b) some component in middle zone (Color figure online).

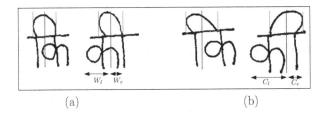

(a)   (b)

Fig. 9. Middle column (in red) lies over (a) some component in middle zone; (b) blank
space in middle zone (Color figure online).

3.3   Segmentation of Lower Modifiers

The proposed method for lower modifier segmentation are shown in Fig. 11. At
first, the components containing the lower modifiers are identified. Thereafter,
they are classified into three classes as given below.

– Middle bar characters (क फ)
– Right bar characters (ख ग घ च ज झ त थ ध न प ब भ म य ल व स)
– No bar characters (छ द र ह)

1	2	3
4	5	6
7	8	9

Fig. 10. 3×3 grid for lower modifier detection.

The steps to determine the classes are:

1. Divide the entire component into 3×3 grid as shown in Fig. 10.
2. Consider block 3 and 6 and check whether more than 90 % of their rows contain object pixels or not. If so, then the component is a right bar character.
3. Consider block 2 and 5 and perform similar check over their rows. If it is satisfied, then it is a middle bar character.
4. If none of the cases are satisfied, then it is a no bar character.

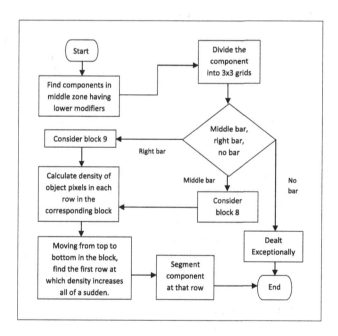

Fig. 11. Flowchart for lower modifier segmentation.

After the final classification, if it is a right bar character then block 9 is considered and if it is a middle bar character then block 8 is considered in the grid. Thereafter, the density of object pixels in each row in the considered block is calculated. Then, moving from top to bottom in that block, the first row at which the density increases all of a sudden is found. This row is omitted to segment the lower modifier from its middle zone component. The no bar characters can be dealt exceptionally.

4 Experimental Results and Discussion

This section presents the experimental results and related discussion of our proposed method.

4.1 Experimental Dataset

Our dataset consists of about 12750 Hindi word samples among which 1200 samples are printed and 11550 samples are handwritten. The handwritten samples are collected from 30 different writers used 10 different types of pens with varying pen width. The samples are scanned at 300 dpi using HP Office Jet 5610 scanner. The implementation has been done on MATLAB (R2010a).

4.2 Character Segmentation Results

Detection of Header Lines: The experimental results are shown in Fig. 12(a). In this figure, the left and right column shows the input samples and the corresponding outputs after the completion of phase 1. It is shown that there is a white single width straight line detected as the header line for each of the inputs. This represents the required row to be removed. Now, we can remove the appropriate number of rows above and below of the obtained row according to the pen width of the written text so as to completely get rid of the header line.

Segmentation of Upper Modifiers: The experimental results are shown in Fig. 12(b). The left and right column in the figure shows the inputs and the corresponding outputs after the completion of phase 2. We can observe that all the upper modifiers along with their middle zone counterpart get totally separated from the rest and are represented individually. For the upper modifiers with their counterparts in the middle zone, appropriate joining has been done and also the extrapolation has been performed if required.

Segmentation of Lower Modifiers: Finally, the lower modifiers are segmented in the last phase of the proposed method. The experimental results are shown in Fig. 12(c). The left and right column in the figure shows the inputs and the corresponding outputs after the completion of third phase. It is observed that all the lower modifiers are detected and segmented from their middle zone component character correctly.

For the above all test cases, we consider the input samples having the shortcomings mentioned earlier to show the efficacy of our proposed method.

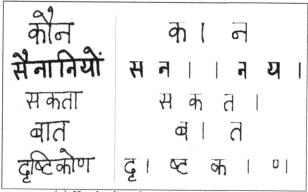

(a) Header line detection and removal.

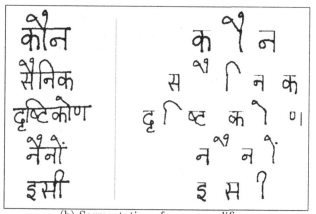

(b) Segmentation of upper modifiers.

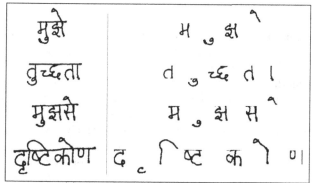

(c) Segmentation of lower modifiers.

Fig. 12. Experimental results of different phases for handwritten Hindi words.

Table 1. Header line detection accuracy.

Method	Total no. of words	No. of header lines detected correctly	Accuracy
Hanmandlu and Agrawal	12750	6739	52.85 %
	1200 (printed)	1168	97.33 %
	11550 (handwritten)	5571	48.23 %
Bansal and Sinha	12750	4022	31.55 %
	1200 (printed)	1168	94.85 %
	11550 (handwritten)	2854	24.7 %
Proposed	12750	12614	98.93 %
	1200 (printed)	1200	100 %
	11550 (handwritten)	11414	98.82 %

Table 2. Upper modifier segmentation accuracy.

Method	Total no. of words	Total no. of upper modifiers	No. of upper modifiers segmented correctly	Accuracy
Hanmandlu and Agrawal	12750	3693	2977	80.61 %
	1200 (printed)	704	640	90.9 %
	11550 (handwritten)	2989	2337	78.18 %
Bansal and Sinha	12750	3693	1648	44.62 %
	1200 (printed)	704	520	73.86 %
	11550 (handwritten)	2989	1128	37.73 %
Proposed	12750	3693	3558	96.34 %
	1200 (printed)	704	704	100 %
	11550 (handwritten)	2984	2854	95.45 %

4.3 Comparison with Other Methods

We compared our results with two existing methods of Bansal–Sinha [5] and Hanmandlu–Agrawal [9]. As the datasets used in these two methods were not available, so to perform the comparative analysis in same platform we prepared our own dataset with reasonable size as discussed in Sect. 4.1. We implemented the other two methods and tested all these methods on our own dataset. Our main objective was to make this comparative analysis unbiased. The accuracy of header line detection, upper modifier segmentation, and lower modifier segmentation are shown in Tables 1, 2, and 3 respectively. We observe that the performance of header line detection is much better than the other two existing methods. This is because of the efficiency of our proposed method to handle large variety of writing styles and skewed header lines as input data. Also for upper and lower modifier segmentation, our algorithm has shown an acceptable improvement in accuracy over the other two existing methods. The overall success rate of our proposed method is 96.93 % which is much better than Bansal–Sinha (48.58 %) and Hanmandlu–Agrawal (72.8 %) methods. During the measuring of accuracy rate, we treated over and under segmentation as an incorrect

Table 3. Lower modifier segmentation accuracy.

Method	Total no. of words	Total no. of lower modifiers	No. of lower modifiers segmented correctly	Accuracy
Hanmandlu and Agrawal	12750	3080	2616	84.94 %
	1200 (printed)	376	320	85.1 %
	11550 (handwritten)	2704	2296	84.91 %
Bansal and Sinha	12750	3080	2143	69.58 %
	1200 (printed)	376	336	89.36 %
	11550 (handwritten)	2704	1807	66.83 %
Proposed	12750	3080	2942	95.52 %
	1200 (printed)	376	360	95.74 %
	11550 (handwritten)	2704	2582	95.49 %

segmentation. This work can be extended later on to make a more generalized method for lower modifier segmentation in case of no bar characters and the segmentation of two characters touch in upper, middle, or lower region. Many a times shadow characters also occur in handwritten text when one totally independent component occurs under some other component. They can also be dealt with in future.

We have also done a comparison in between the above said methods and our proposed method w.r.t. computational time. We used our own test datasets for this experimental analysis. It is noticed that our proposed method is computationally efficient than the other two methods for both the printed and handwritten images.

5   Concluding Remarks

In this paper, we have proposed a character segmentation method based on structure shape of Hindi language. The proposed method has performed significantly well at each level of segmentation to handle large scale shape variation in writing style of Hindi language. The proposed method is tested on handwritten Hindi word images and the results are very promising with an average accuracy rate of 96.93 %. But this method is not performing well for few particular cases as stated earlier. In future, we shall extend our work to improve the accuracy of segmenattion and to make it applicable to character recognition for handwritten Hindi OCR system.

References

1. Bag, S., Harit, G.: Skeletonizing character images using a modified medial axis-based strategy. Int. J. Pattern Recognit. Artif. Intell. **25**, 1035–1054 (2011)
2. Bag, S., Harit, G.: A survey on optical character recognition for Bangla and Devanagari scripts. Sadhana **38**, 133–168 (2013)

3. Bag, S., Bhowmick, P., Behera, P., Harit, G.: Robust binarization of degraded documents using adaptive-cum-interpolative thresholding in a multi-scale framework. In: International Conference on Image Information Processing, pp. 1–6. IEEE Press, New York (2011)

4. Bag, S., Bhowmick, P., Harit, G., Biswas, A.: Character segmentation of handwritten Bangla text by vertex characterization of isothetic covers. In: National Conference on Computer Vision, Pattern Recognition, Image Processing and Graphics, pp. 21–24. IEEE Press, New York (2011)

5. Bansal, V., Sinha, R.M.K.: Segmentation of touching and fused Devanagari characters. Pattern Recognit. **35**, 875–893 (2002)

6. Bishnu, A., Chaudhuri, B.B.: Segmentation of Bangla handwritten text into characters by recursive contour Following. In: International Conference on Document Analysis and Recognition, pp. 236–239. IEEE Press, New York (1999)

7. Casey, R.G., Lecolinet, E.: A survey of methods and strategies in character segmentation. IEEE Trans. Pattern Anal. Mach. Intell. **18**, 690–706 (1996)

8. Garain, U., Chaudhuri, B.B.: Segmentation of touching characters in printed Devnagari and Bangla scripts using fuzzy multifactorial analysis. IEEE Trans. Syst. Man Cybern. Part C **32**, 449–459 (2002)

9. Hanmandlu, M., Agrawal, P.: A structural approach for segmentation of handwritten Hindi text. In: International Conference on Cognition and Recognition, pp. 589–597 (2005)

10. Huang, L., Wan, G., Liu, C.: An improved parallel thinning algorithm. In: International Conference on Document Analysis and Recognition, pp. 780–783. IEEE Press, New York (2003)

11. Otsu, N.: A threshold selection method from gray-level histogram. IEEE Trans. Syst. Man Cybern. **9**, 62–66 (1979)

12. Pal, U., Chaudhuri, B.B.: Indian script character recognition: a survey. Pattern Recognit. **37**, 1887–1899 (2004)

13. Pal, U., Datta, S.: Segmentation of Bangla unconstrained handwritten text. In: International Conference on Document Analysis and Recognition, pp. 1128–1132. IEEE Press, New York (2003)

14. Pal, U., Jayadevan, R., Sharma, N.: Handwritten recognition in Indian regional scripts: a survey. ACM Trans. Asian Lang. Inf. Process. **11**(1), 1–35 (2012)

15. Rosenfeld, A., Kak, A.C.: Digital Picture Processing, 2nd edn., vols. 1 and 2. Academic Press, New York (1982)

16. Sarkar, R., Das, N., Basu, S., Kundu, M., Nasipuri, M., Basu, D.K.: A two-stage approach for segmentation of handwritten Bangla word images. In: International Conference on Frontiers in Handwriting Recognition, pp. 403–408. CENPARMI, Canada (2008)

Retinal Blood Vessel Segmentation and Bifurcation Point Detection

Tapash Dutta[1], Nilanjan Dutta[1],
and Oishila Bandyopadhyay[2]([✉])

[1] Department of Information Technology, Indian Institute of Engineering Science and
Technology, Howrah, India
[2] Advanced Computing and Microelectronics Unit, Indian Statistical Institute,
Kolkata, India
oishila@gmail.com

Abstract. The analysis of retinal blood vessel structure plays an important role in diagnosis of different diseases. Automated extraction of vascular network and identification of bifurcation points can be an important part of computer assisted analysis of retinal vascular disorders. In this paper, we propose an efficient method of automatic blood vessel extraction and bifurcation point detection from retinal images. The proposed method introduces the novel concept of relaxed digital arc for the removal of optic disc region to improve the correctness of the results. Experimental results show the effectiveness of the proposed method. We re-validate the quality of the proposed blood vessel segmentation approach by comparing the segmentation accuracy with existing approaches. The efficiency of bifurcation point detection process is evaluated by comparing manual bifurcation point count with the findings of the proposed approach.

Keywords: Bifurcation points · Optic disc · Retinal blood vessel ·
Digital arc

1 Introduction

Retinal vessel segmentation is an important part of computer aided diagnosis of ophthalmic disorders. Blood vessels and optic disc are the two major components of retinal image analysis. The study of blood vessel morphology can be used as an important indicator for diagnosis of many diseases like diabetes, hypertension, arteriosclerosis, and glaucoma. Identification of bifurcation points in retinal vascular network is essential for prediction of any ophthalmic disorder or use of retinal structure for registration and biometric authentication.

Image segmentation and edge detection play an important role in medical image processing. Advanced edge detection algorithms based on scan line processing and linking of edge points using graph strategy can efficiently detect the edges of the complex images [22,23]. In the last few decades several researchers have worked on retinal blood vessel extraction. Reviews of different

© Springer International Publishing Switzerland 2015
R.P. Barneva et al. (Eds.): IWCIA 2015, LNCS 9448, pp. 261–275, 2015.
DOI: 10.1007/978-3-319-26145-4_19

blood vessel extraction algorithms are discussed by Fraz et al. [8]. The segmentation methods of retinal blood vessel can be classified into two main groups based on scanning and tracking [4]. In tracking based methods, vessel structure segmentation and image feature extraction are performed simultaneously in a single pass. Pixel scanning based segmentation methods extract the image feature point by contrast enhancement operations followed by different thresholding and filtering techniques. In many approaches, matched filter is used for contrast enhancement [3,13]. In some papers, extraction of blood vessel from dark background is performed by local entropy based thresholding methods [13]. Adaptive histogram equalization is also used for contrast enhancement of retina images [6]. Optic disc is an important part of retinal images. Many researchers have developed algorithms for segmentation of optic disc region [14]. In retinal images, vascular bifurcation and crossover junctions are special regions where a single blood vessel splits into two new sub-vessels, or two blood vessels cross each other. Iqbal et al. have used neural network based approach for vascular feature extraction [11]. These vascular features are used for disease diagnosis in medical applications [2] and personal identification in biometric applications [20].

In this paper, we propose an efficient method for extraction of vascular network from retinal images and identification of potential bifurcation points in the vascular network. The proposed method integrates adaptive histogram equalization and matched filter with a novel histogram based segmentation approach for segmentation of vascular network. Global histogram equalization is applied to segment the optic disc region from the retinal image. The novel concept of relaxed digital arc is introduced to remove the optic disc region from vascular network. The segmented vascular network is then used to identify the possible bifurcation points. We have applied the neighborhood pixel masking to identify the bifurcation points and filter the wrong detection. Quality of segmentation is evaluated by comparing the average segmentation accuracy value obtained in the proposed approach against other segmentation approaches. The fundus camera images available in DRIVE database are used for testing the proposed method.

2   Related Definitions

The optic disc segmentation process proposed in this work has utilized the novel concept of relaxed digital arc. A few definitions related to this process are given below:

Chain Code: It is used to encode a direction around the border between pixels. If $G(m, n)$ is a grid point, then the grid point (m', n') is a neighbor of G, provided that $max(|m - m'|, |n - n'|) = 1$. The chain code [9] of G with respect to its neighbor grid point in G_1 can have a value in $0, 1, 2, ..., 7$ as shown in Fig. 1(a).

Digital Curve (DC): A DC C is an ordered sequence of grid points (representable by chain codes) such that each point (excepting the first one) in C is a neighbor of its predecessor in the sequence [21] (see Fig. 1(b)).

Digital Circle: The properties of a digital circle can be defined as follows [1,18]:

(a) Chain code changes in same traversal direction.
(b) Pixels may form few digital straight line segments (DSS) with single chain code.
(c) Difference in chain code for consecutive pixels will never exceed unity.

A digital circle can be divided into eight octants. Each octant in a circle represents a circular arc (as shown in Fig. 1(c)). Hence each circular arc is a part of the digital circle and maintain the properties mentioned above.

Relaxed Digital Arc (RDA): In this work, we introduce the novel concept of relaxed digital circular arc which relaxes the third property of digital arc. So properties of RDA are as follows:

(a) Chain code changes in same traversal direction.
(b) Pixels may form few digital straight line segments with single chain code.
(c) Difference in chain code for consecutive pixels will never exceed 2.

Figure 1(d) and (e) show the digital arc (AB) and relaxed digital arc (CD) respectively with S_1 and S_2 represent the straight and the curved region of the arc.

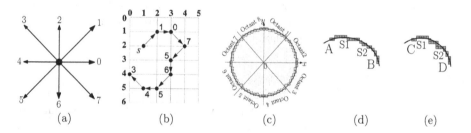

(a)   (b)   (c)   (d)   (e)

Fig. 1. (a) Chain code, (b) Digital curve, (c) Digital circle, (d) Digital arc, (e) Relaxed digital arc.

3   Proposed Method

The proposed method consists of several phases. Figure 2 shows different phases of the proposed approach. In the initial phase, the input retina image undergoes preprocessing to prepare the image for vessel extraction (Fig. 2(b) and (c)). Vessel extraction and bifurcation point detection has different phases. The preprocessed image passes through matched filtering, and thresholding process for segmentation of retinal vascular network (Fig. 2(d),(e), and (f)). Finally, the segmented retinal vascular network undergoes optic disc removal, thinning and masking for detection of potential bifurcation points (Fig. 2(g), (h), and (i)).

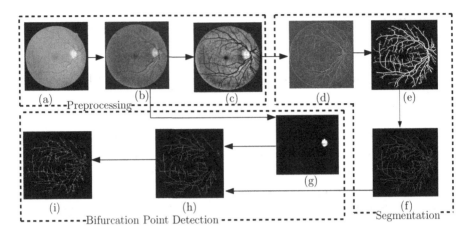

Fig. 2. Proposed Method (a) Input retina image, (b) Adaptive histogram equalization on green channel, (c) Contrast enhancement, (d) Matched filter output, (e) Threshold and noise removal, (f) Morphological thinning, (g) Optic disc region extraction from (b), (h) Optic disc removal from (f), (i) Bifurcation points identified (marked with red points) (Colour figure online).

3.1   Preprocessing

In the preprocessing phase, colour channel selection, adaptive histogram equalization, and contrast enhancement are performed to process the input retinal image for blood vessel segmentation.

3.1.1   Colour Channel Selection

Retina image taken by Fundus camera appears with brighter optic disc region, reddish retina surface and dark vascular network. Segmentation of vascular network from Fundus retina image requires blood vessels to be more prominent. Hence, we have extracted green component of the input RGB retinal image (Fundus images from DRIVE database) and used it for the proposed segmentation and bifurcation point detection approach. Figure 3 shows that green channel retina image (Fig. 3(c)) exhibits greatest contrast between background, vessels, and other foreground objects.

3.1.2   Adaptive Histogram Equalization

The fundus images are often non-uniformly illuminated and they exhibit poor contrast between the background and the blood vessels. The objective here is to enhance the contrast so that vessels can be identified with greater accuracy. Contrast limited adaptive histogram equalization (CLAHE) method [6] has been used with a window size 9×9 to enhance the contrast of the image (as shown in Fig. 4(b)). CLAHE method partitions the image into contextual regions and then

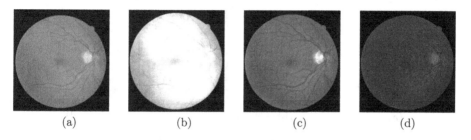

Fig. 3. RGB Component (a) DRIVE Database Image #02, (b) Red Component, (c) Green Component, (d) Blue Component (Colour figure online).

applies histogram equalization to each partition. This results in even distribution of gray values and increases visibility of the low contrast features present in the image.

3.1.3   Contrast Enhancement

To enhance the contrast further, morphological bottom hat transformation [7] is applied on the histogram equalized image. The bottom-hat transformation is used to extract valleys such as dark lines and dark spots. In the proposed method, the bottom hat transformed image (Fig. 5(b)) is subtracted from the adaptive histogram equalized image to generate retina image with more prominent vascular network (Fig. 5(c)).

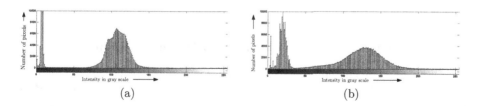

Fig. 4. Adaptive Histogram Equalization (a) Histogram of green channel component (Fig. 3(c)), (b) Histogram after applying contrast limited adaptive histogram equalization (Colour figure online).

4   Extraction of Retinal Vascular Network

Segmentation of retinal blood vessel consists of different phases like matched filtering, thresholding, connected component analysis and removal of outer boundary.

4.1   Linear Segment Detection Using Matched Filter

Matched filter [4] is used to detect piece-wise linear segments of blood vessels in retinal images. As the blood vessels have very low contrast, a two-dimensional

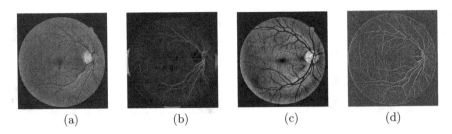

Fig. 5. Contrast Enhancement (a) Adaptive histogram equalized Image, (b) Bottom-hat applied on (a), (c) Contrast enhanced image with prominent vessels, (d) Matched filter output.

matched filter kernel is designed to apply to the retinal image in order to enhance the contrast between blood vessel and background (as shown in Fig. 5(c)). Such a kernel can be expressed as

$$f(x,y) = -exp^{\frac{-x^2}{2\sigma^2}}, \quad for \quad |y| <= L/2 \tag{1}$$

where σ = spread of the intensity profile and L=length of the vessel segment assumed to be a straight line and have a fixed orientation. Here, we have used multi scale matched filtering and the value of σ has been varied in the range 0 to 0.1 with each interval having value of 0.01. As the result shows (Fig. 5(d)), use of multi-scale matched filter enhances the contrast between vessels and the background significantly.

4.2 Threshold Computation from Intensity Distribution

Histogram of an image represents the distribution of image pixels among different intensity levels. The histogram of image I with G intensity levels, is defined as the discrete function $H(r_k) = n_k$ where r_k is the k-th intensity bin in the interval [0,G] and n_k is the number of pixels in the intensity bin r_k [10].

In this paper, we propose a histogram based thresholding approach to extract the vessels from the matched filtered retinal image. After matched filtering, retinal images appear with dark background and white vessel part. The histogram of such gray scale image has maximum number of pixels belonging to the dark background region with lower intensity values and fewer pixels belong to maximum intensity level 255 white pixels. So we compute the pixel number difference between every i-th and $(i+1)$-th intensity bin. The difference in number of pixels among neighboring bins can be represented as

$$\Delta d_i = |d_i - d_{i+1}| \quad for \quad 0 \leq i \leq 255 \tag{2}$$

As the intensity bin with the histogram peak represents the intensity of the background region pixels, we analyze the intensity region between the histogram peak ($index$) and the maximum intensity level ($index1$) for selection of threshold intensity (as shown in Fig. 6(a)). In retinal images, it is noticed that $\Delta d_{min} = 0$ for many high value intensity bins. So the threshold value is computed in a iterative process by comparing every bin count against the mean value (avg) of Δd_{max} and Δd_{min} and replacing the maximum bin index $(index1)$ with index of the mean value (avg). The process terminates when three successive mean value index remain unchanged (see *Procedure threshold()*). Figure 6(a) shows the histogram of matched filtered retinal image. Figure 6(b) shows the retinal image generated after thresholding of matched filtered image (Fig. 5(d)).

Procedure *threshold(Img_{gray})*

1. $(count[], bin[]) \leftarrow generate\_hist(Img_{gray})$
   /*Computes pixel-count for each intensity bin of the gray image*/
2. $m \leftarrow max(count[])$
3. $n \leftarrow sizeof(count[])$
4. $index \leftarrow find\_index(count[], m)$
   /* Computes the index of the value in the given array*/
5. $pos[1], pos[2], pos[3], k \leftarrow 0$
6. $index1 \leftarrow n - 1$
7. *Repeat*
8. $j \leftarrow 0$
9.   for each i in $index$ to $index1$
10. $j \leftarrow j + 1$
11. $\Delta d(j) \leftarrow |(count(i) - count(i + 1)|$
12.   *loop*
13. $\Delta d_{max} \leftarrow max(\Delta d)$
14. $\Delta d_{min} \leftarrow min(\Delta d)$
15. $avg \leftarrow \lceil (\Delta d_{max} + \Delta d_{min})/2 \rceil$
16.   for each i in $index1$ down to $index$
17. $if(avg > count(i))$
18. $index1 \leftarrow i$
19.   end
20.   *loop*
21. $pos(k) \leftarrow index1$
22. $k \leftarrow k + 1$
23. $Until((pos(1) = pos(2))\&(pos(2) = pos(3)))$
24. $i_{thr} \leftarrow bin(pos(1))$
25. **return** i_{thr}

4.3   Connected Component Analysis

Connected component analysis is performed on thresholded image with 8-connected objects. Pixels of the image are grouped in different labels depending on their connectivity. Pixels belonging to background region are labeled with 0.

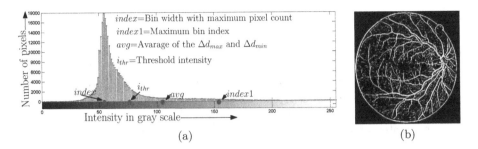

(a)   (b)

Fig. 6. (a) Histogram of Fig. 5(d), (b) Thresholded image.

The proposed method counts the number of pixels in each label to differentiate between vessels and unwanted noise. In the thresholded retinal image, the component with maximum number of pixels represents the image background. As vessels are also connected components, they appear as the components with relatively fewer pixels. Some components with very few pixels are identified as noise or unwanted elements. The proposed method has eliminated the components with fewer pixel count (fewer than 150) to remove the noise and generate the segmented vascular network (Fig. 7(a)).

4.4 Removal of the Outer Boundary

Canny edge detection is used to detect the outer boundary (see Fig. 7(b)) of the segmented retinal image. The boundary image is dialated and subtracted from the segmented vascular network to remove the outer boundary of the segmented retinal image (as shown in Fig. 7(c)).

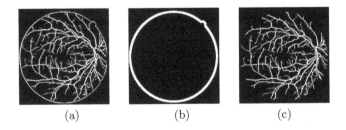

(a)   (b)   (c)

Fig. 7. Outer Boundary Removal (a) Image after noise removal of Fig. 6(b), (b) Outer Boundary of (a), (c) Outer boundary removed from (b).

5 Bifurcation Point Detection

Segmented blood vessel image is used to detect the bifurcation points. The proposed approach detects the candidate bifurcation points by applying 3×3 mask on the thinned segmented vascular network. Removal of optic disc region helps to improve the effectiveness of the proposed bifurcation point identification process. Finally, the small length sub-vessel filtering is performed to identify the potential bifurcation points.

5.1   Thinning

The proposed method have used single pixel wide vessel network for correct identification of bifurcation points. Morphological thinning is applied on the segmented vascular network of retinal image before using it for bifurcation point detection (Fig. 8(b)). The thinned single pixel based vascular network appears with the spurious pixels. These spurious pixels are removed from the thinned vascular network by applying a 3×3 neighborhood filter window (as shown in Fig. 8(c)).

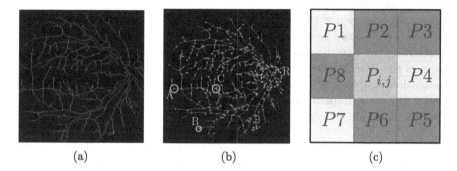

(a)	(b)	(c)

Fig. 8. Thining (a) Thinned vascular Network, (b) Probable bifurcation points detected by 3×3 masking on (a), (c) 3×3 mask pixel combination on marked region 'A' of (b).

5.2   Candidate Bifurcation Points Detection

The thick vessels appear in retinal image bifurcate into thin sub-vessels. Detection of these bifurcation points is important for disease diagnosis and biometric authentication using retinal images. In some images, crossover points among neighboring vessels may also appear as false bifurcation points. So the main challenge in bifurcation point detection is to identify the potential bifurcation points from the group of candidate bifurcation points recognized in the segmented vascular network. The proposed method has applied a 3×3 mask [11] to detect the probable bifurcation points appear in the segmented vascular network. If a pixel represents a candidate bifurcation point, then considering a 3×3 mask over that pixel, at least three pixels withtin the mask will represent vessel pixel (with intensity 1). So, a pixel P_i qualifies to be a bifurcation point if $b(P_i) = \frac{1}{2} \sum_{i=1}^{8} |P_i - P_{i+1}|$ has the value 3.

Applying this property we have identified the candidate bifurcation points (as shown in Fig. 8(b)) on the vascular network. Analysis of the bifurcation points identified by 3×3 mask shows that many points are identified falsely as bifurcation points in the optic disc region of the vascular network as this region has many overlapping vessels. Some bifurcation points have very small sub-vessel length. In some cases, some vessel crossover points are also detected as bifurcation points. Such false bifurcation points are filtered in three phases.

5.3   Optic Disk Removal

In the proposed method, several false bifurcation points are identified in the optic disc region of the vascular network. Those false points can be removed by removing the optic disc region from the segmented vascular network. The proposed method has removed the optic disc region from the contrast enhanced retina image (Fig. 9(a)) by applying intensity thresholding based segmentation, noise removal and relaxed digital arc based circular mask generation.

5.3.1   Segmentation

As the optic disc region is the maximum illuminated area in the retinal image, segmentation of optic disc can be done by intensity thresholding. For intensity thresholding, we have applied global histogram equalization on the green component of the RGB image. Global histogram equalization exhibits the most lucid illumination at optic disc region. So its complement generates the darkest region at optic disc area. Exploiting this fact, the brighter image is subtracted from its complement, to obtain a part of the optic disk (Fig. 9(a)).

5.3.1.1 Noise Removal. After the extraction of optic disc part by intensity thresholding, noise removal process using connected component analysis and length filtering is applied to filter the noise part present in the thresholded optic disc image. To improve the quality of the segmented optic disc portion, region filling is done to fill the thin black region appearing in the segmented optic disc (Fig. 9(b)).

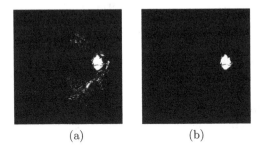

(a)   (b)

Fig. 9. Optic disc segmentation (a) Segmented optic disc, (b) Optic disc after noise removal.

5.3.1.2 Circular mask generation. The contour of the segmented optic disc is analyzed to generate the circular mask to cover the optic disc region in the vascular network. We utilize the properties of digital arc [18,19] and digital circle [1] to generate the circular mask covering the optic disc contour.

Chain code analysis of digital arc: Analysis of digital arc chain code shows that a circular arc can have straight component and curved component. The straight component can have only two chain consecutive codes with one single occurrence and one multiple occurrence (with runs differ in length by at most unity) [19]. At the beginning of the curved region, the chain code with multiple occurrence, appears in single runs and break the straightness. Appearance of a new chain code indicates the start of a new octant. Figure 1(d) shows the segment of a digital arc 'AB' with 'S1' and 'S2' represent the straight component and curved region respectively.

In the proposed method, we analyze the chain code of the optic disc contour and identify the length of different relaxed digital arcs (with consecutive chain code atmost differs by 2). Figure 10(a) shows the relax digital arcs (RDA) identified in the optic disk contour. The RDA with maximum length is considered as the octant of the circular mask (arc $S9$ in Fig. 10(a)).

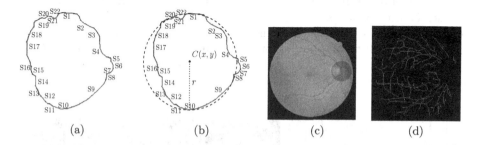

(a)   (b)   (c)   (d)

Fig. 10. Optic disc removal (a) RDAs identified in optic disc contour, (b) Optic disc contour with circular mask, (c) Optic disc encircled in retina image, (d) Vascular network after optic disc removal.

The proposed circular mask generation process includes following steps:

(a) Identify the RDA with maximum length (arc_{max}) from the optic disc contour.
(b) Compute the radius of the arc_{max} (considering arc_{max} as octant of the proposed circular mask).
(c) If the chain code pattern of arc_{max} follows clockwise direction (07/06/ 76/../10 for left to right traversal of the image, like $S1$, $S9$ in Fig. 10(a)), computes the center of the circular mask using arc_{max} end-point positions, and radius of the arc_{max}.
(d) If the chain code pattern of arc_{max} follows counter-clockwise direction (like $S2$, $S4$ in Fig. 10(a)), computes the center of the circular mask in the opposite side of the RDA center (using arc_{max} end-point positions, and radius of the arc_{max}), as couter-clockwise chain code pattern indicates distortion in segmented optic disc contour.

Finally, this circular mask is used to cover the optic disc region in the retinal vascular network. Figure 10(b) and (c) shows the circular mask and Fig. 10(d) shows the vascular network after removal of optic disc region.

5.4   Small Length Sub-vessel Filtering using 5 × 5 Mask

Few bifurcation points detected by 3 × 3 mask have very small length (one pixel) sub-vessels (point 'B' of Fig. 8(b)). These points cannot be considered as potential bifurcation points. We have applied a 5 × 5 mask to filter such bifurcation points. A 5 × 5 mask can be considered as a 3 × 3 mask with a one pixel thick outer layer on it. The pixels in the outer layer are the nearest neighbors of the pixels in the inner 3 × 3 mask. The corner pixels in the 3 × 3 mask have 5 neighboring pixels (e.g. pixel $P1$ has $P10$, $P11$, $P12$, $P25$ and $P24$ as neighbors) in the outer layer and the intermediate pixels of the 3 × 3 mask have 3 nearest neighbor pixels (e.g. $P4$ has $P15$, $P16$ and $P17$ as neighbors) in the outer layer (Fig. 11(a)). For each probable bifurcation point identified by the 3 × 3 mask, the proposed method has checked the the pixel values of the outer layer of 5 × 5 mask. If for each white boundary pixel in the 3 × 3 mask, one neighboring white pixel appears in the outer layer of the 5 × 5 mask, the bifurcation point is considered as the potential bifurcation point. Figure 11(a) shows the pixel distribution of a potential bifurcation point (point 'A' of 8(b)) and Fig. 11(b) shows the pixel distribution of a false bifurcation point with very small sub-vessel length (point 'B' of Fig. 8(b)).

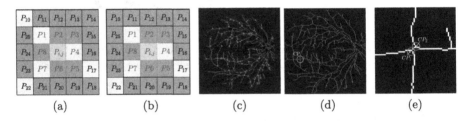

Fig. 11. Bifurcation Point Detection (a) 5×5 mask pixel combination on marked region 'A' of Fig. 8(b), (b) 5 × 5 mask pixel combination on marked region 'B' of Fig. 8(b), (c) Potential bifurcation points after applying 5 × 5 mask and bifurcation point pair filtering, (d) crossover point, (e) false detection of bifurcation point pair.

5.5   False Bifurcation Point Pair Filtering

In many situations, a single crossover point is wrongly detected as a pair of bifurcation points (point 'CP' of Fig. 11(e)) [2]. Such false detection is identified by checking the distance of the nearest bifurcation point for each probable bifurcation point. If the distance is much smaller (less than 3 pixels), those two bifurcation points are removed. Figure 11(c) shows the potential bifurcation points detected after filtering.

Table 1. Vessel Segmentation and Bifurcation Point Detection Accuracy.

Sl No.	Accuracy (%)	Manual Count	Algorithm Count	Sl No.	Accuracy (%)	Manual Count	Algorithm Count
1	95.13	70	71	11	94.77	53	60
2	94.32	65	61	12	95.69	48	52
3	93.08	72	89	13	94.84	60	58
4	95.83	50	48	14	95.26	71	63
5	95.73	39	48	15	94.12	38	33
6	95.05	57	55	16	95.32	85	79
7	94.67	57	54	17	95.21	67	59
8	94.80	22	25	18	95.48	52	47
9	95.92	42	38	19	95.41	76	65
10	95.47	60	55	20	95.55	62	57

Table 2. Comparison of accuracy for Blood Vessel extraction using different approaches.

Serial No.	Algorithm	Average Accuracy
1	Marin [15]	94.52
2	Chaudhuri [3]	87.73
3	Jiang and Mojon [12]	89.11
4	Mendonca [17]	94.63
5	Martinez-Perez [16]	93.44
6	Chinsdikici and Aydin [5]	92.93
7	Staal [24]	94.41
8	**Proposed Method**	**95.26**

6 Experimental Results

The proposed vessel extraction algorithm is tested using the images of the DRIVE database. In order to quantify the accuracy of vessel extraction using proposed method, we have compared the outcome of the proposed method with respective manually segmented test images available in the DRIVE database. The accuracy is checked using the Confusion Matrix [15]. Considering true-positive (TP), true-negative (TN), false-positive (FP) and false-negative (FN) segmented vessels, the accuracy can be represented as $Ac = \frac{TP+TN}{TP+TN+FP+FN}$. The proposed method shows 95.26 % accuracy in vessel segmentation using test images from DRIVE database (shown in Table 1). The accuracy of bifurcation point detection part is evaluated by counting the number of bifurcation points detected manually from the segmented image and those detected using the proposed method. Table 1 also shows the variation in point detection.

In this table, serial number represents DRIVE database test image number, accuracy represents vessel segmentation accuracy, and count represents bifurcation point count.

We have compared the average performance of the proposed approach with other existing retinal blood vessel segmentation algorithms (using same images of

DRIVE database) listed in Table 2. It reveals that the proposed method exhibits better accuracy than existing approaches.

7 Conclusion

In this paper we propose an efficient method for fully automated blood vessel extraction. We have used adaptive histogram equalization, multi-scale matched filtering and an automated thresholding scheme that shows the vessel segmentation accuracy of 95.26 %. The concept of digital arc is utilized to generate the circular mask for optic disc removal from the segmented vascular network. We have identified the potential bifurcation points by applying 3×3 and 5×5 mask on the segmented vascular network. Performance analysis of the experimental findings shows satisfactory results. This work can be extended in future for disease diagnosis and biometric authentication based on retinal images.

Acknowledgements. Authors would like to acknowledge Department of Science & Technology, Government of India, for financial support vide ref. no. SR/WOS-A/ET-1022/2014 under Woman Scientist Scheme to carry out this work. We also acknowledge the use of DRIVE database (http://www.isi.uu.nl/Research/Databases/DRIVE/download.php) images for implementation and testing of the proposed approach.

References

1. Bhowmick, P., Bhattacharya, B.B.: Number-theoretic interpretation and construction of a digital circle. Discrete Appl. Math. **156**, 2381–2399 (2008)
2. Bhuiyan, A., Nath, B., Ramamohanarao, K.: Detection and classification of bifurcation and branch points on retinal vascular network. In: Proceedings of Digital Image Computing Techniques and Applications (DICTA), pp. 1–8 (2012)
3. Chaudhuri, S., Chatterjee, S., Katz, N., Nelson, M., Goldbaum, M.: Detection of blood vessels in retinal images using two-dimensional matched filters. IEEE Trans. Med. Imaging **3**, 263–269 (1989)
4. Chutatape, O., Zheng, L., Krishnan, S.M.: Retinal blood vessel detection and tracking by matched gaussian and kalman filters. In: Proceedings of Engineering in Medicine and Biology Society, pp. 3144–3149 (1998)
5. Cinsdikici, M., Aydin, D.: Detection of blood vessels in ophthalmoscope images using MF/ant (matched filter/ant colony) algorithm. ELSEVIER Trans. Med. Imaging **2**, 85–96 (2009)
6. Fazli, S., Samadi, S., Nadirkhanlou, P.: A novel retinal vessel segmentation based on local adaptive histogram equalization. In: Proceedings of Machine Vision and Image Processing (MVIP), pp. 131–135 (2013)
7. Fraz, M.M., Javed, M.Y., Basit, A.: Evaluation of retinal vessel segmentation methodologies based on combination of vessel centerlines and morphological processing. In: Proceedings of Emerging Technologies (ICET), pp. 232–236 (2008)
8. Fraz, M., Remagnino, P., Hoppe, A., Uyyanonvara, B., Rudnicka, A., Owen, C., Barman, S.: Blood vessel segmentation methodologies in retinal images a survey. Comput. Methods Programs Biomed. **108**(1), 407–433 (2012)

9. Freeman, H.: On the encoding of arbitrary geometric configurations. IRE Trans. Electron. Comput. **EC–10**(2), 260–268 (1961)
10. Gonzalez, R., Woods, R.E.: Digital image processing (2008)
11. Iqbal, M.I., Aibinu, A.M., Nilsson1, M., Tijani, I.B., Salami, M.J.E.: Detection of vascular intersection in retina fundus image using modified cross point number and neural network technique. In: Proceedings of Computer and Communication Engineering, pp. 241–246 (2008)
12. Jiang, X., Mojon, D.: Adaptive local thresholding by verification-based multi-threshold probing with application to vessel detection in retinal images. IEEE Trans. Pattern Anal. Mach. Intell. **25**(1), 131–137 (2003)
13. Kuri, S.K., Patankar, S.S., Kulkarni, J.V.: Optimized MFR & automated local entropy thresholding for retinal blood vessel extraction. In: Proceedings of Electrical and Computer Engineering (ICECE), pp. 141–144 (2012)
14. Lu, S.: Accurate and efficient optic disc detection and segmentation by a circular transformation. IEEE Trans. Med. Imaging **30**(12), 2126–2133 (2011)
15. Marn, D., Aquino, A., Arias, M.E.G., Bravo, J.M.: A new supervised method for blood vessel segmentation in retinal images by using gray-level and moment invariants-based features. IEEE Trans. Med. Imaging **1**, 146–158 (2011)
16. Martinez-Perez, M., Highes, A., Stanton, A., Thorn, S., Chapman, N., Bharath, A., Parker, K.: Retinal vascular tree morphology: a semi-automatic quantification. IEEE Trans. Biomed. Eng. **8**, 912–917 (2006)
17. Mendonca, A., Campilho, A.: Segmentation of retinal blood vessels by combining the detection of centerlines and morphological reconstruction. IEEE Trans. Med. Imaging **9**, 1200–1230 (2006)
18. Pal, S., Bhowmick, P.: Estimation of discrete curvature based on chain-code pairing and digital straightness. In: Proceedings of IEEE International Conference on Image Processing, pp. 1097–1100 (2009)
19. Pal, S., Dutta, R., Bhowmick, P.: Circular arc segmentation by curvature estimation and geometric validation. World Sci. Int. J. Image Graph. **12**, 1250024-1–1250024-24 (2012)
20. Patwari, M.B., Manza, R.R., Rajput, Y.M., Saswade, M., Despande, N.: Personal identification algorithm based on retinal blood vessels bifurcation. In: Proceedings of Intelligent Computing Applications (ICICA), pp. 203–207 (2014)
21. Rosenfeld, A.: Digital straight line segments. IEEE Trans. Comput. **12**, 1264–1269 (1974)
22. Sappa, A.D.: Unsupervised contour closure algorithm for range image edge-based segmentation. IEEE Trans. Image Process. **15**(2), 377–384 (2006)
23. Sappa, A.D., Devy, M.: Fast range image segmentation by an edge detection strategy. In: Proceedings of IEEE International Conference on 3D Digital Imaging and Modeling, pp. 292–299 (2001)
24. Staal, J., Abramoff, M., Niemeijer, M., Viergever, M., van Ginneken, B.: Ridge-based vessel segmentation in color images of the retina. IEEE Trans. Med. Imaging **4**, 501–509 (2004)

Reconstruction of Bicolored Images

Alain Billionnet[1], Fethi Jarray[1,3], Ghassen Tlig[1,3]([✉]),
and Ezzeddine Zagrouba[2]

[1] Cedric-CNAM, 292 Rue St-Martin, Paris, France
[2] Higher Institute of Computer Science, Tunis, Tunisia
[3] Higher Institute of Computer Science, Medenine, Tunisia
alain.billionnet@ensiie.fr, fethi_jarray@yahoo.fr,
ghassen.tlik@gmail.com, ezzeddine.zagrouba@fsm.rnu.tn

Abstract. In this paper, we present an integer programming approach to estimating a discrete bi-colored image from its two-color horizontal and vertical projections. The two-color projections basically refer to the number of pixels per column having colors c_1 and c_2, and likewise for each row as well. The aim of the integer programming approach is to minimize the number of conflict pixels, i.e. the number of pixels that have color c_1 as well as c_2. Since the problem is NP-complete, we give a survey of the literature and we propose a new integer programming formulation of this problem.

Keywords: Discrete tomography · Reconstruction bicolored images · Integer programming

1 Introduction

Discrete tomography deals with the reconstruction of discrete objects from limited number of projections. The projections are quantitative information about the elements of the object that lie on discrete lines parallel to a given set of directions.. Discrete tomography is applicable in many interesting areas such as non-destructive testing, scheduling [7,18], image processing, electron microscopy and radio tomographic imaging.

The first results in the area of discrete tomography were shown independently, in 1957, by Ryser [22] and Gale [15] about the reconstruction of binary matrices from the row sums and column sums. Ryser also provided a polynomial time algorithm for reconstructing binary matrix from two projections. Later, many features about the structural proprieties of the matrices to reconstruct were studied. An overview of the main results in this field are collected in [16,17].

In this paper, we examine a new formulation to reconstruct bicolored images from orthogonal projections. In Sect. 2, we introduce some definitions and preliminary results. In Sect. 3, we present a short survey of the main approaches about the reconstruction of bicolored images. In Sect. 4, we provide an integer programming formulation. In Sect. 5, we discuss the numerical results.

R.P. Barneva et al. (Eds.): IWCIA 2015, LNCS 9448, pp. 276–283, 2015.
DOI: 10.1007/978-3-319-26145-4_20

2   Definitions and Preliminary Results

We consider a black and white image A of size $m \times n$. A can be considered as a binary matrix, $A = [a_{i,j}]$, where the black color is replaced by (1) and the white color is replaced by (0). Let $h_i = \sum_{j=1}^{n} a_{i,j}$ be the number of black cells on row i and $v_j = \sum_{i=1}^{m} a_{i,j}$ is the number of black cells on column j. We denote by $H = (h_1, \ldots, h_m) \in \mathbb{N}^m$ and $V = (v_1, \ldots, v_n) \in \mathbb{N}^n$ the row and column sums of A. For example, the binary image in Fig. 1 has the horizontal and vertical projections $H = (2, 4, 5, 1, 5, 4, 5)$, and $V = (2, 4, 4, 2, 5, 4, 5)$. The class of all binary images having the horizontal projection H and the vertical projection V is denoted by $\mathcal{BM}(H, V)$.

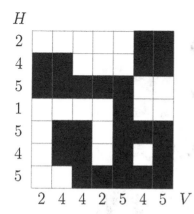
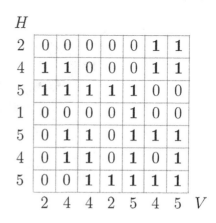

Fig. 1. A binary image and the equivalent binary matrix with horizontal projection $H = (2, 4, 5, 1, 5, 4, 5)$ and vertical projection $V = (2, 4, 4, 2, 5, 4, 5)$.

The reconstruction of a black and white image or a binary matrix is stated as follows:

Reconstruction Binary Image: $RBI(H, V)$
Given: $H = (h_1, \ldots, h_m) \in \mathbb{N}^m$ and $V = (v_1, \ldots, v_n) \in \mathbb{N}^n$.
Goal: Construct an $m \times n$ binary image that satisfies H and V, i.e., row i has exactly h_i black cells and column j has exactly v_i black cells.

$RBI(H, V)$ is solved in polynomial time [22] but the reconstruction is not unique and the reconstruction under additional geometrical constraints is studied in [1,9,13,21]. The most considered constraints are convexity and connectedness [1,5,11,12,23].

Definition 1. *Given a colored image $A = [a_{i,j}]$ of size $m \times n$. The horizontal projection of color c is the vector $H^c = (h_1^c, \ldots, h_m^c) \in \mathbb{N}^m$ where: $h_i^c = |\{j : a_{i,j} = c\}|$. The vertical projection of color c is the vector $V^c = (v_1^c, \ldots, v_n^c) \in \mathbb{N}^n$, for $c \in C$ where $v_j^c = |\{i : a_{i,j} = c\}|$.*

3   Reconstruction of Bicolored Images

In this paper, we are mainly concerned with the problem of reconstructing bicolored images ($c = 2$). We suppose that a cell can be empty, colored by c_1 or colored by c_2 (see Fig. 2). The associated decision problem can be defined as:

Reconstruction Bicolored Image: $RBCI(\mathcal{H}, \mathcal{V})$
Given: Two orthogonal projection for color c_1 and c_2, $\mathcal{H} = (H^{c_1}, H^{c_2})$ and $\mathcal{V} = (V^{c_1}, V^{c_2})$.
Goal: Construct an $m \times n$ bicolored image that satisfies the row and column sums $\mathcal{H}$ and $\mathcal{V}$.

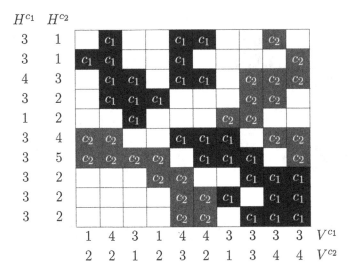

Fig. 2. Bicolored image with its projections $\mathcal{H} = (H^{c_1}, H^{c_2})$ and $\mathcal{V} = (V^{c_1}, V^{c_2})$.

The class of bicolored images having the horizontal projection $\mathcal{H} = (H^{c_1}, H^{c_2})$, and vertical projection $\mathcal{V} = (V^{c_1}, V^{c_2})$ will be denoted by $\mathcal{BCM}(\mathcal{H}, \mathcal{V})$. The reconstruction of bicolored images from their horizontal and vertical projections for each color is NP-complete [14]. Several solvable cases in polynomial time have been studied [2,6,8,10]. In the following, we describe shortly three approximation algorithms from previous works to solve the reconstruction bicolored image, $RBCI(\mathcal{H}, \mathcal{V})$.

Algorithm 1 is a Lagrangean approach published in [19]. Jarray [19] presents a binary quadratic formulation to build an $m \times n$ bicolored images with smoothness proprieties. The objective function minimizes the number of overlapping between colors c_1 and c_2. The constraints ensure that the image solution respects the orthogonal projections of colors c_1 and c_2. Using a Lagrangian relaxation

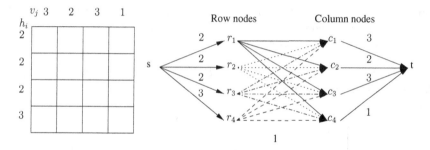

Fig. 3. Reconstruction binary problem $RBI(H, V)$ and the associated max-flow problem, where $H = (2, 2, 2, 3)$ and $V = (3, 2, 3, 1)$.

method, Jarray [19] shows that the problems $RBCI(\mathcal{H}, \mathcal{V})$ can be decomposed into two subproblems of reconstructing binary image where each subproblem is equivalent to a min-cost max-flow problem in a bipartite graph.

Algorithm 2 is an heuristic algorithm published in [3,4]. This method approximates the bicolored image in two steps. In the first step, Ryser's algorithm is used [22] to compute a solution for each color c_1 and c_2 without regarding the overlapping colors. In the second step, the above solutions are merged in unique image, which may have a conflict between colors. For each pixel having a conflict the method searches a convenient switching components that removes the conflicts without changing the projections.

Algorithm 3 is an iterative algorithm published in [20]. Jarray and Tlig [20] propose an approximation based on a max-flow technique. Firstly, an associated max-flow problem is used to compute an initial solution for color c_1 (see Fig. 3). Then, a solution for color c_2 is computed by solving the associated min-cost max flow problem where the cost is the solution of color c_1 obtained in the previous iteration (see Fig. 3). Subsequently, the solution in the previous iteration is used as a cost to determine a solution to solving the associated min-cost max flow problem. This procedure is repeated until the conflict between the solutions of color c_1 and c_2 is resolved or becomes constant.

4   Integer Programming Formulation

Our contribution in this paper is to reconstruct bicolored image by integer programming technique. We introduce for each cell (i, j) two binary variables $x_{i,j}$ and $y_{i,j}$ such that $x_{i,j} = 1$ if the cell (i, j) is colored by color c_1 and $y_{i,j} = 1$ if the cell (i, j) is colored by color c_2. If $x_{i,j} = y_{i,j} = 0$ then the cell (i, j) is said to be empty or uncolored. We call a conflict cell (i, j) a cell colored by both c_1 and c_2, i.e., $x_{i,j} = y_{i,j} = 1$.

Definition 2. *Let X and Y be two $m \times n$ binary matrices. We define the conflict between X and Y as $conf(X, Y) = \sum_{i=1}^{m} \sum_{j=1}^{n} x_{i,j} \, y_{i,j}$, i.e., the number of cells sharing value 1 on both X and Y.*

The reconstruction of a solution of $RBCI(\mathcal{H}, \mathcal{V})$ can be expressed by the following integer programming P

$$P \begin{cases} min \ \sum_{i=1}^{m} \sum_{j=1}^{n} x_{i,j} y_{i,j} \\ s.t. \\ \sum_{j=1}^{n} x_{i,j} = h_i^{c_1} \quad i = 1, \ldots, m; \qquad (1.a) \\ \sum_{i=1}^{m} x_{i,j} = v_j^{c_1} \quad j = 1, \ldots, n; \qquad (1.b) \\ \sum_{j=1}^{n} y_{i,j} = h_i^{c_2} \quad i = 1, \ldots, m; \qquad (2.a) \\ \sum_{i=1}^{m} y_{i,j} = v_j^{c_2} \quad j = 1, \ldots, n; \qquad (2.b) \\ x_{i,j}, y_{i,j} \in \{0, 1\} \quad i = 1, \ldots, m; j = 1, \ldots, n; \end{cases}$$

The constraints ensure that the solution of P satisfies the orthogonal projections of color c_1 and c_2. The objective function minimizes the number of conflicts between X and Y.

Definition 3. *Let X and Y be two binary matrices, $S = X \oplus Y$ is a bicolored image such that the cell (i, j) has the color c_1 (resp. c_2) if $x_{i,j} = 1$ (resp. $y_{i,j} = 1$). For example, $\begin{pmatrix} 1 & 0 \\ 0 & 1 \end{pmatrix} \oplus \begin{pmatrix} 0 & 1 \\ 1 & 0 \end{pmatrix} = \begin{pmatrix} c_1 & c_2 \\ c_2 & c_1 \end{pmatrix}$.*

Let X and Y two $m \times n$ binary matrices satisfying respectively (H^{c_1}, V^{c_1}) and (H^{c_2}, V^{c_2}) with null intersection. The combination of X and Y, $S = X \oplus Y$, gives a bicolored image satisfying the orthogonal projections and the exclusivity condition, where each color is associated with the 1's of one matrix. Even if it is easy to get X and Y independently, it is very hard to get X and Y without conflicts.

The program P can be easily linearized by replacing the quadratic terms $x_{ij} y_{i,j}$ by the variables z_{ij} and by adding the linearization constraints (4.a, 4.b, 4.c and 4.d) to ensure the equivalence between linear program LP and integer program P. We get the following equivalent integer linear program LP:

$$LP \begin{cases} min \ \sum_{i=1}^{m} \sum_{j=1}^{n} z_{i,j} \\ s.t. \\ (1.a), (1.b), (2.a), (2.b) \\ z_{i,j} \leq x_{i,j} \quad i = 1, \ldots, m; j = 1, \ldots, n \qquad (4.a) \\ z_{i,j} \leq y_{i,j} \quad i = 1, \ldots, m; j = 1, \ldots, n \qquad (4.b) \\ z_{i,j} \geq x_{i,j} + y_{i,j} - 1 \quad i = 1, \ldots, m; j = 1, \ldots, n \ (4.c) \\ z_{i,j} \geq 0 \quad i = 1, \ldots, m; j = 1, \ldots, n \qquad (4.d) \\ x_{i,j}, y_{i,j} \in \{0, 1\} \quad i = 1, \ldots, m; j = 1, \ldots, n; \end{cases}$$

Program LP can be directly injected in an IP solver such as Cplex.

Table 1. Numerical comparisons of the reconstruction of small size.

Image	% colored		Program P		Algorithm 1		Algorithm 2		Algorithm 3	
	$\%c_1$	$\%c_2$	$\%Conf$	Time	$\%Conf$	Time	$\%Conf$	Time	$\%Conf$	Time
20×20	34.17	32.01	0	0.08	0	0.79	0	0.06	0	0.00
40×40	33.18	33.59	0	0.19	0.11	2.13	0	0.14	0	0.00
60×60	33.54	32.93	0	0.29	0.12	4.58	0	0.53	0	0.02
80×80	33.14	33.39	0	0.74	1.27	8.72	0	0.85	0	0.04
100×100	33.53	33.16	0	2.44	2.95	15.26	0	0.96	0	0.08

Table 2. Numerical comparisons of the reconstruction of average size.

Image	% colored		Program P		Algorithm 1		Algorithm 2		Algorithm 3	
	$\%c_1$	$\%c_2$	$\%Conf$	Time	$\%Conf$	Time	$\%Conf$	Time	$\%Conf$	Time
120×120	33.51	33.22	0	6.17	4.05	25.53	0	3.64	0	0.10
140×140	33.41	33.18	0	11.23	4.67	41.18	0	10.31	0	0.18
160×160	33.24	33.36	0	30.73	5.28	64.09	0	23.23	0	0.21
180×180	33.21	33.47	0	66.02	5.73	96.10	0	41.68	0	0.28
200×200	33.28	33.43	0	109.21	6.02	139.38	0	66.08	0	0.35
220×220	33.43	33.24	0	148.78	6.32	196.63	0	85.68	0	0.42
240×240	33.31	33.34	0	190.77	6.61	270.71	0	133.43	0	0.49
260×260	33.36	33.38	0	262.26	6.94	365.15	0	168.21	0	0.58
280×280	33.41	33.25	0	371.49	7.28	483.38	0	213.48	0	0.74
300×300	33.35	33.32	0	438.87	7.69	631.23	0	295.25	0	1.01

Table 3. Numerical comparisons of the reconstruction of large size.

Image	% colored		Program P		Algorithm 1		Algorithm 2		Algorithm 3	
	$\%c_1$	$\%c_2$	$\%Conf$	Time	$\%Conf$	Time	$\%Conf$	Time	$\%Conf$	Time
320×320	33.31	33.33	0	536.66	8.11	814.94	0	345.07	0	1.10
340×340	33.41	33.29	0	777.85	8.59	1042.77	0	428.43	0	1.14
360×360	33.31	33.31	0	929.28	9.10	1323.94	0	501.57	0	1.22
380×380	33.31	33.32	0	1353.30	9.69	1668.05	0	612.82	0	1.35
400×400	33.34	33.37	0	1510.99	10.37	2091.19	0	777.47	0	1.83
420×420	33.41	33.28	0	2279.72	11.12	2612.95	0	954.69	0	2.59
440×440	33.34	33.31	0	2718.20	12.00	3259.38	0	1135.71	0	2.87
460×460	33.38	33.26	0	3305.12	13.00	4064.08	0	1336.62	0	3.42
480×480	33.34	33.33	0	4295.35	14.21	5071.09	0	1593.43	0	4.01
500×500	33.33	33.31	0	4884.58	15.07	6343.67	0	1828.30	0	5.02

5 Numerical Results

In this section, we compare the linear integer program P with the Lagrangean approach [19]), say Algorithm 1, the heuristic algorithm [3], say Algorithm 2, and the iterative algorithm [20], say Algorithm 3. We use a large set of random bicolored images with size varying from 10×10 to 500×500. We compare the reconstruction methods from the running CPU time and the number of conflicts.

The results of the computational experiments are summarized in Tables 1, 2 and 3. In these tables, the first column contains the size of the image. The second column displays the ratio of the number of cells colored by color (c_1 and c_2), in the original image. The subcolumn labeled $Conf$ contains the ratio of the number of conflicts provided by each method. The subcolumn labeled time contains the running CPU Time (in seconds) required by each method. We have used three classes of image: small size (see Table 1), average size (see Table 2) and large size (see Table 3).

The first main observation is that *Algorithm 3* is the fastest one and it solves the problem in few seconds and the CPU grows slightly in function of the size.

The second observation is that the Lagrangian relaxation, *Algorithm 1*, cannot solve the problem to the optimality whereas the other approaches get a zero conflicts. Moreover, the CPU is largely high compared to that of *Algorithm 2* and *Algorithm 3*.

The third observation is that the IP approach get an optimal solution with a zero conflict since it is an exact approach. However it needs about the double of CPU consumed by *Algorithm 2*. However, the IP approach can support the integration of new constraints which is not evident in *Algorithm 2* and *Algorithm 3*, since the min-cost max flow problem may not be useful with other constraints.

6   Conclusion

In this paper, we have proposed an exact integer programming formulation to reconstruct bicolored images from their orthogonal projections. Since the objective function is quadratic, we have used the linearization techniques and injected the equivalent program to CPLEX solver. The numerical results show that the IP finds the optimal solution for large size images. The CPU time can be decreased by deeply analyze the polyhedron of IP and adding more valid cuts. As a future work, we plan to try other generation methods to verify how performances change in the analyzed algorithms.

References

1. Barcucci, E., Del Lungo, A., Nivat, M., Pinzani, R.: The reconstruction of polyominoes from their orthogonal projections. Theor. Comput. Sci. **155**, 321–347 (1996)
2. Barcucci, E., Brocchi, S.: Solving multicolor discrete tomography problems by using prior knowledge. Fundamenta Informaticae **125**, 1–16 (2013)
3. Barcucci, E., Brocchi, S., Frosini, A.: Solving the two color problem: an heuristic algorithm. In: Aggarwal, J.K., Barneva, R.P., Brimkov, V.E., Koroutchev, K.N., Korutcheva, E.R. (eds.) IWCIA 2011. LNCS, vol. 6636, pp. 298–310. Springer, Heidelberg (2011)
4. Brocchi, S.: The three color problem solver (2014). http://www.researchand technology.net/discretetomography/3colorproblem/3colorsolver.html

5. Billionnet, A., Jarray, F., Tlig, G., Zagrouba, E.: Reconstructing convex matrices by integer programming approaches. J. Math. Model. Algor. **12**, 329–343 (2013)
6. Brocchi, S., Frosini, A., Rinaldi, S.: A reconstruction algorithm for a subclass of instances of the 2-color problem. Theor. Comput. Sci. **412**, 4795–4804 (2011)
7. Costa, M.C., Jarray, F., Picouleau, C.: An acyclic days-off scheduling problem. 4'OR: Q. J. Oper. Res. **4**(1), 73–85 (2006)
8. Costa, M.C., de Werra, D., Picouleau, C., Schindl, D.: A solvable case of image reconstruction in discrete tomography. Discrete Appl. Math. **148**, 240–245 (2005)
9. Costa, M.C., Jarray, F., Picouleau, C.: Reconstruction of binary matrices under adjacency constraints. Electron. Notes Discrete Math. **20**, 281–297 (2005)
10. Costa, M.C., de Werra, D., Picouleau, C.: Using graphs for some discrete tomography problems. Discrete Appl. Math. **154**, 35–46 (2006)
11. Chrobak, M., Dürr, C.: Reconstructing hv-convex polyominoes from orthogonal projection. Inf. Process. Lett. **69**, 283–289 (1999)
12. Dahl, G., Fatberg, T.: Optimization and reconstruction of hv-convex (0,1)-matrices. Electron. Notes Discrete Math. **12**, 58–69 (2003)
13. Del Lungo, A., Frosini, A., Nivat, M., Vuillon, L.: Discrete tomography: reconstruction under periodicity constraints. In: Widmayer, P., Eidenbenz, S., Triguero, F., Morales, R., Conejo, R., Hennessy, M. (eds.) CALP 2002. LNCS, vol. 2380, pp. 38–56. Springer, Heidelberg (2002)
14. Dürr, C., Guiñez, F., Matamala, M.: Reconstructing 3-colored grids from horizontal and vertical projections Is NP-hard. In: Fiat, A., Sanders, P. (eds.) ESA 2009. LNCS, vol. 5757, pp. 776–787. Springer, Heidelberg (2009)
15. Gale, D.: A theorem on flows in networks. Pac. J. Math. **7**(2), 1073–1082 (1957)
16. Herman, G.T., Kuba, A.: Discrete Tomography: Foundations. Algorithms and Applications. Birkhäuser, Boston (1999)
17. Herman, G.T., Kuba, A.: Advances in Discrete Tomography and Its Applications. Birkhäuser, Boston (2007)
18. Jarray, F.: A 4-day or a 3-day workweeks scheduling problem with a given workforce size. APJOR: Asian-Pac. J. Oper. Res. **26**(5), 1–12 (2009)
19. Jarray, F.: A Lagrangian approach to reconstruct bicolored images from discrete orthogonal projections. Pure Math. Appl. **20**, 17–25 (2009)
20. Jarray, F., Tlig, G.: Approximating bicolored images from discrete projections. In: Aggarwal, J.K., Barneva, R.P., Brimkov, V.E., Koroutchev, K.N., Korutcheva, E.R. (eds.) IWCIA 2011. LNCS, vol. 6636, pp. 311–320. Springer, Heidelberg (2011)
21. Picouleau, C., Brunetti, S., Frosini, A.: Reconstructing a binary matrix under timetabling constraints. Electron. Notes Discrete Math. **20**, 99–112 (2005)
22. Ryser, H.R.: Combinatorial properties of matrices of zeros and ones. Canad. J. Math. **9**, 371–377 (1957)
23. Woeginger, G.J.: The reconstruction of polyominoes from their orthogonal projections. Inf. Process. Lett. **77**, 225–229 (2001)

Combinatorial Exemplar-Based Image Inpainting

Veepin Kumar$^{(\boxtimes)}$, Jayanta Mukherjee,
and Shyamal Kumar Das Mandal

Indian Institute of Technology Kharagpur, Kharagpur 721302, West Bengal, India
jay@cse.iitkgp.ernet.in, {veepinkmr,sdasmandal}@cet.iitkgp.ernet.in

Abstract. We formulate image inpainting as a metric labeling problem. We solve this metric labeling problem via a combinatorial approximation approach. Results show the effectiveness of our method compared to other recent methods.

Keywords: Inpainting · Image restoration · Exemplar · Metric labeling · Combinatorial approximation

1 Introduction

The process of removing any object from an image, and filling it such that it looks like an original image, is known as image inpainting. It also includes filling damaged portions of an image. The inpainting techniques, developed so far, can be broadly classified into structure and texture based. Structure based techniques [1,9] only use local information available at the boundary between source and target region. In these techniques, the geometric structures, like level lines, edges, etc., are extended or extrapolated to fill in the gap of the target region (called 'hole' in this work). The computation usually involves solving partial differential equations or employing variational methods. These techniques give good results when used to inpaint non-textured and smaller regions. However, the inpainted region also gets blurred.

Texture based techniques [4,5] perform relatively better for inpainting textured and large regions. In this case, the objective is set to find the most similar exemplar patches from the source region, iteratively, to fill the target region. The approach suffers from the fact that it may fail to synthesize a geometry, if there is no example of a target patch in the image.

Most of the exemplar based methods do not check for visual coherency of the patches filled in target region with its neighbors. They do not consider whether the filled patch is visually consistent with its surroundings, or it is very different. There have been some combinatorial approaches to solve this difficulty [6,8,11]. The results produced by these techniques are visually better than traditional structure and texture based techniques.

The present work is an improvement over our previous work [7]. In [7], we modified the Criminisi's [4] technique to produce better linear structures. Then, we presented this modified exemplar based image inpainting as a metric labeling

© Springer International Publishing Switzerland 2015
R.P. Barneva et al. (Eds.): IWCIA 2015, LNCS 9448, pp. 284–298, 2015.
DOI: 10.1007/978-3-319-26145-4_21

problem. This metric labeling problem is formulated as the linear programming relaxation of an integer program. This linear programming relaxation is then solved using primal-dual approximation technique [2]. However, the technique suffers from the fact that we can not use its minimum to get the best inpainted image. A user needs to select the visually pleasant image out of a set of candidate images near local optimum point.

In this paper, we first propose a modification to the techniques in [4,7] by improving the quality of inpainiting of linear structures. We call it 'LInear structure retaining Modified Exemplar based inPainting' (LIMEP). After this, we propose a COMbinatorial Exemplar based inPainting (COMEP) technique. We first apply LIMEP on the image to be inpainted, then we apply COMEP on this image. As in [7], in this case also, the problem is formulated as a metric labeling problem. However, we have used a combinatorial approximation approach to solve this metric labeling problem. We observe that at every iteration the algorithm provides a better quality inpainted image (as determined by technique in [10]). Convergence of the algorithm is also guaranteed. The present work is different from the work in [7] in following ways.

- We do not use linear programming based approach, and solve for the max-flow of an augmented graph for performing the task as it was reported in [7].
- We modify the neighborhood criteria for neighboring patches, which ensures better visual consistency inside the hole.
- We use a quality measure [10] for the inpainted images to proceed in the direction of generating good inpainted images. This ensures that results obtained at the end is better than any of the intermediate processed image.

The present approach is simple in conceptualization. It produces better results as compared to those obtained by previous techniques [4,7,8], and gives promising future directions. In the next section we describe our modification to the techniques in [4,7]. After this, we present the proposed combinatorial image inpainting technique. We present typical results of the proposed technique and discuss them in Sect. 4. Conclusions highlighting future research directions that may come out of this work, are drawn in Sect. 5.

2   LIMEP

We have an original image OI as shown in Fig. 1. We want to remove a portion of this image, say Φ and reconstruct it. We call Φ as *target region* and remaining portion $\Omega = OI - \Phi$ as *source region*. In exemplar based approaches, we select a patch along the boundary of source and target region, as shown in Fig. 1. We call this patch as *target patch*. Target patch is selected according to its priority. Then, we find the patch from the *source region* which is most similar to this *target patch*. We call this patch as *source patch*. This procedure is repeated with all the target patches, according to their priority, till whole *target region* is covered. At the end we get the inpainted image.

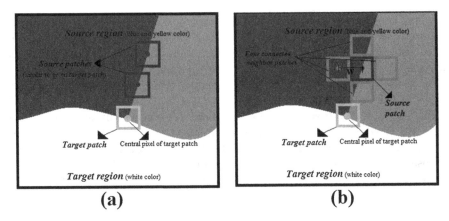

Fig. 1. (a) Diagram showing *source region, target region, target patch,* and two *source patches* similar to this *target patch*. (b) Diagram showing the four connected neighbor patches corresponding to a *source patch* (Color figure online).

In this work, we modify the exemplar based inpainting technique [4,7] to propagate linear structures in a better way. We give priority to *target patches* as introduced by us in [7]. This modified priority '$\tilde{P}(p)$' for each patch, centered at pixel p, is given by:

$$\tilde{P}(p) = C(p)D(p)E(p). \tag{1}$$

where $C(p)$ is called the *confidence term*, $D(p)$ is called the *data term*, and $E(p)$ *edge length term*. The *confidence term* corresponds to the number of known pixels in the *target patch*, the *data term* corresponds to the measure of the edges at the center pixel of the *target patch*, and the *edge length term* corresponds to the number of pixels which are part of an edge in the *target patch*. They are given as below.

$$C(p) = \frac{\sum\limits_{q \in \psi_p \cap (I-\Omega)} C(q)}{\mid \psi_p \mid}, D(p) = \frac{\mid \nabla I_p^\perp . n_p \mid}{\alpha}, E(p) = \frac{\sum\limits_{q \in \psi_p \cap (I-\Omega)} I(q)}{\mid \psi_p \mid}. \tag{2}$$

where $\mid \psi_p \mid$ is the area of the *target patch* denoted by ψ_p, α is the normalization factor, n_p is an orthogonal unit vector to the fill front and $\perp$ denotes the orthogonal operator. For initialization, the confidence i.e. C value of each pixel in the known region of the image is taken to be 1 and that in the unknown region is taken to be 0. $I(q)$ denotes the intensity gradient at any point q and is given as below.

$$I(q) = \mid I_x(q) \mid + \mid I_y(q) \mid . \tag{3}$$

In the above equations, I_x and I_y are, respectively, intensity gradients in x and y directions.

For each *target patch*, denoted by ψ_i centered at pixel p, we find two *source patches*, which are most similar to it. We denote these source patches by ψ_{i1} and

ψ_{i2}. The similarity measure is given by the sum of absolute differences between already known pixels of the *target patch* and the *source patch*, as given below.

$$\Psi_{i1} = \underset{\Psi_{im}\in\Omega}{argmin} \quad d(\Psi_{im},\Psi_i). \tag{4}$$

$$\Psi_{i2} = \underset{\Psi_{in}\in(\Omega-\Psi_{i1})}{argmin} \quad d(\Psi_{in},\Psi_i). \tag{5}$$

where $d(\Psi_{im},\Psi_i)$ denotes the sum of absolute differences between already known pixels of the two patches Ψ_{im} and Ψ_i. It is given as.

$$d(\Psi_{im},\Psi_i) = \sum_{k=1}^{w\times w} \mid \Psi_i(k) - \Psi_{im}(k) \mid . \tag{6}$$

where $w \times w$ is the patch size. Note that this absolute patch difference is the sum of differences between RGB components.

We fill the unknown part of the *target patch* ψ_i from the corresponding parts of the two *source patches* ψ_{i1} and ψ_{i2}. We call the resulting patches as *filled patches*, and denote them by $\hat{\psi}_{i1}$ and $\hat{\psi}_{i2}$, respectively. We check which *filled patch*, of the two, has maximum number of points belonging to edges, in any of the two directions, horizontal and vertical directions. We accept this *filled patch* for inpainting the image. We call this *filled patch* as *inpainted patch*, and denote it by $\hat{\psi}_i$. This ensures that linear structures are retained and propagated better. Mathematically, this operation is given below.

$$\hat{\psi}_i = \begin{cases} \hat{\psi}_{i1}, & \text{if } n_{x1} \geq \{n_{y1}, n_{x2}, n_{y2}\} \text{ or } n_{y1} \geq \{n_{x1}, n_{x2}, n_{y2}\} \\ \hat{\psi}_{i2}, & \text{otherwise.} \end{cases} \tag{7}$$

where n_{x1}, n_{y1}, n_{x2} and n_{y2} are the number of pixels which are part of an edge of the *filled patches* $\hat{\psi}_{i1}$ and $\hat{\psi}_{i2}$ in horizontal and vertical directions, respectively. We use following central differences gradient operator for detecting the edge:

$$L_x = [-1/2 \quad 0 \quad 1/2] \tag{8}$$
$$L_Y = [-1/2 \quad 0 \quad 1/2]' \tag{9}$$

where, L_X is the gradient operator in horizontal direction and L_Y is the gradient operator in vertical direction and $'$ denotes the transpose operator.

After filling the target region with the *inpainted patch*, we again determine the priority of resulting *target patches* at the boundary between *source* and *target* regions. Then, we select the patch with highest priority again to fill it via the same procedure as explained above. We repeat this procedure till no target patch is left to get the final inpainted image.

The proposed modification is based on the fact that edges are perceptually very important in an image. Results demonstrate the effectiveness of our technique in propagating linear structures.

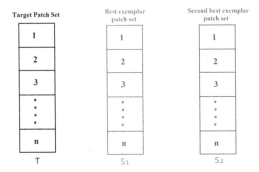

Fig. 2. Diagram showing three patch sets containing (from left to right) target patches, best exemplar patches and second best matching patches.

3   COMEP

In this section we describe our proposed COMEP technique. We have an image OI with a region to be inpainted denoted by Φ, as discussed in the previous section. We apply the LIMEP algorithm to this image. For each *target patch* Ψ_i, we find first two exemplars Ψ_{i1}, Ψ_{i2} which are most similar to Ψ_i, as discussed in previous section. Thus, we construct three patch sets as shown in Fig. 2, namely the set of target patches, the set of best exemplar patches and the set of second best exemplar patches. We denote them by τ, $S1$ and $S2$, respectively. Now, our goal is to decide which patch from sets $S1$ and $S2$ should come at the corresponding position in set τ. Note that some patches from set $S2$ are more preferred to come at corresponding positions of set τ, if they result in larger edges in the resulting *filled patches*, as discussed in the previous section. We do not need to make decision for these patches. Thus, we pose the inpainting problem as a metric labeling problem [3]. Here patches Ψ_i in target region correspond to objects, and the best two exemplars $\Psi_{im}, (m = 1, 2.)$ correspond to its two candidate labels. We refer to them as *label patches*. Whenever we assign a *label patch* to a *target patch*, we incur a cost. This cost has two components - *self cost* and *neighbor cost*. The *self cost* gives a measure of the similarity of the *label patch* with the assigned *target patch*. The *neighbor cost* gives a measure of the similarity of the assigned *label patch* with the neighboring *target patches*. We have considered four connected neighborhood. Distance between the centers of two neighboring patches is w, where $w \times w$ is the patch size as shown in Fig. 1(b).

We denote the total cost of assigning a *label patch* Ψ_{im} to a *target patch* Ψ_i as $T_{\Psi_i, \Psi_{im}}$, and is given below.

$$T_{\Psi_i, \Psi_{im}} = S_{\psi_i, \psi_{im}} + N_{\psi_i \psi_j}. \tag{10}$$

where, $S_{\psi_i, \psi_{im}}$ denotes the cost of assigning *label patch* ψ_{im} to *target patch* ψ_i. It is given by the sum of absolute differences of already filled pixels of the two patches ψ_i and ψ_{im}. $N_{\psi_i \psi_j}$ denotes the neighborhood cost of assigning *label patch* ψ_{im} to *target patch* ψ_i. It is given by the sum of absolute differences of the already filled pixels of the two patches ψ_{im} and ψ_j, for all the four connected neighbor patches ψ_j. Thus, total cost becomes:

$$T_{\Psi_i, \Psi_{im}} = \sum_{k=1}^{w \times w} | \Psi_i(k) - \Psi_{im}(k) | + \lambda \sum_{\psi_j \in N_{\psi_{im}}} \sum_{k=1}^{w \times w} | \psi_i(k) - \psi_j(k) | . \quad (11)$$

where, λ is a constant which determines the weightage between self cost and neighbor cost towards the total cost. We have taken $\lambda = 1$. Now, our objective is to produce a good inpainted image by minimizing this cost for each *label patch*. The computational steps of the algorithm are given below:

1. Assign the best exemplar patch i.e. *label patch* to each *target patch* (we refer to the currently applied *label patch* as the *active label patch*).
2. Calculate the costs of assignment of a *label patch* to a *target patch* for both the *label patches* of all the *target patches*.
3. Compare the costs of both the *label patches* for each *target patch*. For this comparison, the *target patches* are selected based on their priority according to Eq. (1).
4. If a node, say ψ_p, at which cost of assignment of *label patch* ψ_{p2} is less than that of *label patch* ψ_{p1}, then its *label patch* ψ_{p1} is replaced with *label patch* ψ_{p2}. The cost of assignment of a label patch to a target patch is calculated according to Eq. (11). In order to maintain visual consistency among filled patches, the patches with filling priority (refer to Eq. (1)) greater than the patch ψ_p are assigned label ψ_{p2}. Thus, such patches are filled with second best exemplar patch. Now, the priorities of patches corresponding to boundary pixels of the target region which includes the patch ψ_p get changed due to the assignment of the *label patch* ψ_{p2}. So, we again run the base inpainting algorithm (LIMEP) to the remaining target region and calculate 'first two best matching exemplars for each patch to be filled'.
5. At the end of step 4, we get a new inpainted image. We check the quality of this image by the technique "BRISQUE" as proposed in [10]. If the quality is less than the quality of previous image, then we discard step 4 for patch ψ_p, and move on to the next *target patch*. In this technique, distortions in the image are quantified by using the scene statistics of locally normalized luminance coefficients and products of locally normalized luminance coefficients. These coefficients are used to derive the underlying features which are then fed to a support vector machine regressor. The support vector machine regressor takes the feature vector as input and gives the quality score as output. Mathematically, the operation is given by:

$$f(x) = w^T \phi(x) + b \quad (12)$$

where, $f(x)$ corresponds to the quality score, $\phi(x)$ is the feature vector and w, b are the parameters of the support vector machine regressor. The details are available in [10][1].

---

[1] A software release of the technique reported in [10] is available online: http://live. ece.utexas.edu/research/quality/BRISQUE_release.zip.

We keep on repeating steps 1 to 5 for the remaining target region till whole target region is covered. A short block diagram description of COMEP is given in Fig. 3.

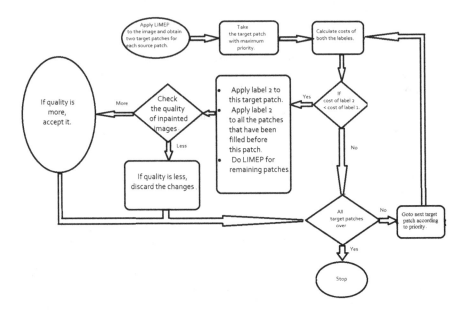

Fig. 3. Block diagram description of COMEP technique.

3.1   Convergence of the Algorithm

We apply LIMEP to the given image and obtain two source patches for each target patch. Thus, we get three patch sets- target patch set (τ), source patch set 1 $(S1)$ and source patch set 2 $(S2)$. Let total number of elements in each set be n. Let the ith element in the set $\tau, S1, S2$ be denoted as $\tau(i), S1(i), S2(i)$ respectively. Note that the elements of τ are sorted in decreasing order according to their priority. i.e. $\tau(1)$ corresponds to the target patch with maximum priority, $\tau(2)$ corresponds to the target patch with second maximum priority and so on. For any i, let the cost of assigning $S1(i)$ and $S2(i)$ to $\tau(i)$ be denoted by $C1(i)$ and $C2(i)$ respectively.

The convergence of algorithm can be described in following points:

1. In COMEP, we first assign $S1(i)$ to each $\tau(i)$ for $i = 1$ to n. Then, starting with the first element of τ we calculate the costs of assigning $S1(i)$ and $S2(i)$ to $\tau(i)$, compare the two costs, till we get an element $\tau(p)$ for which $C2(p) < C1(p)$.

2. Then, we assign $S2(i)$ to $\tau(i)$ for i $= 1$ to p. We apply LIMEP to the remaining target region i.e. target region without patches corresponding to the elements $\tau(i)$ for i $= 1$ to p. We get a new inpainted image. Say this time we get target patch set as $\tau'(i)$ and source patch sets as $S1'(i)$ and $S2'(i)$. Let the total number of elements in each set be m.

3. Now, we check the quality of this new inpainted image with the previous one by using the technique in [10]. If the quality of this inpainted image is more than the previous one, then we apply steps 1 and 2 to the remaining target region with patch sets $\tau'(i), S1'(i)$ and $S2'(i)$.

4. If the quality is less, then we discard the changes made so far. Thus, we again have patch sets τ, $S1$ and $S2$. We move on to next element $\tau(p+1)$ and continue processing for searching the next target patch whose cost for labeling second best exemplar patch is lower. The algorithm stops when whole target region is covered.

4   Results and Discussion

In this section, we show how the proposed LIMEP and COMEP methods perform in the task on image inpainting. We have applied our algorithm for following two purposes:

1. To remove objects from images.
2. To fill scratches in images.

While presenting the results, we compare them with the results obtained by the techniques reported in [4][2], [8][3] and [7]. The technique in [7] was developed by using the technique in [4] while the proposed technique is developed by taking ideas from both [4,7]. We have taken results obtained by technique in [8] for comparison purpose only. Number of iterations required depends upon image size, size of object to be removed, and the patch size. We have taken patches of size 9×9. All experiments were performed on MATLAB R2013a, core i3 processor with 8 GB RAM under Windows 7 64 bit.

We discuss the results for object removal first, and after this we present the results for scratch removal.

4.1   Removal of Objects from Images

In an image, we create a white colored mask around the object to be removed as shown in Figs. 4, 6, 7 and 8. The white region becomes the target region, and rest of the image becomes the source region. We applied our algorithm to such original images with mask. We also applied the techniques in [4,7,8]. We present the output of all these techniques side by side in Figs. 4, 5, 6, 7 and 8 for the purpose of visual comparison. In Fig. 5, zoomed versions of inpainted regions

---

[2] The technique reported in [4] was implemented by us.
[3] We are thankful to Mr. Yunqiang Liu for providing code for the paper [8].

of the set of images of Fig. 4 are shown for better visualization. From Fig. 5, we observe that the techniques in [4, 7, 8] did not construct the wooden stick properly, while the proposed LIMEP technique constructed the wooden stick, but it is thin. The COMEP technique produced the wooden stick properly. In Figs. 6 and 7, we observe that the our results are better than the results produced by techniques in [4, 8], while the final inpainted image obtained from the COMEP technique is comparable to the manually selected visually best inpainted image produced by the primal-dual optimization based inpainting technique [7]. As COMEP does not require any manual intervention for stopping iteration, it is more preferred to [7].

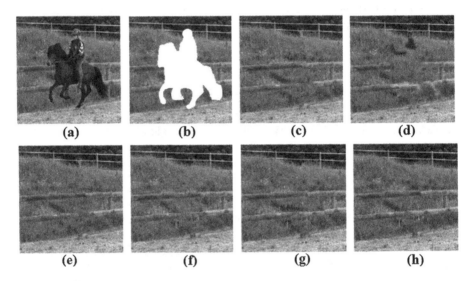

Fig. 4. (a) Original Image. (b) Original Image with mask. (c) Result of [4]. (d) Result of [8]. (e) Result of modified exemplar based inpainting. (f) Result of inpainting via primal-dual optimization [7]. (g) Result of LIMEP (h) Result of COMEP.

From Fig. 8, we observe that the inpainted images produced by both the LIMEP and COMEP techniques are better than those produced by techniques reported in [4, 8]. But, the result obtained by the technique reported in [7] is much better than the image produced by COMEP technique. The failure of our algorithm in this case (Fig. 8) is probably due to the limitation of the quality metric [10] in expressing quality score to the sub-images.

In Table 1 we present the quality score [10] of the inpainted images obtained by the techniques under comparisons. Figure 9 gives the plot of quality score versus the number of iterations, as the COMEP algorithm proceeds, for the four images of Figs. 4, 6, 7 and 8. We observe that the quality score increases as the algorithm proceeds. Note that at iteration 1 the quality score of the inpainted image is the score obtained by the LIMEP technique.

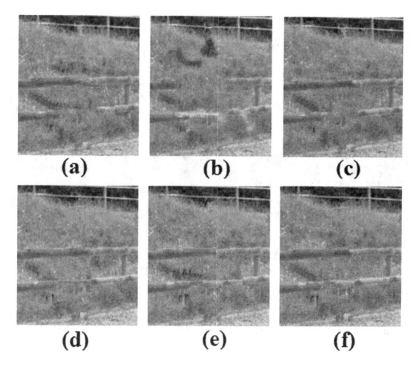

Fig. 5. Zoomed versions of inpainted regions of Fig. 4(c)–(h) for clarity.

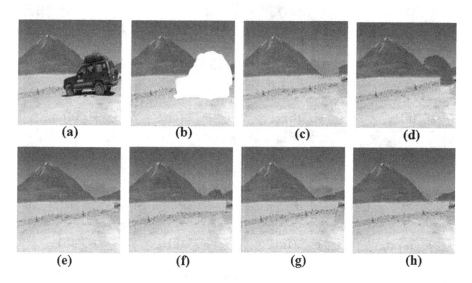

Fig. 6. (a) Original Image. (b) Original Image with mask. (c) Result of [4]. (d) Result of [8]. (e) Result of modified exemplar based inpainting. (f) Result of inpainting via primal-dual optimization [7]. (g) Result of LIMEP (h) Result of COMEP.

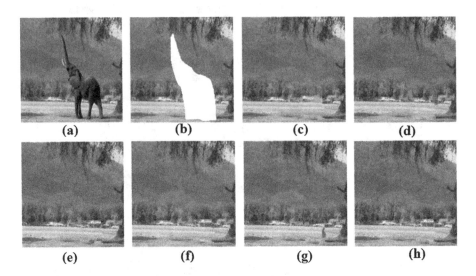

Fig. 7. (a) Original Image. (b) Original Image with mask. (c) Result of [4]. (d) Result of [8]. (e) Result of modified exemplar based inpainting. (f) Result of inpainting via primal-dual optimization [7]. (g) Result of LIMEP (h) Result of COMEP.

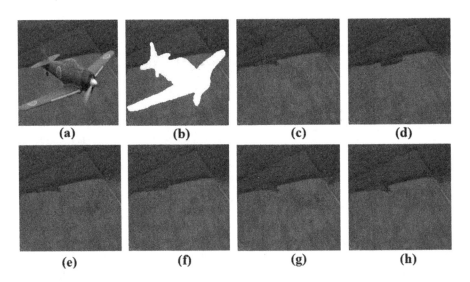

Fig. 8. (a) Original Image. (b) Original Image with mask. (c) Result of [4]. (d) Result of [8]. (e) Result of modified exemplar based inpainting. (f) Result of inpainting via primal-dual optimization [7]. (g) Result of LIMEP (h) Result of COMEP.

Table 1. Object removal results.

		Image in figure number			
		1	2	3	4
Method	Criminisi's [4]	20.85	17.43	19.80	19.26
	Liu's [8]	19.95	18.16	18.61	19.74
	Modified exemplar based [7]	21.20	18.00	20.09	20.16
	Primal-dual [7]	22.23	20.34	**22.58**	23.05
	LIMEP	21.25	19.40	20.29	20.48
	COMEP	**23.68**	**20.58**	22.16	**23.23**

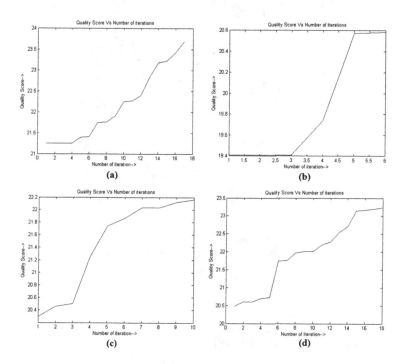

Fig. 9. Diagram showing plot of quality score versus number of iterations for the four images of Figs. 4, 6, 7 and 8 respectively.

4.2 Removal of Scratches from Images

We adapt the above technique to remove scratches from images to get back the original image. Here, the scratch is represented by white color as shown in Fig. 9. Scratch is the target region, and rest of the image becomes the source region. The same algorithm is applied as explained in Sects. 2 and 3. It is observed that the proposed techniques gave better results than techniques in [4,7,8], while in a few cases results (refer to Table 2) of technique in [8] are better. In Table 2 we present the quality score of the inpainted images obtained by our technique, and the techniques in [4,7,8]. As scratches are small in size, the visually good

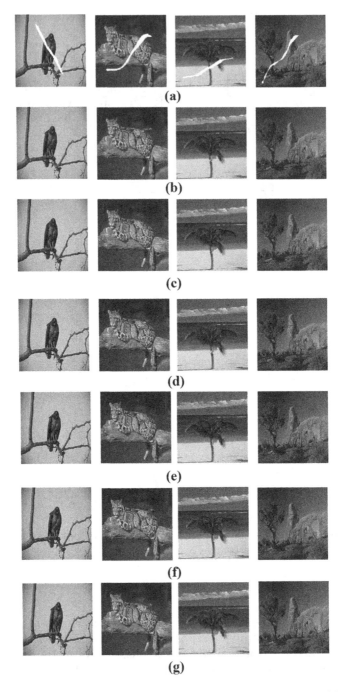

Fig. 10. (a) Original Image with mask. (b) Result of [4]. (c) Result of [8]. (d) Result of modified exemplar based inpainting [7]. (e) Result of inpainting via primal-dual optimization [7]. (f) Result of LIMEP. (g) Result of COMEP.

Table 2. Scratch removal results

		Quality score				PSNR value (in dB)				Execution time (in seconds)			
		Image number				Image number				Image number			
		1	2	3	4	1	2	3	4	1	2	3	4
Method	Criminisi's [3]	40.44	1.16	21.40	6.84	41.53	35.43	40.42	41.53	19.06	35.43	18.09	41.52
	Liu's [7]	40.75	4.06	21.60	6.81	43.25	37.89	43.89	44.67	2.09	3.46	2.03	1.89
	Modified exemplar based [8]	40.52	2.58	21.45	7.04	41.66	36.16	40.42	41.70	21.09	36.16	18.61	41.70
	Primal-dual [8]	41.42	2.81	21.52	7.33	41.75	36.68	41.70	42.14	156.87	145.47	74.21	81.25
	LIMEP	41.41	2.60	21.51	7.28	41.70	36.40	41.49	41.85	21.66	36.78	18.94	42.24
	COMEP	41.43	2.90	21.53	7.34	41.90	36.75	41.88	42.45	291.36	316.78	195.48	285.32

inpainted image should be very similar to the original image. Thus, peak signal to noise (PSNR) ratio is also a good measure of quality of the reconstructed images in this case. In Table 2 we present the PSNR values also for the purpose of comparison. We observe that in terms of PSNR values, the technique in [8] gave best results, while the COMEP technique ranks second.

5 Conclusion and Future Scope of Work

We have proposed a new combinatorial image inpainting technique for removing objects from images. The results have been found to be better than those obtained by the techniques in [4,7,8]. In future we would like to develop a good quality measure for the inpainted images. We would like to extend our technique for handling more labels.

References

1. Bertalmio, M., Sapiro, G., Caselles, V., Ballester, C.: Image inpainting. In: Proceedings of the 27th Annual Conference on Computer Graphics and Interactive Techniques, pp. 417–424. ACM Press/Addison-Wesley Publishing Co. (2000)
2. Chandra, S., Jayadeva, Mehra, A.: Numerical Optimization with Applications. Alpha Science International, Oxford (2009)
3. Chekuri, C., Khanna, S., Naor, J.S., Zosin, L.: Approximation algorithms for the metric labeling problem via a new linear programming formulation. In: Proceedings of the Twelfth Annual ACM-SIAM Symposium on Discrete Algorithms, pp. 109–118. Society for Industrial and Applied Mathematics (2001)
4. Criminisi, A., Pérez, P., Toyama, K.: Region filling and object removal by exemplar-based image inpainting. IEEE Trans. Image Process. **13**(9), 1200–1212 (2004)
5. Efros, A.A., Leung, T.K.: Texture synthesis by Non-parametric sampling. In: Proceedings of the Seventh IEEE International Conference on Computer Vision 1999, pp. 1033–1038. IEEE (1999)
6. Komodakis, N., Tziritas, G.: Image completion using efficient belief propagation via priority scheduling and dynamic pruning. IEEE Trans. Image Process. **16**, 2649–2661 (2007)
7. Kumar, V., Mukhopadhyay, J., Mandal, S.K.D.: Modified Exemplar-Based Image Inpainting via Primal-Dual Optimization. In: Kryszkiewicz, M., Bandyopadhyay, S., Rybinski, H., Pal, S.K. (eds.) PReMI 2015. LNCS, vol. 9124, pp. 116–125. Springer, Heidelberg (2015)

8. Liu, Y., Caselles, V.: Exemplar-based image inpainting using multiscale graph cuts. IEEE Trans. Image Process. **22**, 1699–1711 (2013)
9. Masnou, S.: Disocclusion: a variational approach using level lines. IEEE Trans. Image Process. **11**(2), 68–76 (1981)
10. Mittal, A., Moorthy, A.K., Bovik, A.C.: No-reference image quality assessment in the spatial domain. IEEE Trans. Image Process. **21**, 4695–4708 (2012)
11. Wexler, Y., Shechtman, E., Irani, M.: Space-time completion of video. IEEE Trans. Pattern Anal. Mach. Intell. **29**, 463–476 (2007)

Incremental Updating of 3D Topological Maps to Describe Videos

Guillaume Damiand[1]($\boxtimes$), Sylvain Brandel[1], and Donatello Conte[2]

[1] Université de Lyon, CNRS, LIRIS, UMR 5205, 69622 Villeurbanne, France
`{guillaume.damiand,sylvain.brandel}@liris.cnrs.fr`
[2] Université François-Rabelais de Tours, LI EA 6300, 37200 Tours, France
`donatello.conte@univ-tours.fr`

Abstract. A topological map is an efficient mathematical model for representing an image subdivision where all cells and adjacency relations between elements are represented. It has been proved to be a very good tool for video processing when video is seen as a 3D image. However the construction of a topological map for representing a video needs the availability of the complete image sequence. In this paper we propose a procedure for online updating a topological map in order to build it as the video is produced, allowing to use it in real time.

Keywords: 3D topological maps · Video processing · Combinatorial maps

1 Introduction

Many works have studied models representing partitions of an image. Topological data structures describe images as a set of elements and their adjacency relations. The most famous example is the Region Adjacency Graph (RAG) [19] which represents each region by a vertex, and where neighboring regions are connected by an edge. But the RAG suffers from several drawbacks as it does not represent multiple adjacency or makes no differences between enclosure and adjacency relations. Topological maps [2] have been used to solve these issues. A topological map is a mathematical model that represents an image subdivision. It aims to allow the use of topological and geometrical features of the subdivision in image processing. This kind of features have been used effectively for image segmentation [7] and more recently for video denoising [3].

Approaches for processing and understanding video data are mostly based by statistical representations; they can be broadly divided in two categories: some methods represent videos in a classical way that is an images sequence in which each image is represented by its pixels [11,12], other methods represent videos based on "mid-level" vision, that is using some features to represent and to analyze data [20]. In the last category, motion features, moving objects and trajectories are widely used as representation of video data [13,15]. Structural representation is rarely used for representing videos. In fact, if video metadata

© Springer International Publishing Switzerland 2015
R.P. Barneva et al. (Eds.): IWCIA 2015, LNCS 9448, pp. 299–310, 2015.
DOI: 10.1007/978-3-319-26145-4_22

are stably represented by structured data [9,10,17], structural representation of video content is still an open problem [16], and to the best of our knowledge graph representation was used only for coding videos and not for processing or understanding them.

In our approach a video is seen as a 3D image: a temporal sequence of 2D images is considered as a 3D image in which each voxel is considered as a temporal pixel, described by three coordinates (x, y, t), (x, y) being the spatial coordinates and t being the temporal coordinate. The 3D image is thus represented by a topological map. Then, topological features are proved to be very useful for video analysis and processing [3].

However the main problem in representing a video as a topological map is that for building this kind of data structure the complete video sequence has to be available (to represent it as a 3D image). Therefore algorithms using topological map are not available for real time processing.

The aim of this paper is to propose a method for online updating a topological map as the images of the sequence are produced. We propose two algorithms. The first one allows to remove k slices from the beginning of a given topological map. The second one allows to add k slices after the end of a given topological map. Combining these two operations allows to move a slicing window through a whole video.

The remainder of the paper is organized as follows: in Sect. 2 we introduce preliminary notions on topological maps that will be used afterwards. Section 3 describes the proposed operations by giving the algorithms and illustrating them on examples. In Sect. 4, some results show the complexity of our operations in the context of objects detection for video surveillance applications. Lastly, we conclude and give some perspectives in Sect. 5.

2   Preliminary Notions

A 3D topological map is an extension of a combinatorial map used to represent a 3D image partition. Therefore we first recall the notions on combinatorial maps and then we introduce notions on topological maps that are used in this work.

A combinatorial map is a mathematical model of representation of a space subdivision in any dimension. This is done using abstract elements, called *darts*, and applications defined on these darts, called β_i. The formal definition [6,14] is the following:

Definition 1 (Combinatorial Map). *Let $n \geq 0$. An n dimensional combinatorial map (or n-map) is an algebra $C = (D, \beta_1, \ldots, \beta_n)$ where:*

1. *D is a finite set of darts;*
2. *β_1 is a permutation on D;*
3. *$\forall i, 2 \leq i \leq n, \beta_i$ is an involution on D;*
4. *$\forall i, 1 \leq i \leq n - 2, \forall j, i + 2 \leq j \leq n, \beta_i \circ \beta_j$ is an involution.*

When two darts are linked with β_i, we say that they are i-sewed. An n-map represents each cell of the space subdivision implicitly by a set of darts. Moreover, adjacency and incidence relation between these cells are encoded through β_i operators (see [6,14] for more details).

A 3D topological map is a mathematical model which represents a 3D labeled image. This is a specialization of a 3-map in order to take into account the specific properties of a 3D image regarding any possible 3D subdivision. We can see in Fig. 1 an example of a 3D topological map.

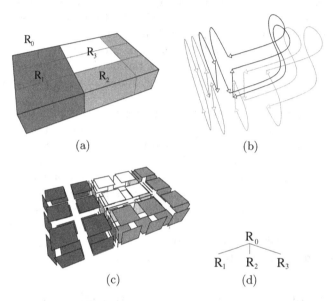

Fig. 1. Example of a 3D topological map. (a) A 3D labeled image. (b) The minimal 3-map. (c) The intervoxel matrix. (d) The region enclosure tree.

A 3D topological map is composed of three parts (see [4] for details):

- a minimal 3-map representing the topology of the image (Fig. 1(b));
- an intervoxel matrix used to retrieve geometrical information associated to the 3-map. The intervoxel matrix is called the embedding of the combinatorial map (Fig. 1(c));
- an enclosure tree of regions (Fig. 1(d)).

The 3-map (Fig. 1(b)) represents each region of the image by describing its surfaces. There is always one external surface, and possibly some internal surfaces, one for each cavity in the region. These surfaces are described in the 3-map through their minimal form (minimal in number of cells, i.e. we cannot remove any cell without changing the topology of the 3-map). This minimality gives interesting properties to the 3-map that will be useful in algorithms:

- each face of the 3-map describes exactly a maximal contact surface between two regions;
- each edge of the 3-map describes exactly a maximal contact curve between several faces.

A last interesting property of the 3-map is to capture the topology of each region. We can thus compute for example Euler characteristics or Betti number of regions and use these characteristics in image and video processing algorithms, as done for example in [3].

The intervoxel matrix is the embedding of the 3-map describing the geometry of the regions in the image (Fig. 1(c)). Each cell of the map is associated with intervoxel elements representing geometrical information of the cell. A face in the combinatorial map is embedded by a set of surfels separating voxels of the two incident regions. The edges, which are the border of faces, are associated to sets of linels. The vertices, which are the border of edges, are embedded by pointels.

The enclosure tree of regions (Fig. 1(d)) represents the enclosure relations. Each region in the topological map is associated to a node in the enclosure tree. The nodes are linked together by the enclosure relation (a region R_i is enclosed in R_j if R_i is completely surrounded by region R_j [5]). To link this tree with the 3-map, each dart d of the map knows its belonging region (called $region(d)$). Each region R knows one of its dart called representative dart (called $rep(R)$).

A 3D topological map can directly represent a 2D video simply by considering the temporal sequence of 2D images as a 3D image: each voxel corresponds to a temporal pixel, described by three coordinates (x, y, t), (x, y) being the spatial coordinates and t being the temporal coordinate.

3   Incremental Updating of 3D Topological Maps

In order to incrementally update a 3D topological map, we introduce in this paper two operations. The first one, given in Algorithm 1, allows to remove the k first slices of a given topological map. The second one, given in Algorithm 2, allows to add k slices after a given topological map. Combining these two operations allows to move a slicing window through a whole video.

3.1   Remove Slices

Algorithm 1 presents the method allowing to remove the first k slices of a given topological map T.

- The first step consists here to split in surfels the surfaces of all the regions intersecting the slices to remove. We use for that the split operation defined in [8]. The test if a region is intersected by the cutting plane is achieved thanks to the minimum and maximum voxel of the region (computed during the extraction of the 3D topological map).

Algorithm 1. Remove slices at the beginning of a 3D topological map

---

Input: T: A 3D topological map;
  k: A number of slices.
Result: T is modified to remove its k first slices.

foreach *region R in* T **do**
 | **if** *the minimum z value in R $\leq k$* **and** *the maximum z value in R $\geq k - 1$*
 | **then**
 | | split the surface of R in surfels;

Create faces F to describe the cutting plane;
2-sewn darts of F with darts of T;
foreach *dart d in* T **do**
 | **if** *the z value of the triplet of d $< k$* **then**
 | | remove the face containing d;

Simplify T;
Recompute the region enclosure tree;

- In the second step, all the faces describing the cutting plane are created, and then 2-sewn to darts around them in the existing 3-map. To retrieve the pair of darts to 2-sew, we use the geometry associated with each dart: two darts must be 2-sewn if they represent the same linel in reverse orientation.
- The third step consists in removing all the faces having one dart belonging to the first k slices. Thanks to the two previous steps, we are sure that these faces totally belong to the slices to remove.
- The last step consists in simplifying the combinatorial map in order to retrieve the minimality property and to recompute the region enclosure tree. We use for that the same algorithms that the ones used for the 3D topological map extraction [4].

We can prove that this algorithm produces the 3D topological map describing the initial 3D image where we have removed its k first slices. Thus this proves directly the consistency of the operation: the modified 3D topological map represents the topology of the updated 3D image.

The remove slices operation is illustrated in Fig. 3. In this example, we consider a 2D topological map since it is really hard to visualize drawings of 3D topological maps, but things are similar in both dimensions by replacing *voxels* by *pixels* and *face* by *edge*. We consider the initial 2D topological map depicted in Fig. 2(b) where we want to remove the three first slices (thus k = 3). In the first step of Algorithm 1, the borders of the two regions R_1 and R_2 are split in linels since some of their pixels belong to the three first slices. We obtain the 2-map shown in Fig. 3(a).

The second step inserts all the red edges in Fig. 3(a) which are new edges describing the cutting plane. These edges are sewn with their neighbor existing edges in the 2-map (illustrated by blue arcs in Fig. 3(b)). The next step (not drawn) consists in removing all the green edges from the 2-map. Indeed, all these

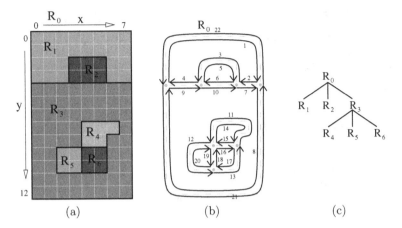

Fig. 2. Example of a 2D topological map. (a) A labeled image. (b) The minimal 2-map. (c) The region enclosure tree.

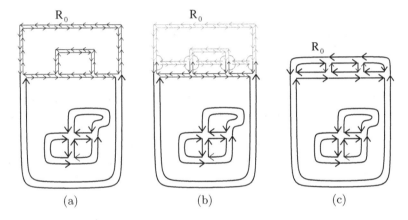

Fig. 3. Illustration of the removing of the three first slices of the 2D topological map depicted in Fig. 2(b). (a) The 2-map obtained after the split in linels of the borders of regions R_1 and R_2. (b) The 2-map obtained after the insertion of edges describing the cutting plane (in red). Then all the green edges are removed since they belong to the removed slices. (c) The final topological map obtained after the simplification of the previous 2-map (Colour figure online).

edges belong to the first three slices. The last step of the algorithm simplifies the map. For that, each degree two vertex is removed. This produces the combinatorial map shown in Fig. 3(c) which is the 2D topological map of the image obtained from the initial labeled image (given in Fig. 2(a)) where the three first slices were removed.

Figure 4 illustrates the three possible configurations of a region around the cutting plane. In these three cases, the surface of the region must be split in

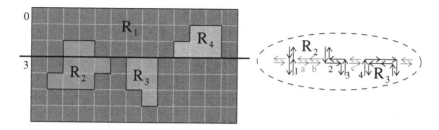

Fig. 4. Illustration of the three possible configurations of a region around the cutting plane $y = 3$. R_1 and R_2 are cut by the plane, i.e. they have pixels before and after it. R_3 touches the plane and have all its pixels after it. R_4 touches the plane and have all its pixels before it.

surfels in order to be correctly 2-sewn with the new faces describing the cutting plane. Other regions are either totally smaller or greater than the cutting plane. Smaller regions are totally removed by the cutting, while greater ones are totally kept. In both cases, there is no need to subdivide their faces. The example of region R_4 illustrates the -1 used in Algorithm 1. Indeed in such a case, the maximal z value of the region is equal to the value of the plane minus 1, and such a region must have its border split in linels to be correctly sewn with the new edges.

The right part of Fig. 4 illustrates the method used to 2-sew correctly the new edges describing the cutting plane with the existing edges. The principle consists in turning around the vertex to sew until finding an existing dart. For dart a, we found dart 1. For dart b, we found dart 2 (because here we 2-sew the extremity of dart b and not their origin). For the origin of dart c we found dart 3 and for its extremity we found dart 4. Note that the 2-sew process is done for one dart per edge since at each step we 2-sew a dart and its β_3.

3.2   Add Slices

In order to add slices at the end of a 3D topological map T, the first step of Algorithm 2 consists in splitting the external surface of T in surfels. This step allows the identification of the new slices with the external boundary of T. Indeed, only faces having the same topology can be identified in combinatorial maps (the two faces must be isomorphic i.e. they must be composed by the same number of darts), and it is simpler to split in surfels the external faces to identify instead of searching to modify each pair of faces in order to have the same topology.

In the second step, we compute the topological map T′ describing the slices to add. During this computation, we do not simplify the external surface of T′. Thus the external boundary of T′ is composed by faces describing surfels.

This allows to directly identify in the third step the right faces of T with the left faces of T′. Retrieving the faces to identify is done by using the geometry of the faces: both faces must have the same geometry but with reverse orientation. Testing if two faces have the same geometry is done by iterating simultaneously through the two cycles of darts, in reverse direction, while comparing the coordinates of the pairs of pointels associated with darts of the first face and darts of the second face.

Then it is enough to merge regions having same labels by removing the face separating them (step four) and to simplify the resulting 3-map (step five) in order to obtain the topological map describing the initial partition added with the k new slices.

Algorithm 2. Add slices at the end of a 3D topological map

Input: T: A 3D topological map;
    I′: A 3D image containing k slices.
Result: T is modified to add the slices in I′ at its end.

foreach *face f of the infinite region of* T **do**
    ⌊ split f in surfels;

T′ ← compute the topological map of I′ without simplifying its external surface;
Identify the right surfels of T with the left surfels of T′;
foreach *face f to the left of* T′ **do**
    ⌊ **if** f *separates two regions with the same label* **then**
        ⌊ remove f;

Simplify T;

As for the remove slices operation, we can prove that this algorithm produces the 3D topological map describing the initial 3D image where we have added the k new slices. Thus this proves directly the consistency of the operation: the modified 3D topological map represents the topology of the updated 3D image.

The add slices operation is illustrated in Fig. 5, once again in 2D in order to simplify the visualisation of combinatorial maps. The topological map depicted in Fig. 3(c) is considered where we want to add the three slices described by the 2D labeled image shown in Fig. 5(b). The first step of the algorithm consists in splitting the infinite face of the topological map to modify. We obtain the 2-map shown in Fig. 5(a). In the second step, the topological map of the new slices is computed while keeping its infinite face subdivided in linels (see Fig. 5(c)).

Now it is possible to identify the bottom linels of Fig. 5(a) and the top linels of Fig. 5(c). Note that during these identifications, the darts of the two identified parts belonging to the infinite regions are removed. Figure 6(a) shows the obtained 2-map. During the next step, all the adjacent regions having the same label are merged by removing the edges separating them. Now it is enough to simplify the map in order to obtain the topological map representing the initial labeled image added by the new slices (shown in Fig. 6(b)).

Note that allowing these two operations to deal with k slices is an optimization comparing to the application of k successive operations for 1 slice. Indeed, each operation make some splits of existing cells following by some merges in

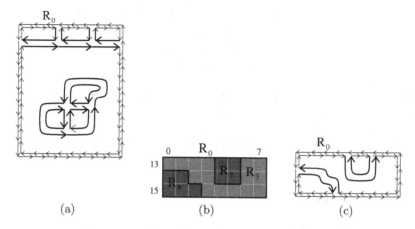

(a)   (b)   (c)

Fig. 5. Illustration of the adding of three new slices at the end of the 2D topological map depicted in Fig. 3(c). (a) The combinatorial map obtained from the initial topological map after the split of its infinite region in linels. (b) The labeled image corresponding to the three new slices. (c) Its corresponding topological map, having its external face subdivided.

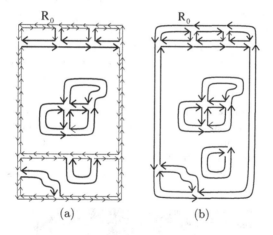

(a)   (b)

Fig. 6. Illustration of the adding of three new slices (cont). (a) The 2-map obtained after the identification of the bottom linels of Fig. 5(a) and the top linels of Fig. 5(c). (b) The final 2D topological map obtained after the merging of regions with same labels and the simplification step.

order to retrieve the minimality property. Dealing with k slices allows to make only one split and one merge instead of k splits and k merges.

The add slices and the remove slices operations can directly be combined in order to propose a sliding window method which allows to iterate through all the frames of a given video. It is enough to call remove_slices(1) then add_slices(1). However, in this case, we can avoid the simplification step and the region enclosure tree computation at the end of the `remove_slices` algorithm which will be done only once at the end of the `add_slices` operation.

Lastly, note that both proposed algorithms can be improved by avoiding some useless splits. For example for the add slice operation, we can only split in surfels the bottom boundary of the initial topological map and not all the infinite face. For the remove slice, we can split in surfels only the part of the concerned regions belonging to the removed slices instead of its entire boundaries. These optimizations will improve the speed of the operations since less cells will be split and then simplified, but the algorithm will become more complex since they are more different cases to consider.

4   Experiments

We have developed both operations `add_slices` and `remove_slices` in 3DTopo Map [1], a software allowing to compute a 3D topological map from a 3D image.

We used the PETS 2010 Dataset [18] which is a well known database containing video sequences of moving people, used for the performance evaluation of tracking and surveillance systems. We use the video of detection masks: white pixels correspond to moving objects and black pixels to background (see Fig. 7).

From this database, we extracted a first 3D topological map containing 65 slices, where each slice corresponds to a frame in the video stream, and each frame has 768×576 pixels. We evaluated the time required by the three operations `add_slices`, `remove_slices` and `add_and_remove_slices` for different k (number of slices to add and/or to remove) starting from 1 and going to 64.

Results are shown in Fig. 8. We can first verify that the time processing for `add_slices` and `add_and_remove_slices` operations is linear regarding the number of added slices. More interestingly, we can see that the time of the `remove_slices` operation decreases linearly regarding the number of removed slices. This can be explained by the number of darts of the obtained 3-map which is smaller after having removed a bigger number of slices.

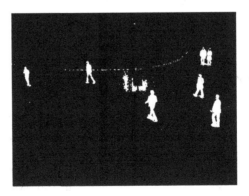

Fig. 7. An image extracted from a video detection masks from the PETS 2010 database (768×576 pixels).

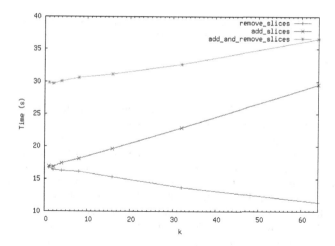

Fig. 8. Times evaluation (in seconds) of `remove_slices`, `add_slices` and `add_and_remove_slices` operations for $k \in \{1, 2, 4, 8, 16, 32, 64\}$ for an initial video sequence containing 65 slices. Each slice corresponds to a frame in the video stream. The resolution of each frame is 768×576 pixels.

5   Conclusion

In this paper, we have presented two operations defined on 3D topological maps. The first operation allows to remove the k first slices of an existing topological map. The second operation allows to add new slices at the end of an existing topological map. By combining these two operations, it is possible to define a sliding window method which allows to iterate through all the frames of a given video.

Now we are working on the use of the new proposed operations in order to define efficient video processing. Our first goal is to update our previous denoising algorithm [3] in order to benefit from the sliding window method. This will allow us to improve the method by combining all the information coming from the different superposed windows. Moreover this will allow us to consider streamed videos.

Another future work is the optimization of our method in order to be able to propose real time processes. For that, we have started to integrate some parallelism in our algorithm. Indeed, several parts can be made in parallel, leading to speed up our method. A second possible optimization is to modify the two algorithms in order to avoid to split faces in surfels. Indeed, this step takes an important time and requires to simplify the map at the end of the algorithms. Modifying this step will give more complex but more efficient algorithms. With these two optimizations we are confident to obtain a method allowing to achieve real-time performance.

Lastly, we plan to propose other video processing algorithms based on 3D topological map and the new sliding window method. Following our first previous results [3], we think that integrating some topological information provided by the 3D topological map can improve several existing video processing algorithms.

Acknowledgment. This work has been partially supported by the French National Agency (ANR), project SoLSTiCe ANR-13-BS02-0002-01.

References

1. 3DTopoMap: Topological 3d image processing software. http://liris.cnrs.fr/guillaume.damiand/carte-topo3D.php?lang=en
2. Bertrand, Y., Damiand, G., Fiorio, C.: Topological map: minimal encoding of 3D segmented images. In: Proceedings of International Workshop on Graph-Based Representations in Pattern Recognition, pp. 64–73, Ischia, Italy, May 2001
3. Conte, D., Damiand, G.: Remove noise in video with 3D topological maps. In: Fränti, P., Brown, G., Loog, M., Escolano, F., Pelillo, M. (eds.) S+SSPR 2014. LNCS, vol. 8621, pp. 213–222. Springer, Heidelberg (2014)
4. Damiand, G.: Topological model for 3d image representation: definition and incremental extraction algorithm. Comput. Vis. Image Underst. **109**(3), 260–289 (2008)
5. Dupas, A., Damiand, G.: First results for 3D image segmentation with topological map. In: Coeurjolly, D., Sivignon, I., Tougne, L., Dupont, F. (eds.) DGCI 2008. LNCS, vol. 4992, pp. 507–518. Springer, Heidelberg (2008)
6. Damiand, G., Lienhardt, P.: Combinatorial Maps: Efficient Data Structures for Computer Graphics and Image Processing. A. K. Peters/CRC Press, Boca Raton (2014)
7. Damiand, G., Resch, P.: Split and merge algorithms defined on topological maps for 3D image segmentation. Graph. Models **65**(1–3), 149–167 (2003)
8. Dupas, A.: Opérations et Algorithmes pour la Segmentation Topologique d'Images 3D. Thèse de doctorat, Université de Poitiers, Novembre 2009
9. Gonno, Y., Nishio, F., Haraoka, K., Yamagishi, Y.: Metadata structuring of audiovisual data streams on MPEG-2 system. In: Metastructures, August 1998
10. Hunter, J., Armstrong, L.: A comparison of schemas for video metadata representation. Comput. Netw. **31**(1116), 1431–1451 (1999)
11. Jain, A.K.: Fundamentals of Digital Image Processing. Prentice-Hall, Englewood Cliffs (1989)
12. Koprinska, I., Carrato, S.: Temporal video segmentation: asurvey. Signal Process.: Image Commun. **16**, 477–500 (2001)
13. Kastrinaki, V., Zervakis, M., Kalaitzakis, K.: A survey of video processing techniques for traffic applications. Image Vis. Comput. **21**, 359–381 (2003)
14. Lienhardt, P.: N-dimensional generalized combinatorial maps and cellular quasi-manifolds. Int. J. Comput. Geom. Appl. **4**(3), 275–324 (1994)
15. Li, X., Zheng, Y.: Patch-based video processing: a variational bayesian approach. IEEE Trans. Circ. Syst. Video Technol. **19**(1), 27–40 (2009)
16. Maugey, T., Ortega, A., Frossard, P.: Graph-based representation and coding of multiview geometry. In: Proceedings of the IEEE International Conference on Acoustics, Speech and Signal Processing (ICASSP), pp. 1325–1329 (2013)
17. Ngo, C.W., Ma, Y.F., Zhang, H.J.: Video summarization and scene detection by graph modeling. IEEE Trans. Circ. Syst. Video Technol. **15**, 296–305 (2005)
18. Pets 2001 dataset. http://www.cvg.rdg.ac.uk/pets2001/
19. Rosenfeld, A.: Adjacency in digital pictures. Inf. Control **26**(1), 24–33 (1974)
20. Wang, J., Adelson, E.: Representing moving images with layers. IEEE Trans. Image Process. **3**(5), 625–638 (1994)

Parallel Strip Segment Recognition and Application to Metallic Tubular Object Measure

Nicolas Aubry[1,2], Bertrand Kerautret[1,2(✉)], Isabelle Debled-Rennesson[1,2], and Philippe Even[1,2]

[1] Université de Lorraine, LORIA, UMR 7503, Vandoeuvre-lès-nancy 54506, France
[2] CNRS, LORIA, UMR 7503, Vandoeuvre-lès-nancy 54506, France
{nicolas.aubry,bertrand.kerautret,isabelle.debled-rennesson,
philippe.even}@loria.fr

Abstract. The segmentation or the geometric analysis of specular object is known as a difficult problem in the computer vision domain. It is also true for the problem of line detection where the specular reflection implies numerous false positive line detection or missing lines located on the dark parts of the object. This limitation reduces its potential use for concrete industrial applications where metallic objects are frequent. In this work, we propose to overcome this limitation by proposing a new strategy which is not based on the image gradient as usually, but exploits the image intensity profile defined inside a parallel strip primitive. Associated to a digital straight segment recognition algorithm robust to noise, we demonstrate the efficiency of our proposed method with a real industrial application.

1 Introduction

The detection or recognition of digital straight segment is always an active research topic where application domains are numerous. As is apparent from the thousand demonstration archive pages of the famous LSD method [15] published on the IPOL journal, the interest of such detection is always meaningful.

Even if recent advances were proposed in the last years, the line segment detection shows always some limits towards object presenting specular reflection. Such physical properties that arise naturally on various shiny materials like plastic or metallic objects, are source of well known difficulties that potentially impact the robustness of other computer vision algorithms. For instance, in the problem of Shape From Shading [12], the specular reflections can dramatically degrade the 3D reconstruction quality. In the image segmentation domain, the specularities are also a source of difficulty since they are at the origin of much more high intensity variations than diffuse reflections. Such difficulties often appear in industrial applications where objects of interest are directly composed of metallic parts.

N. Aubry—This work was supported by the French National Agence of Research and Technology (ANRT).

© Springer International Publishing Switzerland 2015
R.P. Barneva et al. (Eds.): IWCIA 2015, LNCS 9448, pp. 311–322, 2015.
DOI: 10.1007/978-3-319-26145-4_23

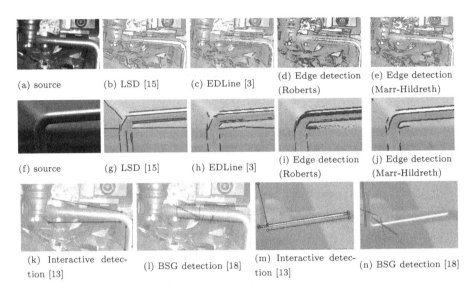

Fig. 1. Experiments of various algorithms for line extraction on source images (a, f). Results of LSD and EDLine algorithms are given in (b, c, g, h). Classic edge detection (from the IPOL online implementation [16]) are given in images (d, e, i, j) and semi automatic algorithms [13,18] were also applied on images (k, l, m, n) with manual markers (resp. resulting lines) represented in blue (resp. red) color (Color figure online).

The Fig. 1 shows an example of industrial application where the objective is to extract the bounding lines of the main tubular object in order to measure its orientation and the length of its straight segments. The LSD method [15] was first experimented on these examples. Based on Gestalt theory and Helmholtz principle, this method has the advantage to not require the tune of any sensible parameters. In the same spirit, the Edge Drawing Line (EDLine) method was also applied on the same images with the author implementation [3]. The experiments were first applied on a configuration where the production engine is presenting numerous metallic pieces and constitutes a difficult and challenging configuration (source image (a)). The second configuration (source image (f)) is in a relative more ideal configuration where the metallic parts of the engine are hidden. As shown on images (b, c) of Fig. 1, these methods produce numerous disconnected segments and false positive parts near specularity. Even on simpler configuration of image (f), the many false positive detections can distort the measures of the tube angle (images (g, h)). Such detections are not convenient to perform simple and robust tube geometric measure for default identification. In complement to the application of these advanced methods, we also apply more elementary contour detection algorithms that confirm that the gradient produced by the specularity are the cause of the numerous false positive line detection. Interactive method based on Hough transform like the one proposed by Even and Malavaud [13] is not efficient to reliably identify the tube with some segments attracted towards specularity and some other ones not always attracted to the tube border

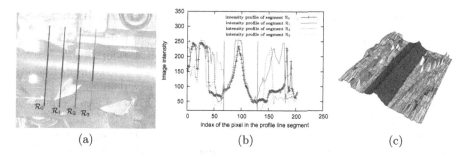

(a)   (b)   (c)

Fig. 2. Illustration of the low evolution of the intensity profiles near the interest area (highlighted in red in image (a)). Our method will exploit the low variation of the intensity profile along the segment $\mathcal{R}_k$. Such variations are visible on image (b) inside the vertical red lines and on the 3D elevation map (c) with the interest area drawn in red (Color figure online).

(see images (k, m) of Fig. 1). A more recent semi automatic method [18] also fails to identify the border of the tube mainly due to the specularities which generate numerous potential candidate contours making wrong the recognition process (images (l, n)).

To overcome the remaining problem dealing with the missing or too numerous gradient information, the contribution of this work is to propose a new strategy by exploiting an intensity profile associated to a parallel strip segment primitive. Following this idea, the recognition process will no more use any gradient information but will be based on the matching of intensity profiles. To illustrate this idea, we have plotted on Fig. 2 the different intensity profiles (image (b)) of four different segments $\mathcal{R}_1$, $\mathcal{R}_2$, $\mathcal{R}_3$ and $\mathcal{R}_4$ (image (a)). On image (b) we can clearly see that the different intensity profiles closely match inside the predefined interest area (denoted later $\mathcal{I}_k$ and highlighted in red on the images (a, b, c)). Such area corresponds, in this example, to the cross section of a metallic tubular object that we want to recognize. The low variations of the intensity profiles on this interest area are also visible on the 3D elevation map (image (c)).

The paper is organized as follows: in the next section, we present the main steps of the new method to recognize a parallel strip from the intensity profiles selection with the help of a digital straight segment recognition robust to noise. Then, in Sect. 3, we propose to exploit this new approach for a concrete industrial application where the object of interest is a metallic specular tube which has to be identified and measured.

2   Recognition of Parallel Strip Segment from Intensity Profiles

Since the proposed method is in the continuity of a previous line detection algorithm [18], we recall in the following section the different steps of the algorithm with its main primitive of blurred segment.

2.1   Previous Work on Digital Straight Segment Recognition

The recognition of a digital straight segment is a classic primitive in the field of discrete geometry which is often used to extract geometric information, like for instance to curvature estimation [17] or to polygonalize a digital contour from its dominant points [5,20]. In order to be robust to noise, the definition of blurred segment [6,11] was proposed to handle non perfect digitization. An example of blurred segment recognition is illustrated on image (a) of Fig. 3. The different points are added incrementally to the primitive until its thickness is not larger than a maximal value. On the example of image (a), starting from point A, 10 points were accepted by the blurred segment illustrated in step (1), 10 more points were also added in step (2) and finally in step (3). The point C can no more be added to the blurred segment since the new thickness will exceed its maximal value.

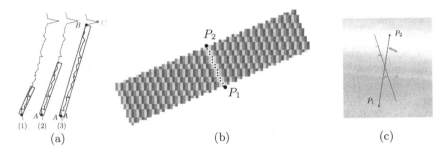

Fig. 3. Illustration of previous works on straight segment recognition, with an example of blurred segment recognition (a), the first step of scanning image pixel for line recognition in gray level [18] (b) and the main limits of this approach (c) (Color figure online).

Initially defined for the recognition of digital straight segment from digital contours or from binary images, we have proposed its extension in order to be able to process gray scale images [18]. The main idea of such algorithm is illustrated on image (b) of Fig. 3 showing the image scanning process designed to find candidate points to be added to the primitive of the blurred segment. Such initial scanning segment given by the two points P_1 and P_2 should be defined manually by the user. The selection of candidate is obtained by using the local maxima of the gradient values and by filtering its gradient direction to prune false candidate due to noise. Even if the proposed algorithm is very fast and suitable to extract segment on shape boundary with single gradient values, it shows some limitations on objects presenting numerous gradient information. As shown in image (c), there exists numerous candidate points with high gradient values (illustrated in green) which are not associated to the same single line. As a consequence, the blurred segment recognition (illustrated in blue) gives a resulting segment with a wrong direction in comparison to the real linear structure oriented horizontally.

2.2   New Approach with Intensity Profile Matching

As presented in the introduction, our main idea relies on the exploitation of the matching of intensity profiles located inside a parallel strip. It is suitable both to diffuse reflectance objects presenting low intensity variations and also to specular/metallic tubular objects. This characteristic is physically due to the specular reflection which presents low intensity variations in the main axis direction of the metallic object.

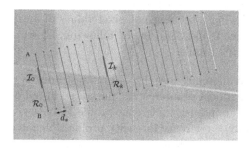

Fig. 4. Example of discrete parallel segments defined as research segments $\mathcal{R}_k$ (Color figure online)

Main Strategy. To apply the segment recognition, we follow the same first steps of the preceding method described in the previous section. In particular, a directional scan is defined in the image starting from two initial points A and B defined manually by the user. Such couple of points provides an initial digital straight segment. This first direction can be defined approximately since we follow several incremental steps in order to refine the final direction. It allows us to generate the other segment in its normal direction which will be the support of the line recognition. Such supports defined later as research segment $\mathcal{R}_k$ are illustrated on Fig. 4. Note that contrary to previous work, we integrate a parameter d_s which corresponds to the distance between two consecutive scans. It helps to balance the precision of the segment recognition with the speed of the algorithm according to the industrial application needs. Then, we analyze each of the segment in order to extract candidate points for the digital straight segment recognition algorithm. As in the previous approach we also exploit the same blurred segment primitive [11] in order to be robust to noise. In particular we use the Alpha-Thick segment from the *DGtal* library framework [1].

The strategy to refine the first manual direction given by the points A and B is also applied in the proposed method. More precisely, we redefine the previous directional scanners in the direction of the first detected blurred segments from the previous candidates.

Selection of Candidates from Intensity Profiles. Contrary to the previous method [18] where the candidate selection was based on the image gradient, we propose here another strategy which relies on the comparison of intensity profiles. In the following, we will use these definitions:

– **Research Segment** $\mathcal{R}_k$: is defined as a digital segment with two predefined initial points A and B for the first research segment $\mathcal{R}_0$. The other research segments $\mathcal{R}_k$ are obtained by translation in the normal direction $\overrightarrow{n}$ of $\mathcal{R}_0$ with a given distance d_s. Equivalently, each points P_i of $\mathcal{R}_0$ are translated into $\mathcal{R}_k$ by the vector: $k * d_s \overrightarrow{n}$. Note that the parameter d_s could be easily set by the user according to awaited recognition precision.
– **Interest Area** $\mathcal{I}_k^{i,j}$: represents a sub sequence of connected pixels of the research segment $\mathcal{R}_k$ starting from index i to j. Such sequence corresponds to the area that is going to be analyzed.
– **Intensity profile** $\mathcal{P}_k$: a sequence of pixel values of an interest area $\mathcal{I}_k$ (i.e. values given between 0 and 255). In the following, $\mathcal{P}_k^l$ will denote the value of lth element of the kth profile.

In a first step, we manually initialize an interest area $\mathcal{I}_0$ on the first research segment $\mathcal{R}_0$ near the object of interest as illustrated in red on Fig. 4. This interest area will produce an intensity reference profile $\mathcal{P}_0$ of the object that we try to detect. More generally, a reference intensity profile is denoted as $\mathcal{P}_r$. Our objective is to find for each research segment $\mathcal{R}_k$ its own interest area $\mathcal{I}_k$ according to a reference profile $\mathcal{P}_r$. For this purpose, the interest area $\mathcal{I}_k^{i,j}$ is incrementally moved from a number of pixels equals to the pixel length of $\mathcal{R}_k - j + i$. For each position the mean squared error (denoted E_r) is computed between the intensity profile $\mathcal{P}_k$ associated to the current interest area $\mathcal{I}_k$ and the reference intensity profile $\mathcal{P}_r$. More formally, if we consider a reference interest area $\mathcal{I}_r^{n,m}$, then the mean squared error E_r of an area of interest $\mathcal{I}_k^{i,j}$ is given by:

$$E_r(\mathcal{I}_k^{i,j}) = \frac{1}{j-i} \sum_{l=i}^{j} (\mathcal{P}_k^l - \mathcal{P}_r^{n+l-i})^2 \tag{1}$$

Such an error computation is illustrated on Fig. 5. For each research segment $\mathcal{R}_k$, the mean squared error is computed for each candidate $\mathcal{I}_k^{i,j}$ (image (a)). Then, from the global plot values, all the candidates with a minimal error (images (b, c)) are retained for the recognition of the parallel strip primitive described in the next section.

2.3   Algorithm

The proposed algorithm relies on the set of research segment $\mathcal{R}_k$ (generated from the first segment $\mathcal{R}_0$) for which we have to compute the interest area $\mathcal{I}_k$ which best matches the reference profile of the $\mathcal{I}_r$.

The general algorithm exploits several primitive or functions:

– `mse(ref, seg)`: computes from Eq. refEqMSE, the mean squared error between the two profiles of the interest area $\mathcal{I}_{ref}^{n,m}$ and $\mathcal{I}_{seg}^{i,j}$.
– `findCandidate(ref, seg)`: finds the interest area $\mathcal{I}_{seg}^{i,j}$ on the segment `seg` which best match the reference interest area $\mathcal{I}_{ref}^{n,m}$. It exploits the previous `mse` method.

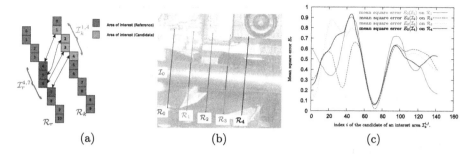

Fig. 5. Illustration of the selection process of the area of interest $\mathcal{I}_k^{i,j}$ (image (a)). The plots of graphic (c) show the different mean squared errors obtained for each point of a research segment $\mathcal{R}_k$ according to the reference area $\mathcal{I}_0$ illustrated on the image (b).

- **extendBlurredSegment**(p): extends the current blurred segment with the point p. It is used to apply the incremental recognition algorithm.
- **recognizeBlurredSegment**$(Seg, StepReference)$: recognizes the two longest blurred segments from the set of research segments $\mathcal{R}_k$. Each blurred segment is recognized from the extremity points of each $\mathcal{R}_k$ by exploiting the previous primitives.

The general algorithm is based on the previous methods. Moreover it applies a refinement of the initial direction given by the user in order to be robust towards the first direction. The process follows these different main steps:

① The function **recognizeBlurredSegment** is applied from the initial list of $\mathcal{R}_k$.
② The new direction is used to define a new list of research segment $\mathcal{R}'_k$. The new direction is defined from the detected blurred segments. More precisely, this direction is given from the direction vector of the detected blurred segments.
③ The function **findCandidate()** is called with reference given by the initial $\mathcal{I}_n^{0,i}$ order to obtain the new reference interest area $\mathcal{I}'_0$. By this way it will allow to find on this new segment the best area associated to the initial area.
④ From these new research segments a new recognition is applied again on the new research segments $\mathcal{R}'_k$.

Here, we exploit a static interest area of reference $\mathcal{I}_k$ defined manually. It is fixed for all the parallel strip recognition process. However, to prevent gradual change of the profile, we can define a dynamic variant where the reference $\mathcal{I}_0$ is updated by using the previous matching interest area $\mathcal{I}_r$ matching. Such a strategy was evaluated and presents the ability to be robust to profile change. However it shows that the reference tends to deviate towards the initial interest area. We choose another strategy by changing the reference only when p matches are achieved. By this way, the algorithm is able to deal with moderate intensity variations and avoids the deviations of previous solution.

Algorithm 1. findCandidate function

---

input : $\mathcal{R}_r$ //Reference research segment
  $\mathcal{R}_i$ //Research segment where interest area must be found

//We define $\mathcal{I}_i^{n,m}$ with $n = 0$ and $m = $ Size of the areas of interest
$(nf, mf) \leftarrow (0, 0)$
min $\leftarrow$ a large number

while $m < $ *of the size of the* $\mathcal{R}_i$ **do**
  mse $\leftarrow$ MSE($\mathcal{R}_r$, $\mathcal{R}_i$)

  if mse $<$ min **then**
    min $\leftarrow$ mse
    $(nf, mf) \leftarrow (n, m)$

  //We define $\mathcal{I}_i^{n,m}$ with $n = n + 1$ and $m = m + 1$

//We define for $\mathcal{R}_i$ his area of interest $\mathcal{I}_i^{nf,mf}$

Algorithm 2. recognizeBlurredSegment function

---

input : Seg //Set of $\mathcal{R}_k$
        stepReference //$\mathcal{R}_r$ change after stepReference iterations
output: blurredSegmentL //Blurred segment that represents an edge
        blurredSegmentR //Blurred segment that represents the other edge of
the tube

$\mathcal{R}_r \leftarrow$ Seg[0]

for $i \leftarrow 0$ **to** *Seg.size* **do**
  findCandidates($\mathcal{R}_r$, $\mathcal{R}_i$)

  // We try to extend the blurredSegmentL with the first point
  // of the area of interest $\mathcal{I}_i^{n,m}$
  blurredSegmentL $\leftarrow$ extendBlurredSegment($P1$)

  // We try to extend the blurredSegmentL with the last point
  // of the area of interest of $\mathcal{I}_i^{n,m}$
  blurredSegmentR $\leftarrow$ extendBlurredSegment($P2$)

  if $P1$ *and* $P2$ *was accepted* **then**
    lastAccepted $\leftarrow \mathcal{R}_i$

  //We change the reference modulo stepReference time
  if ((i mod stepReference) $= 0$) *and* ($P1$ *and* $P2$ *was accepted*) **then**
    $\mathcal{R}_r \leftarrow \mathcal{R}_i$
  else
    $\mathcal{R}_r \leftarrow$ lastAccepted

---

3   Application to Metallic Tube Measuring

In relation to the new proposed method, we focus on the problem of the geometric measure on metallic tubular objects (information about the orientation and length of straight parts) taken in an industrial environment with real time constraints. Such tubular objects represented on Fig. 6(a) are difficult to segment as shown in images (b–h) where six recent segmentation algorithms were applied and failed to segment precisely the metallic tube.

Related to these specularity problems, Chang *et al.* [7] proposed an original approach to estimate the pose of 3D tubular objects. In their work, they propose to exploit the scene image in order to apply a rendering of the piece from the tubular shape model which is supposed to be known. Then from the rendering image, they can search the region matching the initial piece and estimate the initial pose. Such an original work is however not adapted to our main objective to measure shape quality from geometric information since it is only defined to recover position of rigid object without any measure of deformation.

Since this segmentation task on such objects remains a challenge in the computer imagery domain, we follow another strategy by using the new proposed method in order to directly extract the geometric information and exploiting the nature of specular object.

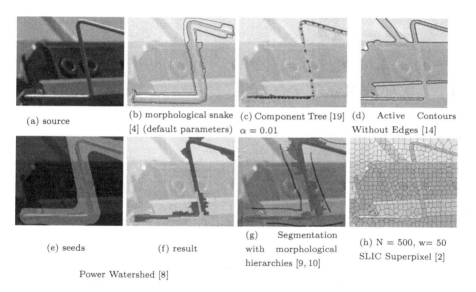

(a) source

(b) morphological snake [4] (default parameters)

(c) Component Tree [19] $\alpha = 0.01$

(d) Active Contours Without Edges [14]

(e) seeds

(f) result

(g) Segmentation with morphological hierarchies [9, 10]

(h) N = 500, w= 50 SLIC Superpixel [2]

Power Watershed [8]

Fig. 6. Experiments of six recent segmentation algorithms on a specular tube (image (a)). Segmentation markers are highlighted in red on images (b, c, e, g) and the segmentation results are represented in blue color on images (b, c, d, f, g, h) (Colour figure online).

Even if the proposed approach needs the set of an initial segment across the object of interest ($\mathcal{I}_0$), we can suppose that it is always available, even approximately, since from the considered industrial application, the main geometric model of the tube is supposed to be known. In the experiments of Fig. 7, we have manually put them with different degrees of precision in order to measure the robustness of the approach. The few parameters of the algorithm were

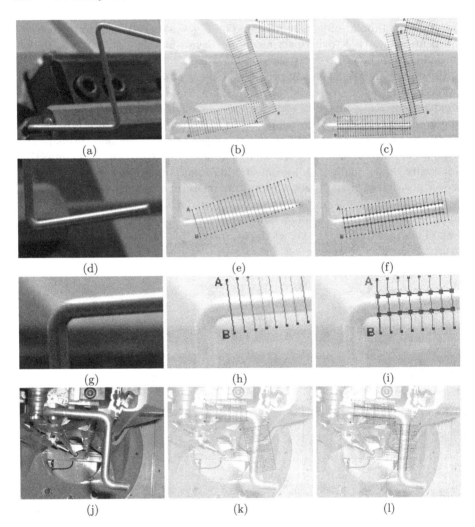

Fig. 7. Results of parallel strip recognition on various images (a, d, g, j). The results obtained after the initialization steps (second row) are given on the third row.

set to 10 for the d_s parameter and to 5 for the maximal width value of the blurred segment parameter (in order to handle some small amount of noise in the recognition process). Different results are presented on the Fig. 7 where various initializations were applied from the two initial points A and B. All the detections were performed in live after the initialization and present fine precision permitting to identify the location and orientation of the different tubular parts given in different configurations.

In order to show the robustness according to the position of the interest area, we have voluntarily initialized this area between the tube and the background.

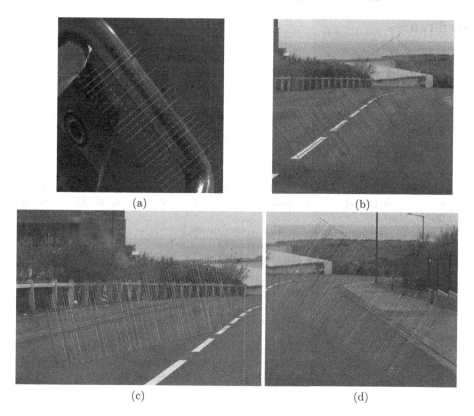

(a)

(b)

(c)

(d)

Fig. 8. Measure of robustness on noisy image with coarse initialization of interest area outside the tube (image (a)). Other experiment to the segment recognition on other type of images containing linear structures (b, c, d).

As it can be shown on the image of Fig. 8 the tubular direction and position is well recovered. Note that the robustness to noise is also visible since this image was taken in low light conditions. Finally, other experiments are presented with others types of image containing linear structures. The different detections well fit the real features associated to road border and center (see images (b, c, d)).

4   Conclusion and Future Work

We proposed a new and original approach to detect parallel strip segment. Instead of basing the detection on image gradient, the intensity profile allows us to be able to detect linear structure even on specular objects. Associated to this work, we have proposed a real industrial application with the recognition of metallic tubular shape. As future work, we planed to extend this same idea to automatically detect all the straight segments of an image in a stand-alone algorithm.

References

1. DGtal: Digital Geometry tools and algorithms library. http://libdgtal.org
2. Achanta, R., Shaji, A., Smith, K., Lucchi, A., Fua, P., Süsstrunk, S.: SLIC super-pixels compared to state-of-the-art superpixel methods. IEEE Trans. Pattern Anal. Mach. Intell. **34**(11), 2274–2282 (2012)
3. Akinlar, C., Topal, C.: EDLines: a real-time line segment detector with a false detection control. Pattern Recogn. Lett. **32**(13), 1633–1642 (2011). http://ceng.anadolu.edu.tr/CV/EDLines/demo.aspx
4. Alvarez, L., Baumela, L., Márquez-Neila, P., Henríquez, P.: A real time morpho-logical snakes algorithm. Image Process. Line **2**, 1–7 (2012)
5. Bhowmick, P., Bhattacharya, B.B.: Fast polygonal approximation of digital curves using relaxed straightness properties. IEEE Trans. Pattern Anal. Mach. Intell. **29**(9), 1590–1602 (2007)
6. Buzer, L.: A simple algorithm for digital line recognition in the general case. Pattern Recogn. **40**(6), 1675–1684 (2007)
7. Chang, J., Raskar, R., Agrawal, A.: 3D pose estimation and segmentation using specular cues. In: CVPR 2009, pp. 1706–1713 (2009)
8. Couprie, C., Grady, L., Najman, L., Talbot, H.: Power watersheds: a new image seg-mentation framework extending graph cuts, random walker and optimal spanning forest. In: IEEE 12th International Conference on Computer Vision, pp. 731–738. IEEE (2009)
9. Cousty, J., Najman, L.: Incremental algorithm for hierarchical minimum spanning forests and saliency of watershed cuts. In: Soille, P., Pesaresi, M., Ouzounis, G.K. (eds.) ISMM 2011. LNCS, vol. 6671, pp. 272–283. Springer, Heidelberg (2011)
10. Cousty, J., Najman, L., Perret, B.: Constructive links between some morphological hierarchies on edge-weighted graphs. In: Hendriks, C.L.L., Borgefors, G., Strand, R. (eds.) ISMM 2013. LNCS, vol. 7883, pp. 86–97. Springer, Heidelberg (2013)
11. Debled-Rennesson, I., Feschet, F., Rouyer-Degli, J.: Optimal blurred segments decomposition of noisy shapes in linear times. Comput. Graphics **30**, 30–36 (2006)
12. Durou, J.D., Falcone, M., Sagona, M.: Numerical methods for shape-from-shading: a new survey with benchmarks. Comput. Vis. Image Underst. **109**(1), 22–43 (2008)
13. Even, P., Malavaud, A.: Semi-automated edge segment specification for an inter-active modelling system of robot environments. Int. Arch. Photogramm. Remote Sens. **33**(B5), 222–229 (2000)
14. Getreuer, P.: Chan-Vese segmentation. Image Process. Line **2**, 214–224 (2012)
15. Grompone von Gioi, R., Jakubowicz, J., Morel, J.M., Randall, G.: LSD: a line segment detector. Image Process. Line **2**, 35–55 (2012)
16. Haldo, S., Juan, C.: A review of classic edge detectors. Image Process. Line **5**, 90–123 (2015). doi:10.5201/ipol.2015.35
17. Kerautret, B., Lachaud, J.O.: Curvature estimation along noisy digital contours by approximate global optimization. Pattern Recogn. **42**(10), 2265–2278 (2009)
18. Kerautret, B., Even, P.: Blurred segments in gray level images for interactive line extraction. In: Wiederhold, P., Barneva, R.P. (eds.) IWCIA 2009. LNCS, vol. 5852, pp. 176–186. Springer, Heidelberg (2009)
19. Naegel, B., Passat, N.: Interactive segmentation based on component-trees. Image Process. Line **4**, 89–97 (2014)
20. Nguyen, T.P., Debled-Rennesson, I.: A discrete geometry approach for dominant point detection. Pattern Recogn. **44**(1), 32–44 (2011)

Analysis and Performance Evaluation of ICA-Based Architectures for Face Recognition

Anu Singha, Mrinal Kanti Bhowmik$^{(\boxtimes)}$,
Prasenjit Dhar, and Anjan Kumar Ghosh

Department of Computer Science and Engineering,
Tripura University (A Central University), Suryamaninagar 799022, India
{anusingh5012,prasenjitdhar.cse}@gmail.com,
mkb_cse@yahoo.co.in, anjn@ieee.org

Abstract. Prediction of the best ICA architecture for face recognition systems is somewhat complicated. This paper shows how the recognition performance of both architectures depends on the nature of feature vectors rather than several criteria such as different databases, number of subjects, and number of principle components. The investigation finds that Architecture-II yields the better performance than Architecture-I based on face feature vectors. The experiments are done on different face datasets like FERET, ORL, CVL, and YALE.

Keywords: ICA · Architecture-I · Architecture-II · Performance evaluation · Analysis

1 Introduction

In image analysis and understanding, face recognition have been a challenging and quite attractive key area of research. It is usually used in security systems and can be compared to other biometrics such as eye iris recognitions or fingerprint. The recognition task has been done by selecting proper subspace projection to get facial features followed by classification in the space of compressed features. There are varieties of techniques employed for selecting subspace projection which projects consider face images as the points in high-dimensional spaces and reduce the dimension to find a more meaningful description. The central issue is how to determine and define image appearance in a high-dimensional image space to a low-dimensional subspace. The most noticeable method in this category is Principle Component Analysis (PCA) [14], which is concerned only second-order dependencies between variables. For past one and half decades, a generalized method of PCA, Independent Component Analysis (ICA) has received spacious notice. ICA technique is a relatively new invention which has been mainly used to Blind Signal Separation (BSS), though it has been successfully applied to the face recognition problem too [5]. ICA is concerned with high-order dependencies between variables in addition to the

© Springer International Publishing Switzerland 2015
R.P. Barneva et al. (Eds.): IWCIA 2015, LNCS 9448, pp. 323–336, 2015.
DOI: 10.1007/978-3-319-26145-4_24

second order. PCA makes the data uncorrelated while ICA makes the data uncorrelated as well as unit variance i.e. as independent as possible. There are at least two benefits for face recognition using ICA: first, the high order dependencies among data may contain more information that is useful for face recognition than the second-order statistic representations. Secondly, ICA finds the directions such that the projection of the data into those directions has maximally "non-Gaussian" distributions.

2   Literature Review

The literature review of ICA on the subject is very contradictory. Bartlett et al. [1] were among the first to apply ICA to face identification task. They have used the Infomax algorithm [9] to employ ICA and recommended two ICA based architectures. Both architectures were evaluated on a subset of the FERET database along with PCA, and claims that the two ICA based architectures were equally powerful and both outperformed the PCA. Liu and Wechsler [11] also used FERET database to study the comparative assessment of ICA performance through Comon [4] ICA algorithm, and claims ICA outperform PCA. Guo et al. [8] also present the process of facial expression recognition based on ICA model. Their experimental results have shown that ICA is a more effective facial expression recognition method than that based on PCA and 2DPCA. Kishor et al. [10] proposed a new face recognition technique based on Independent Component Analysis of GaborJet (GaborJet-ICA). They transformed this GaborJet feature vector into the basis space of PCA, and prove that the difference in performance is insignificant between GaborJet-ICA and GaborJet-PCA. While other researchers reported differently. Socolinsky et al. [15] has reported that ICA performs better in case of visible images, and PCA performs better in case of infrared images. Draper et al. [6] again tested ICA architectures and PCA on the FERET database to come out from these conflicting results. The analysis has shown that ICA architecture-II provides the best results, followed by PCA with L1 or Mahalanobis distance metrics. Recommends the FastICA algorithm for ICA architecture-II, although the difference between FastICA and Infomax is not large. In recognizing facial actions, the recommendation is reversed: found the best result using Infomax to implement ICA architecture-I. Jian Yang et al. [18] also re-evaluated ICA architectures and PCA on the FERET, ORL, AR face databases, and claims as similar to Drapper et al. [6]. They construct two PCA baseline algorithms to re-evaluate ICA-based architectures, but observed no significant performance difference between ICA-I (II) and PCA-I (II).

The performance analyses of the two ICA architectures depended on the property of "intra feature correlation" and "inter feature correlation" of the feature vector of face images. The term "intra feature correlation" refers to the relationship of a feature vector with itself, and the term "inter feature correlation" refers to the relationship of a feature vector with other feature vectors. The relationship within the intra feature vectors should be very strong i.e. variance is one, and the relationship between the inter feature vectors should be poor i.e.

uncorrelated. The poor inter feature relationship decreases the correlation value making it more independent (as a contribution part) which gives better accuracy rate in classification of face images in a small dataset. From the several literature reviews, it has also been observed that many authors evaluated ICA according to some criteria, for example image pre-processing, ICA pre-processing steps, effect of different ICA algorithms, distance metrics, different types of images, and so on. As a consequence, it has been complicated to predict the best ICA architectures for a particular domain. So an analysis is carried out on feature vectors based on the correlation coefficient property, and it is found that Architecture-II is more potential than Architecture-I for performance analysis of face feature vectors.

3   Independent Component Analysis (ICA) and Its Two Architectures

Basically ICA is a method for extracting statistically independent components from a mixture of them [3]. The performance of ICA varies on the databases, the number of images, and the number of subspace dimensions reduced. Typically, the performance depends even more on the two ICA-based architectures of face representations namely Architecture-I and Architecture-II.

3.1   Architecture-I: Statistically Independent Basis Faces

In this architecture, ICA can be represented to treat face images as random variables and pixels are trials for the variables [2]. The approach illustrated in Fig. 1. Organizes each face image in the database as a lengthy vector with size of dimensions in the image, into a matrix X where each row vector is a different image. It makes sense to talk about independence of images. The ICA algorithms learn the weight matrix W, which is then projected onto the input images X to produce the independent basis images in the rows of S.

To illustrate the mathematical basis for Architecture-I, the following steps have been described. In the first step, form an image data matrix $X_{N \times M} = [x_1, x_2, ..., x_N]^T$ of a given set of N training samples $x_1, x_2, ..., x_N$ in R^M. Then center the data matrix X in a trail space R^M by subtracting the mean vector μ^I from each trail, and get the centered matrix $\bar{X}_{N \times M}$. In step 3, whiten the centered data matrix using PCA elements $U_{m \times m}$ and $V_{N \times m}$, and obtain the whitening matrix as

$$H_{N \times m} = VU^{-\frac{1}{2}} \tag{1}$$

Where U is the diagonal matrix of m largest positive eigenvalues, and V is a matrix of orthonormal eigenvectors of corresponding m largest positive eigenvalues. PCA enhances the performance of ICA by throwing away small-negative eigenvalues before whitening, and reduce computational complexity by minimizing pair-wise dependencies. So the data matrix $\bar{X}$ can be whitened using the transformation

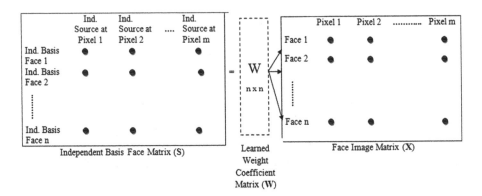

Fig. 1. Block diagram of finding statistically independent basis images.

$$\tilde{X}_{m \times M} = H^T \bar{X} \qquad (2)$$

The main purpose of whitening is to make its components as uncorrelated and unit variances, such that $E\{\tilde{X}\tilde{X}^T\} = I$ [9]. Fourth, process the ICA on $\tilde{X}$ whitened matrix to generate a square learned weight matrix $W^I_{m \times m}$ by a given ICA algorithm. As a fifth step, produce the space S^I with m independent basis images in its rows by projecting weight matrix onto the centered whitened matrix as

$$S^I_{m \times M} = W^I \tilde{X} \qquad (3)$$

At last, the compressed representation of images i.e. feature vectors space X^f of face image matrix X is given by

$$X^f_{N \times m} = \bar{X}(S^I)^T \qquad (4)$$

Each row of X^f that represents the feature vectors, is used to represent the image matrix X for recognition purposes.

3.2 Architecture-II: Statistically Independent Coefficients

According to ICA definition, the coefficient matrix should be orthogonal. But in practice, it might be non-orthogonal. Apart from FastICA, many ICA algorithms such as Infomax, Comons give results in a non-orthogonal coefficient matrix [19]. So the basis images obtained in Architecture-I are statistically independent, but the coefficient matrix that represents input face images in the subspace defined by the basis face images is not statistically independent. Conversely, in Architecture-II, ICA is used to find a set of statistically independent coefficients to represent a face image and the resulting basis images may be statistically dependent. So the input face data matrix X is transposed from Architecture-I i.e. the pixels are variables and the images are trails [2], as shown in Fig. 2.

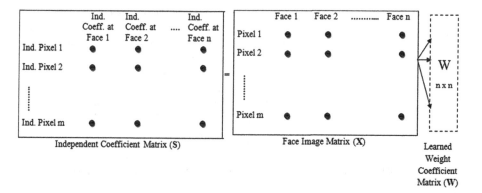

Fig. 2. Block diagram of finding statistically independent coefficients.

Now, each row of the learned weight coefficient matrix W is the basis images, and the statistically independent coefficients that comprise the input images are recovered in the columns of S. It makes sense to talk about independence of pixels.

The illustration of the mathematical basis for Architecture-II is roughly analogous to Architecture-I except (i) starting image data matrix will be in the form of transpose of $X_{N \times M}$, say $Y_{M \times N} = [y_1, y_2, ..., y_N]$, (ii) Architecture-I centering the data matrix by removing the mean of each image, and Architecture-II centering the data matrix by removing the mean image of all image samples. In this architecture, the independent coefficients are recovered in the columns of S^{II} as

$$S^{II}_{m \times M} = W^{II} H^T \qquad (5)$$

Each column of Y^f that represents the feature vectors is given by

$$Y^f_{m \times N} = S^{II} \bar{Y} \qquad (6)$$

4 System Overview

Figure 3 shows a block diagram of a generic human face analysis model for ICA. In image pre-processing stage, the database face images are manually cropped, resized, and finally the histogram equalization for image enhancement is processed. In the second stage, the two most important pre-processing steps that are centering and whitening to simplify and minimize the complexity of the problem for the actual ICA algorithms are carried out. While PCA is used to reduce the dimensions. Then the ICA algorithms for maximizing non-gaussianity as a measure of statistical independence are applied. In the fourth stage, the feature vectors of all the database images are extracted by projecting onto independent source outputs of ICA algorithms. At last, classification of the extracted features has been done using support vector machines, and the accuracy estimated.

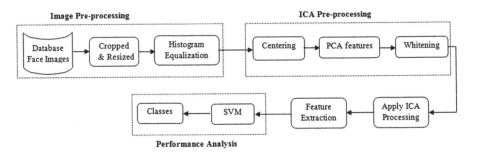

Fig. 3. Block diagram of ICA based feature extraction and classification.

5 Analysis

The analysis has been done in two steps. First, a statistical estimation has been carried out on the feature vectors, and secondly, a performance evaluation over the several databases has been conceded against different factors.

5.1 Numerical Analysis of Feature Vectors of Both Architectures

Table 1 represents some column feature vectors (FV 1 to FV 4) of dimension 15×1 taken from both architectures. In the case of intra feature vector analysis, it has been observed that the variations between the values of the feature vector (say, FV 1 column vector itself) of Architecture-I are very large than the variations of Architecture-II. Similarly it is tough to measure the correlation property of the intra feature vector analysis. Therefore, in Table 2, the values of correlation coefficient of intra and inter feature vectors are reported. In Table 2, rows 1, 5, 8, 10 show the correlation coefficient values of intra feature vectors from both architectures which give us the variance values of 1 indicating a very strong relationships. In the case of inter feature vector analysis, it has also been observed the similar variations (like values within the intra feature vectors) between the values of the feature vectors (FV1-FV2, FV1-FV3, FV1-FV4, FV2-FV3, FV2-FV4, FV3-FV4) from both architectures. For example, the data have been plotted between the feature vector of 1 and 4 in Fig. 4. Although the values between the feature vectors of Architecture-I contain large variation which is shown in Fig. 4a indicating that the points are scattered along the regression line and closer except one single point which indicates a stronger correlation. In Fig. 4b, some points are close enough towards regression line while rest of data points scattered in a wider band from regression line. The correlation coefficient value of 0.010 in Fig. 4b is showing a very weaker positive correlations in Architecture-II, where value of 0.074 in Fig. 4a is showing a comparably stronger positive relationship in case of Architecture-I. Other comparative inter correlation coefficient results between the feature vectors listed in Table 2. From rows 2, 3, 4, 6, 7, 9 it has been noticed that the correlation coefficient values of Architecture-II are less than Architecture-I. In case of Architecture-II, some coefficient values (row

Table 1. Some (*feature vectors*) from both architectures. FV indicates Feature Vectors.

Architecture-I				Architecture-II			
FV 1	FV 2	FV 3	FV 4	FV 1	FV 2	FV 3	FV 4
373.926	313.148	855.538	827.955	0.297	0.119	0.340	0.121
280.486	−14.316	253.632	−19.910	0.174	−0.521	−0.133	0.088
38.602	55.394	6.900	75.637	0.0687	−0.145	−0.0202	−0.151
−246.216	−301.438	250.502	−12.590	−1.180	−0.919	0.173	−0.885
−119.288	−50.0679	−96.324	−81.723	−1.224	−0.813	0.714	1.332
−568.983	−471.674	−332.363	−355.828	−0.950	−0.780	1.177	1.211
269.229	397.313	147.475	33.0255	0.116	−0.350	0.945	0.0500
−456.105	−440.801	361.348	386.277	−3.387	−4.0462	0.1225	−0.506
−192.065	−43.943	−184.124	47.544	0.424	−0.140	0.219	0.558
194.332	134.580	7.094	128.992	−0.173	0.0621	−0.980	−0.550
110.288	51.262	−39.467	−152.434	0.551	−0.438	−1.160	−2.915
−100.337	−130.741	017.498	−59.829	0.571	−0.594	0.676	0.350
−600.982	−727.88	−262.249	−245.191	−0.358	0.370	1.860	2.397
243.287	32.925	−405.614	−536.619	−0.274	0.654	0.350	−0.0311
66.1689	10.324	−14.044	−20.276	0.232	−0.264	−0.892	−0.300

Table 2. Comparisons of the one feature vector with other feature vectors corresponding to Table 1

Sl. No.	FV	FV	Correlation type	Correlation coefficient value (R^2) of Architecture-I	Correlation coefficient value (R^2) of Architecture-II
1	FV 1	FV 1	intra	1	1
2	FV 1	FV 2	inter	0.908	0.684
3	FV 1	FV 3	inter	0.101	0.035
4	FV 1	FV 4	inter	0.074	0.010
5	FV 2	FV 2	intra	1	1
6	FV 2	FV 3	inter	0.113	0.002
7	FV 2	FV 4	inter	0.111	0.031
8	FV 3	FV 3	intra	1	1
9	FV 3	FV 4	inter	0.865	0.673
10	FV 4	FV 4	intra	1	1

4 and 6) of inter feature vectors are close to zero (0.010 and 0.002 respectively). Therefore, it has been found that Architecture-II has a better independedness property through correlation coefficient than Architecture-I which may lead better classification performance.

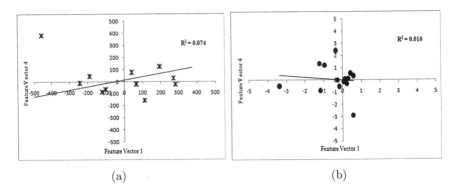

Fig. 4. Correlation between two feature vectors [1 and 4] from (a) Architecture-I, (b) Architecture-II.

5.2 Performance Analysis of ICA Architectures

In this section, the experiments of ICA based two architectures are performed using four face databases: the FERET, ORL, CVL, and YALE databases. From these databases, experiments have done in frontal faces with different expressions, and illuminations.

Database Organization: The FERET [13] has total five probe sets of frontal (pose angle of zero degree) images namely fa, fb, ba, bj, bk. The number of subjects of "f" series does not match with the subjects of "b" series. So, for experiment analysis only "b" series has been taken where each of the 200 subjects has 3 images belonging to probe sets ba, bj, bk respectively. Probe set ba consists of 200 images of 200 subjects, and also set bj, bk consists of 200 images of 200 subjects each. The ORL database [17] consists of 10 different images of each of 40 distinct subjects. For experiments, only 3 frontal position images have been taken from each of the 40 subjects of varying lighting and facial expressions, and these 3 images are put to the manual sets namely set 1, set 2, set 3 backed by the idea to keep same number of images of each subject and same number of sets from several databases for the comparative study. Another face database called FRI CVL [16] consists of 7 different images of 114 number of unique people consists of 108 male and 6 female. The images were taken at different conditions: profile left/right, 45 degrees left/right, frontal, frontal smile, and frontal smile with teeth. As for comparative study, 3 frontal images are taken to the manual sets: set 1 consists of frontal smile images, set 2 consists of frontal smile with teeth, and set 3 consists of only frontal images. In the same way, the Yale database [7] is also prepared.

The purpose of the experiments is to compare the performance of two ICA based architectures for face recognition. To observe the recognition performance, 3 training sample set has been prepared and the recognition rate of each samples averaged. All these training samples are shows in Table 3. In training sample 1,

Table 3. Image sets used for training and testing.

Training sample	Condition		Description of sets	
	Training set	Testing set		
Sample 1	Set 1 and Set 2	Set 3	Set 1	Frontal regular facial expression. In FERET database, set 1 is indicated by ba
Sample 2	Set 2 and Set 3	Set 1	Set 2	Alternative facial expression to set 1. In FERET database, set 2 is indicated by bj
Sample 3	Set 3 and Set 1	Set 2	Set 3	This also contains frontal image taken under different lighting. In FERET database, set 3 is indicated by bj

the set 1/ba and set 2/bj are used for training purpose, and set 3/bk has been used for testing purpose. Similarly in training sample 2, the set 2/bj and set 3/bk are used for training purpose, and set 1/ba has been used for testing purpose. In training sample 3, the set 3/bk and set 1/ba are used for training purpose, and set 2/bj has been used for testing purpose.

Experiments: In the experiments, the face portion of each original image is manually cropped and resized to an image of 60×70 resolutions using bilinear interpolation. The resulting image is then pre-processed using histogram equalization method. Figure 5 shows some sample images after pre-processing.

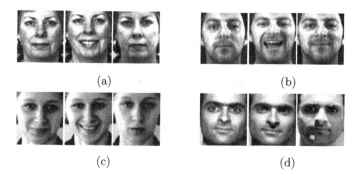

(a)   (b)

(c)   (d)

Fig. 5. Sample images of one subject from the (a) FERET (b) ORL (c) CVL (d) YALE databases.

Based on the first investigation, the assumption is that Architecture-II will produce better result than Architecture-I. In this correspondence, there are three

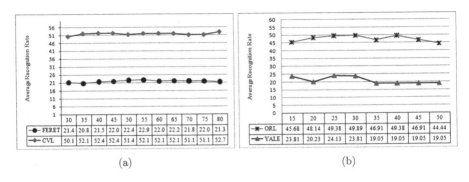

Fig. 6. Recognition rate (*Average*) of architecture I against number of (*Principle Components (PCs)*) over ((*a*) *FERET, CVL (b) ORL, YALE*) databases.

experimental analyses have been done. Experiment 1 concerned on keeping same number of subjects and same number of independent components for each database, where experiment 2 is involved in keeping different number of subjects and same number of independent components. Liu [12] has shown that the selection of number of principle components has a significant effect on the performance of ICA based face recognition. In this regards, the third experiment has been drawn which is based on keeping different number of subjects along with different number of independent components (or principle components) for each database. So an analysis has conceded to find a number of independent components (ICs) for each database which maximizes the performances of ICA architectures. In selection of ICs approach, the numbers of principle components (PCs) vary from 30 to 80 with an interval of 5 in case of FERET and CVL database, and from 15 to 50 with an interval of 5 for ORL and YALE database. Hold the different ranges of PCs for the databases because number of subjects is not equal for all databases. Figure 6 shows the average recognition rates of Architecture-I versus the variation of the PCs. Finally an optimal number of PCs 55, 80, 30, 25 for FERET, CVL, ORL, and YALE databases respectively is chosen. These optimal PCs have been carried out in case of Architecture-II also.

The estimation of a weight matrix is prepared through the FastICA algorithm with the contrast function $G(u) = -\exp(\frac{u^2}{2})$. After feature extraction, SVM multi-classification strategy has been taken. For training support vector machines, polynomial kernel with degree 1 is used and 10-fold cross validation has been done to select proper parameters for kernel function.

Analysis and Observation

1. *Experiment 1: Same number of subjects and same number of ICs:* In this experiment, numbers of 20 subjects are taken from each database i.e. FERET, ORL, CVL, and YALE. From these 20 subjects, a total of 60 face images has been collected for each database. So, each set of training samples consists of 40 training images and 20 testing images. For each face images 15

features are extracted for recognition task. Table 4 lists the recognition rate of each training sample sets and two architectures over four databases. Average recognition rate of two architectures is also listed in this table. Table 4 shows us that Architecture-II significantly outperforms the Architecture-I in all three training samples no matter what database is used by Architecture-I. Also, in terms of the average recognition rate holding the same situation, but in CVL database the performance of Architecture-I is slightly better than other databases although less than Architecture-II with difference of accuracy between two architectures is 20 %. FERET and ORL giving us almost similar results with difference of accuracy between two architectures are 38.33 % and 35 % where YALE database with highest difference of 40 %. All these results are giving contradictory verdict with Bartlett et al. [1] where they had concluded that two ICA representations were equally powerful for face recognition. This analysis is consistent with Draper et al. [6] where concluded that Architecture-II is better than Architecture-I. At this instant, a question is: what are the causes that make notable distinction between the architectures performance? First reason is the architectural representation where Architecture-I centering the data matrix by removing the mean of each image, and Architecture-II centering the data matrix by removing the mean image of all training samples. Secondly, nature of the feature vectors of both architectures i.e. compactness of the feature values such that low compact features will enhance result than high compact features. It is clearly observed that for all four databases Architecture-II perform better than Architecture-I, no matter which databases are used.

Table 4. Keeping same number of subjects and same number of independent components from several databases.

Databases	FERET		ORL		CVL		YALE	
Methods	ICA I	ICA II	ICA I	ICA II	ICA I	ICA II	ICA I	ICA II
Training sample 1	50	75	40	75	55	80	50	45
Training sample 2	35	80	35	65	55	80	30	85
Training sample 3	45	90	50	90	55	65	20	90
Average	**43.33**	**81.66**	**41.66**	**76.66**	**55**	**75**	**33.33**	**73.33**

2. *Experiment 2: Different number of subjects and same number of ICs:* To verify whether the conclusion of experiment 1 depends on the varying number of subjects, the tested two architectures by taking different number of subjects for each database. In this experiment, numbers of 180 subjects and total of 540 face images are taken from FERET database, 103 subjects and total of 309 face images from CVL database, 28 subjects and total of 84 face images from YALE database, 27 subjects and total of 81 face images from ORL database are taken respectively. So, each set of training samples consists of 360, 206, 56, and 54 training images respectively for FERET, CVL,

YALE, ORL databases and 180, 103, 28, 27 testing images respectively for FERET, CVL, YALE, ORL databases. Like experiment 1, 15 features for each face images from all the databases are extracted. Table 5 lists the recognition rate and average recognition rate of each training sample set and two architectures over four databases. As similar with experiment 1, Table 5 also shows us that Architecture-II notably outperforms the Architecture-I in all three training samples no issue of how many images are intended for the experiment. But the recognition rate is become somehow lesser as compare to first experiment which has been more highlighting in FERET database. The explanation behind this the number of features is fewer as compared to the number of images. In terms of the average recognition rate, ORL and CVL database giving almost similar recognition rate with difference of accuracy between two architectures are 22.22 % and 23.31 %. The YALE database giving good recognition rate in case of Architecture-II but highest difference of accuracy along with Architecture-I, is 45.23 %. It seems that the performance varying when different number of subjects is considered, but it does not affect performance of the Architecture-II against Architecture-I.

Table 5. Keeping different number of subjects and same number of independent components from several databases.

Databases	FERET		ORL		CVL		YALE	
Methods	ICA I	ICA II	ICA I	ICA II	ICA I	ICA II	ICA I	ICA II
Training sample 1	14.44	41.67	48.15	70.37	41.75	67.96	39.29	39.29
Training sample 2	21.11	49.44	40.74	59.26	48.54	77.67	21.43	85.71
Training sample 3	17.78	43.33	48.15	74.07	47.57	62.14	10.71	82.14
Average	**17.78**	**44.81**	**45.68**	**67.9**	**45.95**	**69.26**	**23.81**	**69.04**

3. *Experiment 3: Different number of subjects and different number of ICs:* The choice of an optimum number of principle components has a significant effect on the recognition rate of ICA architectures. The number of ICs which maximizes the performances of ICA architectures is 55, 80, 30, and 25 for database FERET, CVL, ORL, and YALE respectively as shows in Fig. 6. So, this experiment is basically based on the different number of ICs along with different number of subjects for each database. The collection of number of subjects is analogous to previous analysis i.e. experiment 2. Table 6 lists the recognition rate and average recognition rate of each training sample set and two architectures over four databases. Although optimum number of PCs effects the recognition rate of ICA architectures as compared to previous experiment, this experiment also notify that the Architecture-II superior than the Architecture-I. The average recognition performances of experiment 2 and 3 are somehow reduced as compared to first experiment except few cases. So the number of subjects or images could also effect in the performances.

Table 6. Keeping different number of subjects and different number of independent components from several databases.

Databases	FERET		ORL		CVL		YALE	
Methods	ICA I	ICA II	ICA I	ICA II	ICA I	ICA II	ICA I	ICA II
Training sample 1	26.67	56.11	44.44	77.78	54.37	79.61	39.29	50
Training sample 2	17.22	72.78	44.44	66.67	51.46	85.71	21.43	85.71
Training sample 3	25.00	66.11	59.26	88.89	52.43	78.57	10.71	78.57
Average	**22.96**	**65**	**49.38**	**77.78**	**52.75**	**81.30**	**23.81**	**71.42**

In a word, the strong and weakly compact correlations between feature vectors of two architectures clearly define which architecture is better than the other. There is no effect of different databases, number of subjects, and number of principle components in the performances of both architectures against one another.

6   Conclusion

Evaluation between ICA architectures is difficult since there are lot of considerations must be taken into account such as differences in representation of architectures, database complication, optimum parameters selection for ICA processing and classification etc. This paper is to investigate the relationship of the feature vectors, and experiments based on several factors. As a conclusion, it has been seen that Architecture-II is outperforms than Architecture-I. In this process, it has been possible to verify the results of similar claims with few researchers and diverse views in case of other researchers.

Acknowledgments. The work presented here is being conducted in the Biometrics Laboratory and Bio-Medical Infrared Image Processing Laboratory of Department of Computer Science and Engineering of Tripura University (A Central University), Tripura, Suryamaninagar-799022. The research work was supported by the Grant No. 12(2)/2011-ESD, Dated 29/03/2011 from the DeitY, MCIT, Government of India and also supported by the Grant No. BT/533/NE/-TBP/2013, Dated 03/03/2014 from the Department of Biotechnology (DBT), Government of India. The authors would like to thank Prof. Barin Kumar De, Department of Physics, Tripura University (A Central University) and Dr. Debotosh Bhattacharjee, Associate Professor, Department of Computer Science and Engineering, Jadavpur University for their kind support to carry out this research work.

References

1. Bartlett, M.S., Lades, H.M., Sejnowski, T.J.: Independent component representations for face recognition. In: Proceedings of the SPIE Symposium on Electronic Imaging: Science and Technology; Conference on Human Vision and Electronic Imaging III, San Jose, California (1998)

2. Bartlett, M.S., Movellan, J.R., Sejnowski, T.J.: Face recognition by independent component analysis. IEEE Trans. Neural Netw. **13**, 1450–1464 (2002)
3. Bell, A., Sejnowski, T.: An information maximization approach to blind separation and blind deconvolution. J. Neural Comput. **37**, 1129–1159 (2007)
4. Comon, P.: Independent component analysis: a new concept? Signal Process. **36**, 287–314 (1994)
5. Deniz, O., Castrillon, M., Hernandez, M.: Face recognition using independent component analysis and support vector machines. Pattern Recogn. Lett. **24**, 2153–2157 (2001)
6. Draper, B.A., Baek, K., Bartlett, M.S., Beveridge, J. R.: Recognizing faces with PCA and ICA. Comput. Vis. Image Underst. 91, 115–137 (2003)
7. Georghiades, A.S., Belhumeur, P.N., Kriegman, D.J.: From few to many: illumination cone models for face recognition under variable lighting and pose. IEEE Trans. PAMI **23**, 643–660 (2001)
8. Guo, X., Zhang, X., Deng, C., Wei, J.: Facial expression recognition based on independent component analysis. J. Multimedia **8**, 402–409 (2013)
9. Hyvarinen, A., Oja, E.: Independent component analysis: algorithms and applications. Neural Netw. **13**, 411–430 (2000)
10. Kinage, K.S., Bhirud, S.G.: Face recognition using independent component analysis of GaborJet (GaborJet-ICA). In: IEEE International Colloquium on Signal Processing and Its Applications (CSPA), Malacca City, pp. 1–6 (2010)
11. Liu, C., Wechsler, H.: Comparative assessment of independent component analysis (ICA) for face recognition. In: International Conference on Audio and Video Based Biometric Person Authentication, Washington (1999)
12. Liu, C.: Enhanced independent component analysis and its application to content based face image retrieval. IEEE Trans. Syst. Man Cybern. B Cybern. **34**, 1117–1127 (2004)
13. Phillips, P.J., Wechsler, H., Huang, J., Rauss, P.J.: The FERET database and evaluation procedure for face-recognition algorithms. Image Vision Comput. **16**, 295–306 (1998)
14. Sirovich, L., Kirby, M.: Low-dimensional procedure for characterization of human faces. J. Opt. Soc. Am. A **4**(3), 519–524 (1987)
15. Socolinsky, D.A., Selinger, A.: A comparative analysis of face recognition performance with visible and thermal infrared imagery. In: Proceedings of the International Conference on Pattern Recognition, Quebec City (2002)
16. Solina, F., Peer, P., Batagelj, B., Juvan, S., Kovac, J.: Color-based face detection in the "15 seconds of fame" art installation. In: Mirage 2003, Conference on Computer Vision/Computer Graphics Collaboration for Model-based Imaging, Rendering, Image Analysis and Graphical Special Effects, pp. 38–47. INRIA Rocquencourt, France, Wilfried Philips, Rocquencourt, INRIA (2003)
17. The AT&T face database. http://www.uk.research.att.com/facedatabase.html
18. Yang, J., Zhang, D., Jing-Yu, Y.: Constructing PCA baseline algorithms to reevaluate ICA-based face-recognition performance. IEEE Trans. Syst. Man Cybern.-Part B Cybern. **37**, 1015–1021 (2007)
19. Zhao, W., Chellappa, R., Rosenfeld, A., Phillips, P.: Face recognition: a literature survey. Technical report, University of Maryland, College Park, MD (2002). Technical report, Global Grid Forum (2002)

Optimization of Low-Dose Tomography via Binary Sensing Matrices

Theeda Prasad[1]([⊠]), P.U. Praveen Kumar[1], C.S. Sastry[1], and P.V. Jampana[2]

[1] Department of Mathematics,
Indian Institute of Technology Hyderabad, Telangana, India
{ma13p1004,csastry}@iith.ac.in, praveen577302@gmail.com
[2] Department of Chemical Engineering,
Indian Institute of Technology Hyderabad, Telangana, India
pjampana@iith.ac.in

Abstract. X-ray computed tomography (CT) is one of the most widely used imaging modalities for diagnostic tasks in the clinical application. As X-ray dosage given to the patient has potential to induce undesirable clinical consequences, there is a need for reduction in dosage while maintaining good quality in reconstruction. The present work attempts to address low-dose tomography via an optimization method. In particular, we formulate the reconstruction problem in the form of a matrix system involving a binary matrix. We then recover the image deploying the ideas from the emerging field of compressed sensing (CS). Further, we study empirically the radial and angular sampling parameters that result in a binary matrix possessing sparse recovery parameters. The experimental results show that the performance of the proposed binary matrix with reconstruction using TV minimization by Augmented Lagrangian and ALternating direction ALgorithms (TVAL3) gives comparably better results than Wavelet based Orthogonal Matching Pursuit (WOMP) and the Least Squares solution.

Keywords: Discrete tomography · Compressive sensing · WOMP · Binary sensing matrix · TVAL3

1 Introduction

Computed Tomography (CT) is a technique for reconstructing the cross-section of an object from measurements that are essentially the line integrals of it. The general image reconstruction in CT is a mathematical process that generates an image from X-ray projection data acquired at different angles around the object. As X-rays are harmful to human bodies, the basic objective in CT in medical use is to obtain high quality images from projection data with as little of radiation dosage as possible [11,19]. This objective was realized in several frameworks ([19,26] and references therein). There is, however, another dimension to the CT reconstruction provided by the notion of *sparsity*, which is not yet exploited properly. The emerging theoretical developments, by the name of compressive

© Springer International Publishing Switzerland 2015
R.P. Barneva et al. (Eds.): IWCIA 2015, LNCS 9448, pp. 337–351, 2015.
DOI: 10.1007/978-3-319-26145-4_25

sensing (CS) have potential in exploiting inherent sparsity in CT images, and resulting in low-dose and stable reconstruction methods.

In the recent CS based CT reconstruction methods [5,26], the property of CT images being sparse in transform domains such as wavelets, frames was used and reconstruction based on convex optimization was proposed. Unlike the existing methods, the present work, however, aims at analyzing the structure of the underlying matrix. The underlying matrix is binary (with elements being equal to 0 or 1) and we demonstrate empirically that the row restricted matrix satisfies sparse recovery properties. As a result, one may be able to determine the data acquisition geometry (i.e., a particular and restricted set of projection samples) for low-dose reconstruction. We believe that this analysis helps in providing theoretical guarantees for faithful CT image reconstruction.

The current work is an attempt towards giving a handle on the data acquisition geometry for low-dose reconstruction in sparsity framework. The paper is organized as follows: Sects. 2, 3 and 4 give an account of related work, brief introductions to CT imaging and basics of the compressive sensing respectively. The proposed reconstruction method via binary sensing matrix is presented in Sect. 5. The experimental results are discussed in Sect. 6. Finally, Sect. 7 gives the concluding remarks.

2   Related Work

In the current literature, CS based techniques are employed to perform CT image reconstruction from incomplete datasets. CS theory allows a sparse signal to be accurately reconstructed from samples far less than what is required by the Shannon/Nyquist sampling theorem [20,21]. The key to the success of CS is the sparsity of a signal under study. In general, an object is not sparse and often times a sparsifying transform can be used to convert it into a domain in which the signal has a sparse representation. One common sparsifying transform is the discrete gradient transform (DGT) whose coefficients can be summed up to form the so-called total variation (TV).

Inspired by CS theory, various TV minimization algorithms were suggested to solve the few-view, limited-angle, and interior problems. For example, Chen et al. proposed a prior image constrained compressed sensing (PICCS) algorithm for dynamic CT application [6]. Yu and Wang proved that a piecewise constant interior region of interest (ROI) can be uniquely reconstructed by a TV minimizing technique [23,24]. Xu et al. extended this CS based interior tomography formulation into a Statistical Iterative Reconstruction (SIR) framework [22]. Ritschl et al. proposed an improved TV method within the Adaptive Steepest Descent-Projection Onto Convex Sets (ASD-POCS) framework for clinical applications [18].

As stated already, current work, however, attempts to study the structure of underlying Radon transform matrix and its compliance with the sparse recovery properties. The study helps one answer the question "what is the data-acquisition geometry such that the sparse recovery properties are satisfied for faithful low-dose reconstruction".

Although these TV-based algorithms are successful in a number of cases, the power of the TV minimization constraint is still limited. First, the TV constraint is a global requirement, which can not directly reflect structures of an object. Second, the DGT operation can not distinguish true structures and image noise. Consequently, images reconstructed with the TV constraint may lose some fine features and generate a blocky appearance in incomplete and noisy cases. In order to overcome the above limitations, in this paper, we employ TV minimization by Augmented Lagrangian and ALternating direction ALgorithms (TVAL3) for reconstruction of tomographic image. This method gives better reconstruction results compared to the state-of-the-art TV minimization methods. We also observe that TVAL3 works well for limited number of rays that are acquired from the CT scanner and the reconstructed image is free from streak artifacts (caused due to scatter) which is generally observed [14] in the reconstructed image using traditional filtered backprojection.

3 Introduction to Computed Tomography

The parallel-beam CT scanning system uses an array of equally spaced unidirectional sources of focused X-ray beams. The basic principle of CT measurement [12] is shown in Fig. 1. The X-ray source, together with primary collimators, provides a fine beam of radiation (ideally an infinitesimally narrow ray) that passes through the object, the intensity of the beam is then measured by a detector. The integral attenuation for each ray position τ is given as:

$$R(\tau) = -\int_{r=0}^{r_p} \mu(\tau, r)dr = \log \frac{I_m}{I_0}, \qquad (1)$$

where I_m is the intensity measured by the detector and is dependent on the initial ray intensity I_0 (i.e. $I_m = I_0 \exp(-\int_{r=0}^{r_p} \mu(r)dr)$), and r is the radial distance along the ray from the source at $r = 0$, limited by the radial distance r_p of the projection plane.

The whole measuring arrangement, including the frame enabling the mentioned linear movement, can be rotated as seen in the Fig. 1. This way, we may obtain a projection for any angle θ of the measurement coordinates (τ, r) with respect to the object coordinates (x, y). It is also possible to obtain the projections for a continuum of θ, so that Eq. (1) can be rewritten as

$$R(\tau, \theta) = \int_{x_{min}}^{x_{max}} \int_{y_{min}}^{y_{max}} \mu(x, y)\delta(x \cos \theta + y \sin \theta - \tau)dxdy, \qquad (2)$$

where the δ-function selects the ray point set, the limits of x and y are given by the object size. Equation (2) is called *Radon transform*, in the continuous-space formulation, obviously, the task of reconstruction of the original image $\mu(x, y)$ from its projection representation $R(\tau, \theta)$ is the problem of finding the inverse Radon transform.

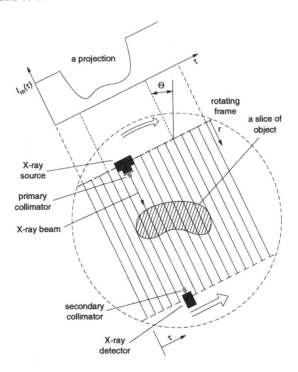

Fig. 1. Principle of measurement of projections basic rectangular arrangement ([12])

To reconstruct CT image, two major categories of methods exist, namely Analytical Reconstruction and Iterative Reconstruction. Methods based on *Filtered Backprojection* (FBP) are one type of analytical reconstruction. This method is currently used in clinical CT scanners because of its computational efficiency and numerical stability [17]. Despite being computationally expensive, iterative reconstruction methods have potential for low-dose reconstruction [25]. One of the commonly known methods is *Algebraic Reconstruction Technique* (ART) which is iterative in nature. There are different variants of ART [13], viz. Simultaneous Algebraic Reconstruction Technique (SART) [1], Multiplicative Algebraic Reconstruction Technique (MART) [2], etc.

4   Compressive Sensing

Compressed sensing is a signal processing technique for efficiently acquiring and reconstructing a signal, by finding solutions to under-determined linear systems [8]. Consider a full rank matrix $A \in \mathbb{R}^{m \times n}$ with $m < n$ then the linear system of equations $y = Ax$ where $y \in \mathbb{R}^m$ and $x \in \mathbb{R}^n$ has infinitely many solutions. Among all, the solution with specific property $J(x)$ may be obtained from the following optimization problem:

$$(P_J) : \min_x J(x) \quad \text{subject to } y = Ax.$$

Considering $J(x)$ as $\|x\|_2$, one obtains the following pseudo-inverse solution:

$$\hat{x} = A^\dagger y,$$

where $A^\dagger = A^T(AA^T)^{-1}$. This solution is unique as the function $\|x\|_2$ is strictly convex, and is not in general sparse.

Definition 1. *A vector $x \in \mathbb{R}^n$ is k-sparse if it has k non-zero co-ordinates. i.e., $\|x\|_0 := |\{i|x_i \neq 0\}| = k < n$.*

By considering $J(x)$ as $\|x\|_0$, one may obtain sparse solution from

$$(P_0) : \min_x \|x\|_0 \text{ subject to } y = Ax.$$

The (P_0) problem is a non-convex optimization problem and finding a solution to it is NP-hard. Since (P_0) problem is intractable, several approaches [10] were proposed to approximate (P_0) based on greedy and convex relaxation methods. Among all greedy methods, Orthogonal Matching Pursuit (OMP) is the most popular.

With $\|x\|_1$ in place of $\|x\|_0$, the convex relaxation problem can be posed as

$$(P_1) : \min_x \|x\|_1 \text{ subject to } y = Ax.$$

The $\|\cdot\|_1$ norm in (P_1) tends to provide sparse solution. One way of establishing equivalence between both the problems is through coherence parameter which is defined as follows:

Definition 2. *The mutual coherence μ_A of a matrix A is the largest absolute inner-product between different normalized columns of A. i.e.,*

$$\mu_A = \max_{1 \leq i,j \leq m, i \neq j} \frac{|a_i^T a_j|}{\|a_i\|\|a_j\|}, \text{ where } a_i \text{ is the } i^{th} \text{ column of } A.$$

For any matrix A of order $m \times n$, the mutual coherence is bounded by $\sqrt{\frac{n-m}{m(n-1)}} \leq \mu_A \leq 1$. The lower bound on mutual coherence μ_A is called Welch bound. The following Theorem 1 [9] relates the equivalence between P_0 and P_1 problems via mutual coherence.

Theorem 1. *Let A be an $m \times n$ matrix and let $0 \neq x \in \mathbb{R}^n$ be a solution of (P_0) satisfying*

$$\|x\|_0 < \frac{1}{2}(1 + (\mu(A))^{-1}).$$

Then x is the unique solution of (P_0) and (P_1).

The Restricted Isometry Property (RIP) is another sufficient condition that ensures the equivalence between P_0 and P_1 problems.

Definition 3. *An $m \times n$ matrix A is said to satisfy the Restricted Isometry Property (RIP) of order k with constant $\delta_k \in (0, 1)$ if*

$$(1 - \delta_k)\|x\|_2^2 \leq \|Ax\|_2^2 \leq (1 + \delta_k)\|x\|_2^2 \ \ \forall x \in \mathbb{R}^n \ \text{with} \ \|x\|_0 \leq k. \qquad (3)$$

Theorem 2. *[4] Suppose an $m \times n$ matrix A satisfies RIP of order $2k$ with constant*

$$\delta_{2k} < \sqrt{2} - 1,$$

then P_0 and P_1 have same k-sparse solution if P_0 has a k-sparse solution.

5   CS in CT: Proposed Method

In this section, we discuss the structure of underlying Radon matrix and an empirical analysis of its compliance with RIP, followed by a convex optimization technique for CT image reconstruction.

5.1   Radon Transform via Binary Matrices

Consider a ray, corresponding to some view θ_i and radial parameter τ_j (Fig. 1). In discrete setting, the Radon measurement may be rewritten as

$$R(\theta_i, \tau_j) = \sum_l I_{i,j}(l)p_l, \qquad (4)$$

where $I_{i,j}(l) = \begin{cases} 1 & \text{if } (\theta_i, \tau_j) \text{ ray hits } l^{th} \text{ pixel} \\ 0 & \text{else.} \end{cases}$

In formulating (4), we consider nearest neighbour interpolation of pixels that fall in the path of the ray. The number p_l stands for the pixel value of lth pixel.

The above equation may rewritten as

$$R(\theta_i, \tau_j) = [I_{i,j}(1) \dots I_{i,j}(N)][p_1 \dots p_N]^T, \qquad (5)$$

where N is related to the size of the image. For all θ_i and τ_j, proceeding this way, one obtains a matrix system

$$y = Ax, \qquad (6)$$

where y contains Radon measurements (i.e. $R(\theta_i, \tau_j)$) in vector form. Accordingly, A is a binary matrix (whose elements are $I_{i,j}(l)$) and x is the vector whose elements p_l. The size of A is dictated by the number of radial and angular sampling parameters and the size of the image to be reconstructed. Suppose A is the matrix corresponding to full set of measurements y, and RA the row restriction of A corresponding to the restricted measurement set Ry (here R may be treated as a restriction matrix). In low-dose CT, as we deal with a small set of projection samples, the system in (6) becomes under-determined, admitting thereby infinitely many solutions in general. The inherent sparsity present in CT images (as detailed in Table 1) makes CS a natural choice [17] for recovering the underlying image.

Table 1. (%) Measure of natural sparsity in standard Shepp-Logan Phantom image. Here $\|x\|_{0,\epsilon} = |\{i | x_i > \epsilon\}|$ and W represents wavelet transform matrix based on 'db8' [7].

Threshold value (ϵ)	x_ϵ in Spatial domain (i.e. $x_\epsilon = \frac{\|x\|_{0,\epsilon}}{N^2} \times 100$)	$x_{W\epsilon}$ in Wavelet domain (i.e. $x_{W\epsilon} = \frac{\|Wx\|_{0,\epsilon}}{N^2} \times 100$)
10^{-1}	41.68	11.39
10^{-2}	41.82	11.81
10^{-4}	41.82	11.81

5.2 On the RIP Compliance of Radon Matrix

In view of tomographic image possessing natural sparsity, the RIP concept is potentially useful in designing the data acquisition mechanism. This is because the sensing matrix provided by the projection set may be designed so that δ_{2k} is minimized, where k is the expected sparsity of tomographic image, which determines the number of measurements to be used for faithful recovery. As the row restricted binary sensing matrix (RA) is not known to satisfy RIP, we obtain k numerically, such that $\delta_{2k} < \sqrt{2} - 1$ (as in Theorem 2), by looking at the distribution of the quantity:

$$\sigma_k(\alpha) = \|RA\alpha\|_2^2, \text{ for } \alpha \in \mathbb{S}^{n-1}, \|\alpha\|_0 \leq k, \tag{7}$$

where $\mathbb{S}^{n-1}$ is unit sphere in $\mathbb{R}^n$. From the values of σ_k, we estimate δ_k as detailed below:

For each k, we considered 1000 vectors, $\alpha \in \mathbb{S}^{n-1}$, such that $\|\alpha\|_0 \leq k$. Here n is related to the size of the image to be reconstructed. From the values of $\sigma_{\{k,\max\}}$ and $\sigma_{\{k,\min\}}$, we estimate δ_k from

$$\delta_k = \max\{1 - \sigma_{\{k,\min\}}, \sigma_{\{k,\max\}} - 1\}. \tag{8}$$

We estimated the values of δ_k for different *sparsity* (k) levels and for four different down-sampling factors (i.e., for different row restriction matrices R generated with uniform distribution). We observed that $k = 21$, $\delta_{2k} < \sqrt{2} - 1$ holds for a matrix of size 2048 × 4096, 1365 × 4096, 1024 × 4096. This observation justifies that the binary sensing matrix provided by the Radon transform appears to possess sparse recovery properties, albeit for low sparsity levels (i.e., k is small). Though the natural sparsity of CT images (Table 1) does not appear to match the k values obtained by our RIP analysis, however, in simulations it was observed that the reconstruction was of good quality even for lower values of k. Since each row of binary matrix corresponds to a pair of angular and radial parameters, the resulting set of rows provides a kind of handle on the data acquisition geometry for faithful reconstruction. To the best of our knowledge, verifying the RIP-compliance of the matrix of Radon Transform analytically is an open problem.

5.3   Reconstruction via an Optimization Method

Motivated by the empirical analysis connecting sparsity level in image to be reconstructed and the number of measurements needed, one may consider the following optimization technique for recovering x.

$$\min_x \|x\|_{TV}, \text{ subject } to \ y = Ax. \tag{9}$$

where the TV norm of x is defined as

$$\|x\|_{TV} = \|\nabla x\|_1 = \sum_{a,b} \sqrt{(x_{a,b} - x_{a-1,b})^2 + (x_{a,b} - x_{a,b-1})^2}.$$

6   Experimental Results

We carried out experiments to reconstruct the tomographic image (Shepp-Logan) by using binary sensing matrix (measurement matrix) and the measurement vector (projection data). To begin with, we focused on the construction of binary sensing matrix. For constructing it, we considered the following radial and angular sampling [16].

Radial Sampling

$$t_m = \delta_r(p - (n/2)), \tag{10}$$

where $p = \{0, 1, \cdots, n-1\}$, δ_r is the length of detector.

Angular Sampling

$$\theta_l = \frac{q + 0.5}{n}\pi, \tag{11}$$

where $q = \{0, 1, \cdots, n-1\}$. As explained in Sect. 5, we obtained A with following property:

$$A \in \{0, 1\}^{n \times n} \text{ with entries } A_{i,j} = \begin{cases} 1 & \text{for } j \in G_i \\ 0 & \text{for } j \notin G_i \end{cases}$$

the matrix whose entries in the ith row correspond to the detector (sensor) locations for the ith measurement. And $G_i \subset \{1, \cdots, n\}$ corresponds to the set of all detector locations selected for the i-th measurement.

We conducted experiments for R (row restriction matrix) of several sizes, viz. 4096×4096, 2048×4096, 1365×4096 and 1024×4096, obtained as row submatrices of full Radon matrix via uniform distribution (i.e. the corresponding number of measurements are 4096, 2048, 1365 and 1024 respectively). Using solvers viz., WOMP [20], TVAL3 [15] and standard Least Squares, we reconstructed the tomographic image. The algorithms were numerically implemented in the MatLab environment on a machine having 4.0 GB RAM and processor speed of 2.6 GHz. Figure 2 is the standard Shepp-Logan phantom image which

Fig. 2. Standard Shepp-Logan Phantom image.

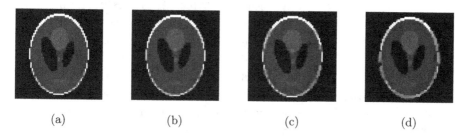

(a)   (b)   (c)   (d)

Fig. 3. Reconstruction of CT image using Total-Variation Augmented Lagrangian Method with measurement matrix of 4 different sizes: (a) 4096×4096 (b) 2048×4096 (c) 1365×4096 (d) 1024×4096.

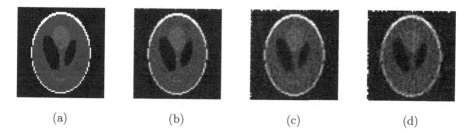

(a)   (b)   (c)   (d)

Fig. 4. Reconstruction of CT image using Least Squares with measurement matrix of 4 different sizes: (a) 4096×4096 (b) 2048×4096 (c) 1365×4096 (d) 1024×4096.

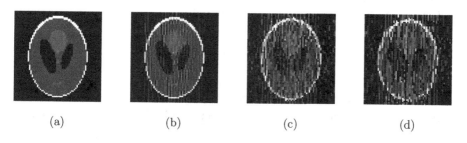

(a)   (b)   (c)   (d)

Fig. 5. Reconstruction of CT image using WOMP with measurement matrix of 4 different sizes: (a) 4096×4096 (b) 2048×4096 (c) 1365×4096 (d) 1024×4096.

Table 2. Performance evaluation of reconstructed tomographic image using solvers viz. TVAL3, Least Squares and WOMP for different dimension of the binary sensing matrix.

Solvers	Dim. of Binary Sensing Matrix	MSE	PSNR (dB)	SSIM	Comp. Time (secs)
Total Variation Augmented Lagrangian (TVAL3) [15]	4096 × 4096	8.775e-4	30.57	0.9795	7.5676
	2048 × 4096	1.977e-3	27.04	0.9552	3.8765
	1365 × 4096	4.112e-3	23.86	0.9060	2.3527
	1024 × 4096	5.521e-3	22.58	0.8724	1.9151
Least Squares	4096 × 4096	1.677e-6	57.75	0.9997	24.4377
	2048 × 4096	4.607e-3	23.37	0.8634	5.9497
	1365 × 4096	1.458e-2	18.36	0.7732	3.1495
	1024 × 4096	2.245e-2	16.48	0.7028	2.3498
Wavelet based Orthogonal Matching Pursuit(WOMP)	4096 × 4096	3.545e-6	54.50	0.9995	3364.82
	2048 × 4096	0.0068	21.69	0.7830	2234.57
	1365 × 4096	0.0425	13.72	0.4646	565.73
	1024 × 4096	0.0615	12.11	0.3810	167.04

is used for comparison purpose. From Figs. 3, 4 and 5 one can visually observe that TVAL3 gives comparably better reconstruction results than standard Least Squares and WOMP based solvers.

In order to evaluate quantitatively the accuracy of reconstruction results, we computed Mean Squared Error (MSE) and Peak Signal to Noise Ratio (PSNR) to measure the similarity between the ground truth and the reconstructed tomographic image. MSE is widely used to evaluate image quality, and is defined as

$$MSE = \frac{\sum_{i=1}^{m} \sum_{j=1}^{n} (x_{i,j} - \widehat{x}_{i,j})^2}{m \times n}, \tag{12}$$

where $x_{i,j}$ is the pixel value of ground truth image and $\widehat{x}_{i,j}$ is the pixel value of reconstructed image.

The corresponding PSNR value is computed as:

$$PSNR = 10 * \log_{10} \frac{(\max(x))^2}{MSE}. \tag{13}$$

Table 2 reports MSE and PSNR errors along with the execution time needed by each of the solvers. The structural similarity (SSIM) index is highly effective way of measuring the structural similarity between two images [21]. Suppose ρ and

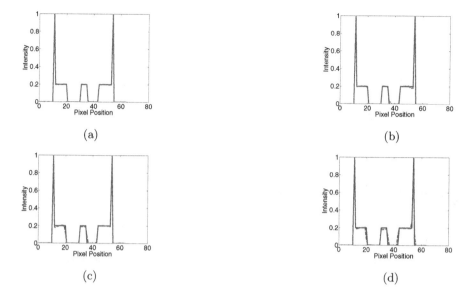

Fig. 6. Comparison of pixel-intensity profiles of ground truth phantom image (continuous line) with reconstructed images (dashed line) using TVAL3 with different dimension of binary sensing matrices: (a) 4096×4096 (b) 2048×4096 (c) 1365×4096 (d) 1024×4096.

t are local image patches taken from the same location of two images that are being compared. The local SSIM index measures three similarities of the image patches: the similarity of luminance $l(\rho, t)$, the similarity of contrast $c(\rho, t)$, and the similarity of structures $s(\rho, t)$. The local SSIM [21] is defined as

$$S(\rho, t) = l(\rho, t) \cdot c(\rho, t) \cdot s(\rho, t),$$

$$S(\rho, t) = \left(\frac{2\mu_\rho \mu_t + C_1}{\mu_\rho^2 + \mu_t^2 + C_1} \right) \left(\frac{2\sigma_\rho \sigma_t + C_2}{\sigma_\rho^2 + \sigma_t^2 + C_2} \right) \left(\frac{2\sigma_{\rho t} + C_3}{\sigma_\rho \sigma_t + C_3} \right), \quad (14)$$

where μ_ρ and μ_t are local means, σ_ρ and σ_t are local standard deviations, and $\sigma_{\rho t}$ is cross-correlation after removing their means. C_1, C_2 and C_3 are stabilizers. The SSIM score of the entire image is then computed by pooling the SSIM map, i.e. by simply averaging the SSIM map. SSIM is highly effective for measuring image quality. Higher SSIM value indicates better image quality.

To better compare the reconstruction results, we plotted the horizontal intensity profile for the chosen row index (here 32nd row was selected). The continuous line corresponds to ground truth image and the dashed line (- - -) corresponds to reconstructed image with different measurement matrices. Figures 6, 7 and 8 are the line intensity profiles of TVAL3, Least Squares and WOMP respectively.

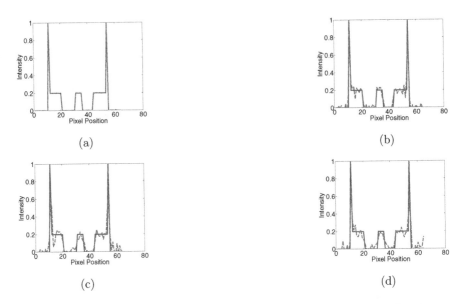

Fig. 7. Comparison of pixel-intensity profiles of ground truth phantom image (continuous line) with reconstructed images (dashed line) using Least Squares with different dimension of binary sensing matrices: (a) 4096 × 4096 (b) 2048 × 4096 (c) 1365 × 4096 (d) 1024 × 4096.

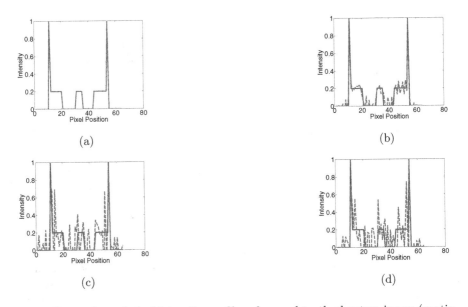

Fig. 8. Comparison of pixel-intensity profiles of ground truth phantom image (continuous line) with reconstructed images (dashed line) using WOMP with different dimension of binary sensing matrices: (a) 4096 × 4096 (b) 2048 × 4096 (c) 1365 × 4096 (d) 1024 × 4096.

7   Conclusions

In the present work, we have formulated the problem of reconstruction of low-dose tomography in terms of a matrix system involving a binary matrix. With a view to gaining handle on the data acquisition geometry, we have verified empirically the compliance of the associated binary matrix with sparse recovery properties. Our experimental results have demonstrated that the binary matrix as given by the data acquisition system satisfies RIP for lower sparsity levels. The reconstructions carried out by three different solvers indicate that TVAL3 gives relatively the reconstruction of better quality. The analytical justification of sparse recovery properties of the matrix of Radon transform, nevertheless, is important in proving the efficacy of CS based ideas in CT. Our future efforts shall attempt to address this issue.

Acknowledgments. One of the authors (CSS) is thankful to CSIR (No. 25(219)/13/EMR-II), Govt. of India, for its support.

A   Appendix: Wavelet Based Orthogonal Matching Pursuit (WOMP)

The conventional form of orthogonal matching pursuit proposed by Troop et al. [12] is a greedy method which builds up the support set of the reconstructed sparse vector iteratively by adding one index to the current support set at each iteration. The input parameters for the conventional OMP algorithm are the measurement matrix (binary matrix) and the measurement vector. Here, we modified the existing OMP algorithm by incorporating the sparsifying transform (i.e. wavelet transform) to further sparsify the binary matrix. We call the modified algorithm as WOMP, which is given below:

Algorithm

Input Parameters: measurement matrix A, wavelet matrix W, measurement vector b, and the error threshold ϵ_0 **Initialization:** Initialize $k = 0$, and set

- The initial solution $(Wx)^0 = 0$.
- The initial residual $r^0 = b - (AW^T)(Wx)^0 = b$.
- The initial solution support $S^0 = Support\{(Wx)^0\} = \phi$

Main Iteration: Increment k by 1 and perform the following steps:

- **Sweep**: Compute the errors $\epsilon(j) = \min_{z_j} \|(a_j w_j^T)z_j - r^{k-1}\|_2^2$ for all j using the optimal choice $z_j^* = (a_j^T w_j^T)^T r^{k-1}/\|a_j w_j^T\|_2^2$.
- **Update Support**: Find a minimizer, j_0 of $\epsilon(j) : \forall j \notin S^{k-1}, \epsilon(j_0) \leqslant \epsilon(j)$, and update $S^k = S^{k-1} \cup \{j_0\}$.
- **Update Provisional Solution**: Compute $(Wx)^k$, the minimizer of $\|(AW^T)(Wx) - b\|_2^2$ subject to $Support\{(Wx)\} = S^k$
- **Update Residual**: Compute $r^k = b - (AW^T)(Wx)^k$.
- **Stopping Rule**: If $\|r^k\|_2 < \epsilon_0$, stop. Otherwise, apply another iteration.

Output: The WOMP solution is $(Wx)^k$ obtained after k iterations.

References

1. Andersen, A.H., Kak, A.C.: Simultaneous algebraic reconstruction technique (SART): a superior implementation of the ART algorithm. Ultrason. Imaging **6**(1), 81–94 (1984)
2. Badea, C., Gordon, R.: Experiments with the nonlinear and chaotic behaviour of the multiplicative algebraic reconstruction technique (MART) algorithm for computed tomography. Phy. Med. Biol. **49**(8), 1455–1474 (2004)
3. Boyd, S., Vandenberghe, L.: Convex Optimization. Cambridge University Press, New York (2004)
4. Candes, E.J.: The restricted isometry property and its implications for compressed sensing. C. R. Math. **346**, 589–592 (2008)
5. Candes, E.J., Romberg, J.: Practical signal recovery from random projections. In: Proceedings of the SPIE Conference on Wavelet Applications in Signal and Image Processing XI, vol. 5914 (2005)
6. Chen, G.H., Tang, J., Leng, S.: Prior image constrained compressed sensing (PICCS): a method to accurately reconstruct dynamic CT images from highly undersampled projection data sets. Med. Phys. **35**(2), 660–663 (2008)
7. Daubechies, I.: Ten Lectures on Wavelets. Society for Industrial and Applied Mathematics, Philadelphia (1992)
8. Donoho, D.L.: Compressed sensing. IEEE Trans. Inf. Theory **52**(4), 1289–1306 (2006)
9. Elad, M. (ed.): Sparse and Redundant Representations: from Theory to Applications in Signal Processing. Springer, New York (2010)
10. Foucart, S., Rauhut, H.: A Mathematical Introduction to Compressive Sensing. Birkhauser, Basel (2013)
11. Frush, D.P., Donnelly, L.F., Rosen, N.S.: Computed tomography and radiation risks: what pediatric health care providers should know. Pediatrics **112**, 951–957 (2003)
12. Jan, J.: Medical Image Processing, Reconstruction and Restoration: Concepts and Methods. CRC Press, Boca Raton (2005)
13. Kak, A.C., Slaney, M.: Principles of Computerized Tomographic Imaging. Society for Industrial and Applied Mathematics, Philadelphia (2001)
14. Kudo, H., Suzuki, T., Rashed, E.A.: Image reconstruction for sparse-view CT and interior CT - introduction to compressed sensing and differentiated backprojection. Quant. Imaging Med. Surg. **3**(3), 147–161 (2013)
15. Li, C., Yin, W., Jiang, H., Zhang, Y.: An efficient augmented Lagrangian method with applications to total variation minimization. Comput. Optim. Appl. **56**(3), 507–530 (2013)
16. Natterer, F.: The Mathematics of Computerized Tomography. Society for Industrial and Applied Mathematics, Philadelphia (2001)
17. Pan, X.C., Sidky, E.Y., Vannier, M.: Why do commercial CT scanners still employ traditional, filtered back-projection for image reconstruction? Inverse Prob. **25**(12), 1230009 (2009)
18. Ritschl, L., Bergner, F., Fleischmann, C., Kachelrieß, M.: Improved total variation-based CT image reconstruction applied to clinical data. Phys. Med. Biol. **56**(6), 1545–1561 (2011)
19. Sastry, C.S., Das, P.C.: Wavelet based multilevel backprojection algorithm for parallel and fan beam scanning geometries. Int. J. Wavelets Multiresolut. Inf. Process. **4**(3), 523–545 (2006)

20. Tropp, J.A., Gilbert, A.C.: Signal recovery from random measurements via orthogonal matching pursuit. IEEE Trans. Inf. Theory **53**(12), 4655–4666 (2007)

21. Wang, Z., Bovik, A.C., Sheikh, H.R., Simoncelli, E.P.: Image quality assessment: from error visibility to structural similarity. IEEE Trans. Image Proc. **13**(4), 600–612 (2004)

22. Xu, Q., Mou, X., Wang, G., Sieren, J., Hoffman, E., Yu, H.: Statistical interior tomography. IEEE Trans. Med. Imaging **30**(5), 1116–1128 (2011)

23. Yu, H.Y., Wang, G.: Compressed sensing based interior tomography. Phys. Med. Biol. **54**(9), 2791–2805 (2009)

24. Yu, H.Y., Yang, J.S., Jiang, M., Wang, G.: Supplemental analysis on compressed sensing based interior tomography. Phys. Med. Biol. **54**(18), N425–N432 (2009)

25. Zhang, H., Huang, J., Ma, J., Bian, Z., Feng, Q., Lu, H., Liang, Z., Chen, W.: Iterative reconstruction for X-ray computed tomography using prior-image induced nonlocal regularization. IEEE Trans. Biomed. Eng. **61**(9), 2367–2378 (2014)

26. Zhou, W., Cai, J.-F., Gao, H.: Adaptive tight frame based medical image reconstruction: a proof-of-concept study for computed tomography. Inverse Prob. **29**, 125006 (2013)

Knot Detection from Accumulation Map by Polar Scan

Adrien Krähenbühl[1,2]($\boxtimes$), Bertrand Kerautret[1,2], and Fabien Feschet[3]

[1] Université de Lorraine, LORIA, UMR 7503, 54506 Vandoeuvre-lè-Nancy, France
{adrien.krahenbuhl,bertrand.kerautret}@loria.fr
[2] CNRS, LORIA, UMR 7503, 54506 Vandoeuvre-lès-Nancy, France
[3] IGCNC - EA 7282, Université Clermont Auvergne, Université d'Auvergne,
63000 Clermont-Ferrand, France
fabien.feschet@u-auvergne.fr

Abstract. This paper proposes to improve the approach presented in Krähenbühl *et al.* [11] to build automatic methods for the wood knot detection from X-Ray CT scanner images. The major drawbacks of the previous method mostly depends on the variety of the distribution of knots and their geometric shapes. Our aim is to extend the robustness by performing the accumulation process of Z-Motion differently and by suppressing the whorl distribution constraint. This is achieved both through a polar Z-Motion accumulation and an aggregation process of connected components related to maxima localization in the accumulation space. The experimental results are in favor of an increase in the robustness while being more sensitive to small and isolated knots. This opens the way to a method fully independent of wood species.

Keywords: Wood knots · X-Ray CT scanners · Accumulation map

1 Introduction

The knots are the part of a tree branch located inside the trunk. They are the larger objects of the trunk structure, originating in the pith and growing toward the bark. The knots are largely studied in agronomic research in order to precisely model the global tree growth [3,5,13,15]. Moreover, informations about knots interest the sawmills with the objective to predict the plank quality. Agronomic researchers and sawmills both use X-Ray CT scanners to obtain images of the internal structure of trunks (see samples in Fig. 1). The major issue consists in precisely segmenting knots in these acquisitions. Unfortunately, the presence of wet sapwood, wood part where the sap passes through, with the same density as knots, is a major problem.

Various approaches were proposed to address this segmentation problem. For instance, Aguilera *et al.* [2] propose to segment knots helped by active contours and Wells *et al.* [16] use morphological operators. More recently, Johannson *et al.* [6] proposed a knot segmentation method based on a sub-sampling of original image in concentric surfaces on which they detect knot sections as ellipses.

© Springer International Publishing Switzerland 2015
R.P. Barneva et al. (Eds.): IWCIA 2015, LNCS 9448, pp. 352–362, 2015.
DOI: 10.1007/978-3-319-26145-4_26

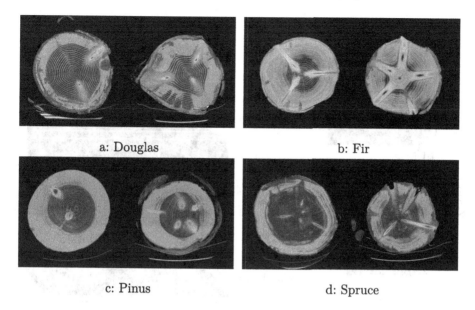

Fig. 1. Original scanned slices of the four wood species used for experimentations. For each species, a slice from the bottom part (on the left) and the top part (on the right) of the tree. All these trunks have a wet sapwood (light grey ring near from the bark). We can see the various knot aspects depending on the species (Color figure online).

All these works are efficient in the context of dry sapwood. The last seems to be adapted to wet sapwood but lacks in implementation details do not allow to reproduce results.

From our side, we have proposed an approach to identify knot areas around each knot, based on the Z-Motion notion [10] (see Sect. 2.2). This approach was adapted to trees containing a whorl distribution of knots. From on this detection, we are able to apply an individual segmentation for each knot in its supporting area [11]. However, the knot areas detection step suffered from some limitations. It is mainly adapted to big knots and cannot avoid the possible presence of two knots in a same detected knot area leading to confusions in the precise characterization of knots. Moreover, isolated knots are rarely identified and thus are missed in the detection process. It must also be added that the original method lacks robustness in some wood species where the whorl distribution of knots is not evident.

We propose in this paper, a new approach to detect knot areas around each individual knot. It combines the Z-Motion with a polar consideration of the tree centered in the pith. Knot areas are then detected from the local maxima of Z-Motion accumulation, by using a connected component merging. This new method allows to easily count the number of knots and to locate them with an high precision. All knots can be identified without size or whorl distribution constraint. Moreover, this method is easy to implement with few parameters

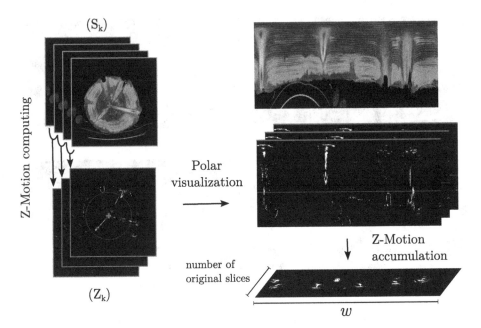

Fig. 2. Generation of a Z-Motion accumulation image. The pith location (in red) allows to accumulate the Z-Motion of each angular sector of each slice on a disc centered on the pith of r_{max} radius (the yellow circle and line) (Color figure online).

which are directly linked to physical properties of wood. Our implementation is based on the IPOL demonstration engine and can be tested on-line [7].

The remainder of this paper is organized in two parts. The next section presents the new method with theoretical points and implementation details. Then, the last section is devoted to the comparisons between the previous and the new approach with visualizations of located knots in the initial tree coordinate system (Fig. 2).

2   Method

The method proposed in this paper locates knot areas from 3D density images in merely three steps:

1. pith computing;
2. generation of the Z-Motion accumulation map;
3. supporting knot areas identification.

2.1   Pith Computing

The pith computing step must provide the trunk pith coordinates for each slice. For that, we use the detection pith algorithm proposed by Boukadida *et al.* [4]

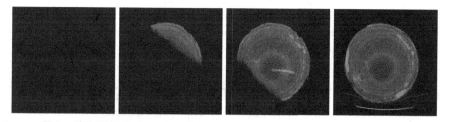

a: Slice 0: invalid b: Slice 19: invalid c: Slice 33: invalid d: Slice 76: valid

Fig. 3. Difference between valid and invalid slices due to an inclined tree cut.

which was already put to good use in a previous work [11]. This algorithm provides two informations: the interval of valid slices and the pith location for each valid slice. Indeed, the original image contains some empty slices or slices containing a partial trunk section due to the tree cut (see Fig. 3). In the remainder of this paper, I will denote the original image restricted to the subset of valid slices. The pith coordinates for a slice S_k of I will be denoted $p_k = (x_k, y_k)$.

More precisely, the pith detection is based on a Hough transform applied on each slice from the higher contours detected by a Sobel filter. It assumes that the higher contours are located on the annual tree rings concentric to the pith. Moreover, these steps are applied on the complete first slice then iteratively in a sub-window on the next slices, centered on the found pith location of the previous slice. More details about the method implementation are available in [4] with a link to the *ImageJ* plugin proposed by authors.

2.2   Generation of the Z-Motion Accumulation Image

This second step uses the Z-Motion notion that we introduced in [10]. The Z-Motion allows to identify the presence of knots by considering them as the biggest objects in motion inside the tree. The Z-Motion can be computed for each pixel (x, y) of a slice S_k as the absolute difference between two consecutive slices:

$$Z_k(x, y) = |S_k(x, y) - S_{k-1}(x, y)|$$

To generate the Z-Motion accumulation image, we consider the division of each slice S_k into a set of angular sectors Ω_a centered on the pith and restricted to a radius r_{max}. The number of angular sectors Ω_a is a parameter denoted w.

From this division, we build the accumulation image A where each row corresponds to a slice of I and each column corresponds to an angular sector. The value of A in (k, a) is computed by summing the Z-Motion values inside the Ω_a sector in the S_k slice helped by the following formula:

$$A(k, a) = \sum_{(x,y) \in S_k \cap \Omega_a} Z_k(x, y) > z_{min}$$

The z_{min} parameter allows to not consider the low Z-Motion values induced by other structures than knots. The details of this step are given in Algorithm 1 with the two main parameters rMax and zMin.

Algorithm 1. Computing of the Z-Motion accumulation image (accImage, A) according to all angular sectors of each slice.

Input	: image	// The 3D wood trunk image (I)
	pith	// Pith with the 2D coordinates of each slice
	rMax	// Maximum radius of analysis around the pith
	nbAngularSectors	// Angular resolution (in sector number) (w)
	zMin	// Minimum Z-Motion considered (z_{min})
Ouput	: accImage	// Z-Motion accumulation image (A) with a size
		// of nbAngularSectors × image.depth pixels

```
1  foreach slice in image do
2      k = index of slice
3      c = pith[k]
4      foreach p in slice do
5          r = dist(p,c) // Euclidean distance between p and c coordinates
6          if r < rMax then
7              a = ⌊orientation of c⃗p × nbAngularSectors/2π⌋ // Angular sector index
8              zMotion = abs( slice[p.x][p.y] - image[k-1][p.x][p.y] )
9              if zMotion > zMin then
10                 accImage[k][a] += zMotion - zMin
```

2.3 Generation of Knot Supporting Areas

As previously described in the introduction, our method is based on the Z-Motion accumulation in a polar space and on the detection of the support area of each knot. However, since the precise geometric distribution of knots depends on wood species, the approach must deal with arbitrary knot distributions. It must also be noted that the Z-Motion can occur around knots and does not only correspond to knots. However, maximums of Z-Motion accumulation indeed correspond to knots. To correctly extract the supporting area of each knot, the Z-Motion accumulation image is binarized through a thresholding process of parameter t_{bin}. In the binary image, knots are preserved by definition. The position of each accumulation maximum indicates the knot locations. Hence, computing the supporting area of each knot consists in an aggregation process of connected components around those maximums. Therefore, the areas are linked to the connected components of the binary image and each supporting area can potentially contain several connected components.

Our process can be summarized as follows. After the connected component detection, we iteratively merge the detected connected components in the order of the ascending Z-Motion maximums. To avoid ambiguous merging, we also

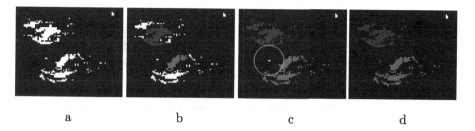

a   b   c   d

Fig. 4. Example of connecting component merging to define knot areas as a set of connected components, from a sub-image of Spruce. Image a represents the binarized image of the Z-Motion accumulation image. Image b shows the two connected components corresponding to an accumulation peak. In c the blue circle represents the distance d_{max} around the last component. In d, the last component is merged to the nearest supporting area, the green one (Color figure online).

Algorithm 2. Detection of the supporting areas.

Input	: imageAcc	// The 2D Z-Motion accumulation image
	tBin	// Threshold value to binarize imageAcc
	dMax	// Maximum distance between two connected
		// components of the same supporting area
Ouput	: supportingAreaList	// List with the supporting area of each knot
Variable	: thImage	// Binarized version of imageAcc using tBin
	ccList	// List of connected components of thImage
	ccProcessedList	// List of processed connected components
	sortedPixels	// List of sorted imageAcc pixels

```
 1  supportingAreaList = emptyList()
 2  ccProcessedList    = emptyList()
 3  Compute thImage as the binarized image of imageAcc using tBin as threshold
 4  Compute ccList as the list of connected components of thImage
 5  Sort all pixels of imageAcc in decreasing order into sortedPixels
 6  while ccProcessedList.size() != ccList.size() do
 7  │   p = sortedPixels.popFirst()
 8  │   cc = connected component containing p
 9  │   if cc not in ccProcessedList then
10  │   │   ccProcessedList.append(cc)
11  │   │   Identify ccMin as the nearest component of cc in supportingAreaList
12  │   │   d = minDist(cc, ccMin)
13  │   │   if d < dMax then
14  │   │   │   Merge cc to ccMin into supportingAreaList
15  │   │   else
16  │   │   │   Insert cc into supportingAreaList
```

introduce one parameter: the maximum distance d_{max} between two distinct connected components belonging to the same knot. Each new considered connected component is merged with the nearest supporting area if the distance is lower than d_{max}. Otherwise, this component belongs to the support area of a new knot (see Fig. 4). The distance computation relative to d_{max} is done by a distance transform based on a Fast Marching Method [14]. Finally, we obtain a set of supporting areas (Φ_p). The resulting Algorithm 2 is given with the two main parameters tBin and dMax, respectively corresponding to t_{bin} and d_{max}.

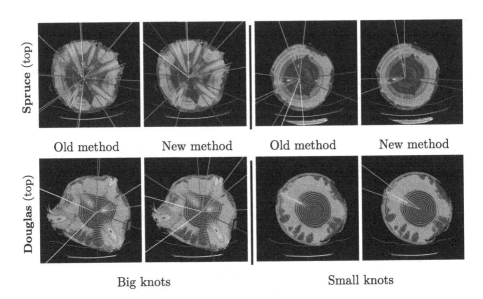

Fig. 5. Comparison between previous and new knots areas drawn on four initial slices, for a Spruce in the first row and a Douglas in the second row.

3   Experimentations

In order to compare the knot areas obtained with the previous method and the new one, we defined a common knot area description. We chosen the extremal coordinates of a supporting area Φ as the bounding box from $(\Phi.k_{min}, \Phi.a_{min})$ to $(\Phi.k_{max}, \Phi.a_{max})$. In this way, we obtain a definition of knot areas equivalent to the one in [10].

In Fig. 6, we can visually compare the knot areas obtained by the two approaches and represented on the Z-Motion accumulation image. The comparison is done on the 8 samples presented in Fig. 1. The knot areas obtained by the previous approach were defined in two steps: first the slice intervals (the red lines) then the angular intervals in each slice interval (colored lines perpendicular to the red ones). On the right, each supporting area detected by the new method is represented by a colored bounding box drawn on top of its detected connected components.

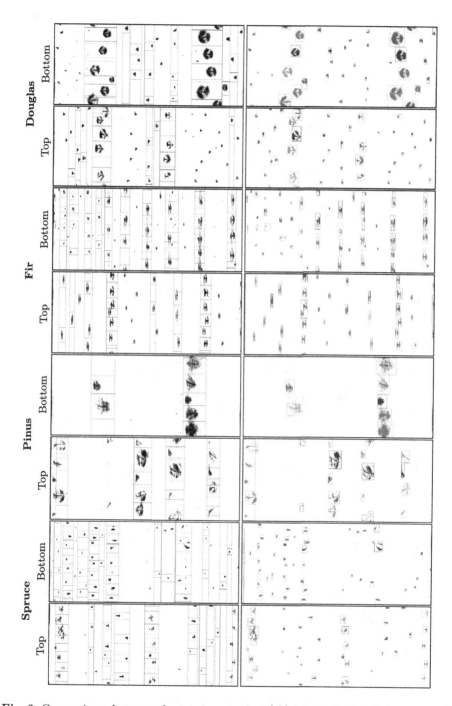

Fig. 6. Comparisons between the previous method [10] (on the left) and the proposed method (on the right) by considering 2 samples of 4 tree species.

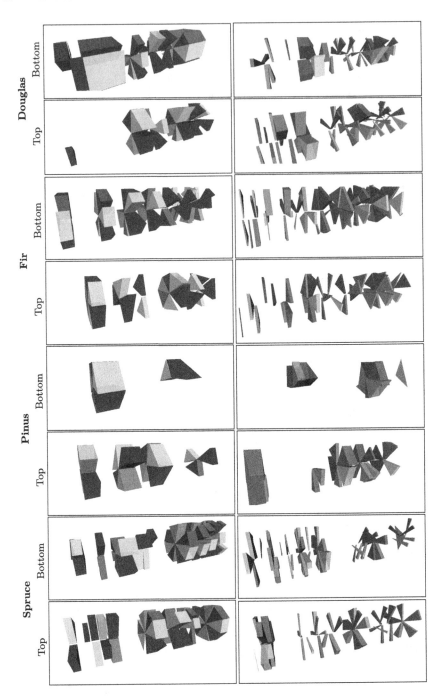

Fig. 7. 3D visual comparisons of locations and dimensions of knot areas inside the original trunk system. We see the knots areas obtained by the previous method (on the left) and by the proposed method (on the right).

Figure 5 focuses on two slices of the Spruce and Douglas samples (top parts). We see that the new knot areas are well estimate and more precise than with the previous method. On the bottom right slice, we can see a knot area detected by the new method previously not detected. On the top left and right, we can some previous knot areas without the presence of knot. This is due to a wide whorl with a small gap between the knot that not appear on this slice. Globally, all the new knot areas are smaller and more precise around each knot.

The previous knot areas were generated and exported by TKDetection [8] (release 3.0) with the default parameters proposed in [11]. The supporting areas and the corresponding new knot areas were generated by a $C++$-based implementation [7] using the DGtal library [1]. The illustrations on the right of Fig. 6 use the parameters $w = 500$, $z_{min} = 200$, $t_{bin} = 100$ and $d_{max} = 20$. The computing of r_{max} parameter is detailed in [11]. It corresponds to the lower euclidian distance from the pith to the bark by following 20 directions around the pith on each slice. The bark limit is located in each of these directions as the first pixel from the pith with a value corresponding to the air density.

Figure 7 visualize the same knot areas than in Fig. 6 in the trunk coordinate system. Colors are identical to a direct matching of the knot locations around the pith.

The new approach corrects two major defects from the previous one:

1. it allows to identify all knots, even the small and/or isolated ones ;
2. it clearly detects more precise areas.

The results show that the proposed approach is robust to specificities of images and, consequently, to tree species.

4   Conclusion

The method proposed in this paper was designed to increase the robustness of the knot detection, especially for small and isolated knots, and to remove the whorl distribution constraint. The first experiments have shown that both goals have been reached. Moreover, the method is easy to implement, and some parameters can be fixed depending on wood species. Automatic determination of those parameters will be the subject of future works. Since the detected supporting areas of knots are compatible with the segmentation methods given in [9,11,12], the presented algorithm can serve as input of those segmentation algorithms leading to a complete characterization of detected knots. It must also be noted that our supporting areas of the knots are more precise than the original algorithm due to the merging process of close connected components. As an extension of this work, we will also decompose the trunks into concentric volumetric sets to obtain several accumulation locations from the pith to the bark. This should provide a way to estimate the 3D main orientations of a knot by defining a procedure to follow this knot through concentric volumes.

References

1. DGtal: Digital Geometry tools and algorithms library. http://libdgtal.org
2. Aguilera, C., Sanchez, R., Baradit, E.: Detection of knots using x-ray tomographies and deformable contours with simulated annealing. Wood Res. **53**, 57–66 (2008)
3. Baño, V., Arriaga, F., Guaita, M.: Determination of the influence of size and position of knots on load capacity and stress distribution in timber beams of pinus sylvestris using finite element model. Biosyst. Eng. **114**(3), 214–222 (2013)
4. Boukadida, H., Longuetaud, F., Colin, F., Freyburger, C., Constant, T., Leban, J.M., Mothe, F.: Pithextract: a robust algorithm for pith detection in computer tomography images of wood - application to 125 logs from 17 tree species. Comput. Electron. Agric. **85**, 90–98 (2012)
5. Funck, J., Zhong, Y., Butler, D., Brunner, C., Forrer, J.: Image segmentation algorithms applied to wood defect detection. Comput. Electron. Agric. **41**(1–3), 157–179 (2003). developments in Image Processing and Scanning of Wood
6. Johansson, E., Johansson, D., Skog, J., Fredriksson, M.: Automated knot detection for high speed computed tomography on pinus sylvestris l. and picea abies (l.) karst. using ellipse fitting in concentric surfaces. Comput. Electron. Agric. **96**, 238–245 (2013)
7. Kerautret, B.: Knot detection from accumulation map by polar scan: Online demonstration (2015). http://ipol-geometry.loria.fr/kerautre/ipol_demo/KnotDetectIPOLDemo/
8. Krähenbühl, A.: TKDetection (2012). https://github.com/akrah/TKDetection/
9. Krähenbühl, A., Kerautret, B., Debled-Rennesson, I.: Knot segmentation in noisy 3D images of wood. In: Gonzalez-Diaz, R., Jimenez, M.-J., Medrano, B. (eds.) DGCI 2013. LNCS, vol. 7749, pp. 383–394. Springer, Heidelberg (2013)
10. Krähenbühl, A., Kerautret, B., Debled-Rennesson, I., Longuetaud, F., Mothe, F.: Knot detection in X-Ray CT images of wood. In: Bebis, G., Boyle, R., Parvin, B., Koracin, D., Fowlkes, C., Wang, S., Choi, M.-H., Mantler, S., Schulze, J., Acevedo, D., Mueller, K., Papka, M. (eds.) ISVC 2012, Part II. LNCS, vol. 7432, pp. 209–218. Springer, Heidelberg (2012)
11. Krähenbühl, A., Kerautret, B., Debled-Rennesson, I., Mothe, F., Longuetaud, F.: Knot segmentation in 3D CT images of wet wood. Pattern Recognit. **1**, 1–17 (2014)
12. Krähenbühl, A., Roussel, J.R., Kerautret, B., Debled-Rennesson, I., Mothe, F., Longuetaud, F.: Segmentation robuste de nœuds partir de coupes tangentielles issues d'images tomographiques de bois. In: Actes de la conférence RFIA 2014, June 2014
13. Moberg, L.: Models of internal knot properties for picea abies. For. Ecol. Manage. **147**(2–3), 123–138 (2001)
14. Sethian, J.A.: Fast marching methods. SIAM Rev. **41**, 199–235 (1998)
15. Todoroki, C., Lowell, E., Dykstra, D.: Automated knot detection with visual post-processing of douglas-fir veneer images. Comput. Electron. Agric. **70**(1), 163–171 (2010)
16. Wells, P., Som, S., Davis, J.: Automated feature extraction from tomographic images of wood. In: Image Computing: Techniques and Applications (DICTA), pp. 56–62. No. 1, Melbourne, Australie, December 1991

Author Index

Printed in the United States
By Bookmasters